AF676578

Molecular Research for Cereal Grain Quality

Molecular Research for Cereal Grain Quality

Editors

Jinsong Bao
Jianhong Xu

MDPI

Basel • Beijing • Wuhan • Barcelona • Belgrade • Novi Sad • Cluj • Manchester

Editors

Jinsong Bao
Hainan Institute
Zhejiang University
Sanya
China

Jianhong Xu
Hainan Institute
Zhejiang University
Sanya
China

Editorial Office
MDPI
St. Alban-Anlage 66
4052 Basel, Switzerland

This is a reprint of articles from the Special Issue published online in the open access journal *International Journal of Molecular Sciences* (ISSN 1422-0067) (available at: www.mdpi.com/journal/ijms/special_issues/Cereal_Seed_Quality).

For citation purposes, cite each article independently as indicated on the article page online and as indicated below:

Lastname, A.A.; Lastname, B.B. Article Title. *Journal Name* **Year**, *Volume Number*, Page Range.

ISBN 978-3-0365-9055-4 (Hbk)
ISBN 978-3-0365-9054-7 (PDF)
doi.org/10.3390/books978-3-0365-9054-7

© 2023 by the authors. Articles in this book are Open Access and distributed under the Creative Commons Attribution (CC BY) license. The book as a whole is distributed by MDPI under the terms and conditions of the Creative Commons Attribution-NonCommercial-NoDerivs (CC BY-NC-ND) license.

Contents

About the Editors

Jinsong Bao

Dr. Jinsong Bao is currently a professor in the College of Agriculture and Biotechnology at Zhejiang University, China. He received his B.S. (1993) and M.S. (1996) degrees in horticulture from Zhejiang Agricultural University, and his Ph.D. degree in biophysics from Zhejiang University in 1999. His research interests focus on molecular genetics of rice quality, more specifically in the areas of starch chemistry and quality, nutritional quality, and molecular breeding. He has edited two books, including *Rice: Chemistry and Technology* (4th edition), and published more than 150 peer-reviewed research articles and 17 book chapters in these areas. He has received two professional awards for his achievements in the genetic study and molecular improvement of rice quality from the Zhejiang Provincial Government. He has also been awarded the "Young Scientist Research Award" from AACCI and the Zhejiang Young Science and Technology Award from the Zhejiang Association of Science and Technology. He has traveled and worked extensively in Hong Kong and USA. He teaches undergraduate and graduate courses in mutational genetics and developmental biology in plants. He is a member of the Editorial Boards of the *International Journal of Molecular Sciences*, the *Journal of Cereal Science*, *Cereal Chemistry*, *Starch*, *Genes & Genomics*, *Rice Science*, *Frontiers in Nutrition* and *Applied Science*.

Jianhong Xu

Dr. Jian-Hong Xu is a professor at Zhejiang University. In 2001, he won a Japanese government scholarship to study in Japan and received his PhD degree from the University Tokyo in 2004. Then, he conducted his postdoctoral research at Dr. Messing's Lab at Rutgers University. He joined Zhejiang University in 2011. His research mainly focuses on plant molecular biology and genomics, including the molecular mechanism of seed quality formation, epigenetics (DNA methylation and demethylation) and small RNAs, molecular evolution of genes and genomes, identification and transposition mechanism of transposable elements and pollen sterility, and transition mechanism of rice photoperiod-and-thermo-sensitive genic male sterility via methods of comparative genomics, bioinformatics, evolutionary biology, molecular biology, and transgenes. His work has been published in journals with a high impact factor, such as *PNAS*, *Genome Research*, *Molecular Biology and Evolution*, *PLoS Genetics*, *Molecular Plant* and *Plant Journal*.

International Journal of *Molecular Sciences*

Editorial

Molecular Research for Cereal Grain Quality

Jinsong Bao * and Jian-Hong Xu

Hainan Institute, Zhejiang University, Yazhou Bay Science and Technology City, Sanya 572025, China; jhxu@zju.edu.cn
* Correspondence: jsbao@zju.edu.cn

Cereals such as wheat (*Triticum aestivum* L.), rice (*Oryza sativa* L.), and maize (*Zea mays* L.) provide key sources of dietary energy for human beings. Deficiency in their production will bring serious food security problems. Due to technical advances, for example, the utilization of heterosis in the development of hybrid cereals, many countries have achieved self-sufficiency in cereal production. However, due to increases in the human population and urbanization and a decline in arable land, the higher production demand remains challenging for many nations. On the other hand, with the increase in living standards thanks to economic development, our desire for a better life requires more production of high-quality cereal foods [1,2].

Cereal grain quality is governed by all the features and characteristics of the grain and its products to meet the demands of end users, which includes milling efficiency, processing quality, grain shape and appearance, ease of cooking, palatability, and nutrition [3]; it mainly reflects the physical and chemical properties of the grain. Physically, the grain's shape and size affect the quality of appearance and also the yield, while chalkiness (especially for rice) affects its appearance and processing quality. Chemically, the major constituents of cereals, i.e., starch, protein, and lipids, affect cooking and eating quality, while protein, lipids, and other micronutrients affect nutritional quality [3]. All the physical and chemical properties are considered to be complex traits that are affected by both genetic and environmental factors [3]. However, molecular mechanisms underlying grain quality formation are poorly understood, which may constrain our ability to produce high-quality cereal grain. This Special Issue aims to provide a forum on the most recent advances in the application of molecular tools to understand the mechanism for improving any cereals' grain quality. A total of 13 papers were collected for this Special Issue [1,2,4–14], mainly covering rice and wheat crops.

Rice is one of the most important staple food crops in the world and feeds more than half of the world's population. The grain quality of rice generally includes the milling, appearance, cooking and eating, and nutritional qualities [3]. Liu et al. [4] cloned a novel quantitative trait locus (QTL), *GLW7.1* (*Grain Length, Width and Weight 7.1*), which encodes the CCT motif family protein GHD7. It was hypothesized that GHD7 participates in GA biosynthesis to increase grain size and is regulated by the GID1-GA-DELLA module as the feedback of the pathway. The near-isogenic line constructed with the dominant allele showed reduced chalkiness, improved cooking and eating quality, and increased grain length [4].

Rice cooking and eating quality is especially important as it directly affects consumer taste preferences and market values. The eating quality can be indirectly predicted with a series of starch physicochemical property evaluations [15]. Starch, including amylose and amylopectin, is synthesized with the actions of a series of enzymes, such as ADP-glucose pyrophosphorylase (AGPase), granule-bound starch synthase (GBSS), soluble starch synthases (SSs), starch branching enzymes (BEs), starch debranching enzymes (DBEs), and phosphorylases [5]. Amylose is synthesized by GBSS I, encoded by the *Waxy* gene (*Wx*). Apparent amylose content (AAC) is an important indicator to evaluate the eating quality of rice grains [15]. Starch gelatinization temperature (GT) is another important

check for updates

Citation: Bao, J.; Xu, J.-H. Molecular Research for Cereal Grain Quality. *Int. J. Mol. Sci.* **2023**, *24*, 13687. https://doi.org/10.3390/ijms241813687

Received: 29 August 2023
Accepted: 4 September 2023
Published: 5 September 2023

Copyright: © 2023 by the authors. Licensee MDPI, Basel, Switzerland. This article is an open access article distributed under the terms and conditions of the Creative Commons Attribution (CC BY) license (https://creativecommons.org/licenses/by/4.0/).

predictor to determine the cooking quality of rice. GT is mainly controlled by the *SSIIa/ALK* (*alkaline degenerate*) gene [1,15]. Fragrance is appealing to consumers, which is controlled by the recessive gene *fgr/OsBADH2* (*betaine aldehyde dehydrogenase 2*) [6,7,15]. Pan et al. [1] analyzed the allelic diversification of the *Wx* and *ALK* loci in indica restorer lines developed over the last 50 years. The proportion of the Wx^a allele decreased, while that of the Wx^b allele increased, leading to a decrease in AAC and an increase in cooking and eating quality in current rice cultivars. *ALK* had no significant effect on taste value, and there was no strict requirement for the GT of high-quality rice, resulting in no selection pressure for *ALK* [1]. CRISPR/Cas9 technology was used to edit the 5′ UTR of the *Wx* gene of three lines with undesirably high AAC to create a batch of soft rice (low-AAC rice) breeding materials. The edited rice lines had an AAC of 16–18% and gel consistency of 77–80 mm, suggesting that the eating quality was successfully improved [2]. Lu et al. [6] obtained a high-quality chromosome-level genome assembly (~378.78 Mb) for a new fragrant japonica cultivar "Changxianggeng 1813", with 31,671 predicted protein-coding genes. They found that it was the *badh2-E2* type of deletion (a 7 bp deletion in the second exon) that causes fragrance in this rice. The identification of many single-nucleotide polymorphisms, insertion-deletion polymorphisms (InDels), and large structure variants (SVs, >1000 bp) will be useful for genomics-assisted breeding in fragrant japonica rice. Tian et al. [7] edited *Wx* and *OsBADH2* simultaneously using CRISPR/Cas9 system to produce both homozygous two-line male sterile mutant lines and homozygous restorer mutant lines with free Cas9. The obtained mutants had a much lower AAC while having a significantly higher 2-acetyl-1-pyrroline aroma content. Based on this, a fragrant glutinous hybrid rice was developed without much effect on most agronomic traits.

Heading date may affect starch structure and grain yield due to different temperatures during seed development. Crofts et al. [8] constructed near-isogenic rice lines with *ss2a Hd1* (*heading date1*), *ss2a Hd1 hd1*, and *ss2a hd1* genotypes. They found that the *ss2a Hd1* line showed the highest plant biomass but with varied grain yield across different years. The *ss2a Hd1 hd1* line showed a higher total grain weight than *ss2a hd1*. The *ss2a hd1* line produced the lowest number of premature seeds and showed higher GT and lower AAC than *ss2a Hd1*, suggesting that *Hd1* is a candidate gene for developing high-yielding rice cultivars with the desired starch structure. Seed development and grain quality may also be affected by epigenetic regulation; DNA methylation is one of the main epigenetic modifications. Irshad et al. [9] generated a null mutant of a rice DNA demethylation gene, *Repressor Of Silencing 1a* (*OsROS1a*), with an in-frame deletion of the complete loss function of the Per-CXXC domain using CRISPR/Cas9 technology. The *osros1a* mutant showed longer and narrower grains, and seeds were deformed and contained an underdeveloped and less-starch-producing endosperm with slightly irregularly shaped embryos. The grains of the mutant were slightly opaque with rounded starch granules. RNA-Seq results indicated that the key genes for starch synthesis (*OsSSIIa* and *OsSSIIIa*) and cellulose synthesis (*CESA2*, *CESA3*, *CESA6*, and *CESA8*) and genes encoding polysaccharides and glutelin were downregulated in the mutant endosperm. Furthermore, 378 differential alternative splicing (AS) genes were identified in the mutant, suggesting that *OsROS1a* has an impact on AS events. Further analyzing the generated mutants, they produced a frameshift mutation to truncate the Pem-CXXC and RRMF domains of *OsROS1a*; this mutant had shrink spikelets, smaller anthers, and pollen grains and was not stained by iodine staining, showing a significant reduction in total soluble sugar and starch contents as compared to the wild type, which caused complete male sterility [10].

The enzymes to synthesize starch usually form a multienzyme complex, but whether these complexes change during seed development is not fully understood. Ying et al. [5] revealed that most of the enzymes except for SSIVb were eluted from GPC, first in smaller-molecular-weight fractions at the early developing stage, and then transferred to higher-molecular-weight fractions at the later stage in both WT and a *BEIIb* mutant (*be2b*). However, the inverse elution pattern of SSIVb may be attributed to its vital role in the initiation step of starch synthesis. The number of protein complexes was markedly decreased in *be2b* at all development stages as compared to those in BEIIb. Although SSIVb could partially compensate for the role of BEIIb in protein complex formation, it was difficult to form a larger protein complex containing over five proteins in *be2b*. These findings unraveled a dynamic change in the protein complex during seed development, which enables a deeper understanding of the complex mechanism of starch biosynthesis and quality improvement in rice.

Storage proteins represent the second-largest storage substance in cereal grains and play an important role in determining the nutritional as well as cooking and eating quality of cereals. In cereal endosperm cells, seed storage proteins are synthesized on the endoplasmic reticulum (ER), where they proteins are translocated to the lumen. The localization of mRNAs plays an essential role in governing gene expression and protein targeting and thus determines cell fate, development, and polar growth. Zhang et al. [11] reviewed the current knowledge of the mechanisms and functions of mRNA localization to the ER in cereal endosperm cells. mRNA targeting to ER subdomains is driven by specific RNA zipcodes and requires a set of trans-acting RBPs that recognize and bind these zipcodes and recruit other factors to mediate active transport. A more detailed network of cotransported mRNAs and the mechanism of assembly and remodeling of multi-RNA-binding protein (RBP) complexes to recognize and bind target mRNAs deserve further investigation.

The storage of rice is also an important part of its production and transactions, and only with good storage performance can its commercial value be maintained in commodity transactions. Zhu et al. [12] found that under high-temperature storage conditions (35 °C), the indica–japonica hybrid rice Yongyou 1540 was not significantly worse in terms of fatty acid value, whiteness value, and changes in electron microscope profile. Metabolomics analysis identified 19 key differential metabolites, in which the lipid metabolites related to palmitoleic acid were found to affect the aging of rice. In addition, two substances, guanosine 3′,5′-cyclophosphate and pipecolic acid, were beneficial to enhancing the resistance of rice to harsh storage conditions, thereby delaying the deterioration of, and maintaining, its quality.

Wheat grain quality mainly includes processing and nutritional quality. The processing quality mainly depends on the content and characteristics of storage proteins, which are important indicators for market value and consumer acceptance. Cao et al. [13] identified two novel high-molecular-weight glutenin subunits (HMW-GS) 1Ax2.1* at *Glu-A1* and 1By19* at *Glu-B1* from German spelt wheat with 2478 bp and 2163 bp encoding 824 and 720 amino acid residues, respectively. The specific single-nucleotide-polymorphism-based markers for 1Ax2.1* and 1By19* genes were developed and validated by using a wide range of wheat accessions, which provide new gene resources and molecular markers for improving wheat's breadmaking quality.

Spring cold stress (SCS) causes a serious threat to wheat reproductive tissues and grain production, which is a major constraint in achieving high grain yield and quality in winter wheat. Su et al. [14] reviewed the physiological and molecular mechanisms involved in wheat floret and spikelet SCS tolerance and summarized QTLs that regulate SCS to identify candidate genes for breeding. To sustain grain setting and quality under SCS, it is necessary to breed novel wheat cultivars using novel SCS-tolerant QTLs or genes with regards to floret and spikelet development in new breeding strategies and uncover the fundamental resistance mechanism.

All these papers provide new scientific insights into molecular research for cereal grain quality. However, most of them focus on the grain quality of rice and wheat crops, so information on other important cereals, including maize, oat, barley, millets, sorghum, rye, etc., is still limited. We hope to collect papers with a broad coverage of the grain quality features of more cereal crops in future Special Issues.

Funding: This work was financially supported by the Hainan Provincial Natural Science Foundation of China (323MS066).

Conflicts of Interest: The authors declare no conflict of interest.

References

1. Pan, L.-X.; Sun, Z.-Z.; Zhang, C.-Q.; Li, B.; Yang, Q.-Q.; Chen, F.; Fan, X.-L.; Zhao, D.-S.; Lv, Q.-M.; Yuan, D.-Y.; et al. Allelic diversification of the *Wx* and *ALK* loci in indica restorer lines and their utilisation in hybrid rice breeding in China over the last 50 years. *Int. J. Mol. Sci.* **2022**, *23*, 5941. [CrossRef] [PubMed]
2. Yang, J.; Guo, X.; Wang, X.; Fang, Y.; Liu, F.; Qin, B.; Li, R. Development of soft rice lines by regulating amylose content via editing the 5′UTR of the *Wx* gene. *Int. J. Mol. Sci.* **2022**, *23*, 10517. [CrossRef] [PubMed]
3. Bao, J.S. Genes and QTLs for rice grain quality improvement. In *Rice—Germplasm, Genetics and Improvement*; Yan, W.G., Bao, J.S., Eds.; InTech: London, UK, 2014; pp. 239–278.
4. Liu, R.; Feng, Q.; Li, P.; Lou, G.; Chen, G.; Jiang, H.; Gao, G.; Zhang, Q.; Xiao, J.; Li, X.; et al. *GLW7.1*, a strong functional allele of *Ghd7*, enhances grain size in rice. *Int. J. Mol. Sci.* **2022**, *23*, 8715. [CrossRef] [PubMed]
5. Ying, Y.; Xu, F.; Zhang, Z.; Tappiban, P.; Bao, J. Dynamic change in starch biosynthetic enzymes complexes during grain-filling stages in BEIIb active and deficient rice. *Int. J. Mol. Sci.* **2022**, *23*, 10714. [CrossRef] [PubMed]
6. Lu, R.; Liu, J.; Wang, X.; Song, Z.; Ji, X.; Li, N.; Ma, G.; Sun, X. Chromosome-level genome assembly of a fragrant japonica rice cultivar 'Changxianggeng 1813' provides insights into genomic variations between fragrant and non-fragrant *japonica* rice. *Int. J. Mol. Sci.* **2022**, *23*, 9705. [CrossRef] [PubMed]
7. Tian, Y.; Zhou, Y.; Gao, G.; Zhang, Q.; Li, Y.; Lou, G.; He, Y. Creation of two-line fragrant glutinous hybrid rice by editing the *Wx* and *OsBADH2* genes via the CRISPR/Cas9 system. *Int. J. Mol. Sci.* **2023**, *24*, 849. [CrossRef] [PubMed]
8. Crofts, N.; Hareyama, K.; Miura, S.; Hosaka, Y.; Oitome, N.F.; Fujita, N. Effect of heading date on the starch structure and grain yield of rice lines with low gelatinization temperature. *Int. J. Mol. Sci.* **2022**, *23*, 10783. [CrossRef] [PubMed]
9. Irshad, F.; Li, C.; Wu, H.-Y.; Yan, Y.; Xu, J.-H. The function of DNA demethylase gene *ROS1a* null mutant on seed development in rice (*Oryza Sativa*) using the CRISPR/CAS9 system. *Int. J. Mol. Sci.* **2022**, *23*, 6357. [CrossRef] [PubMed]
10. Xu, J.-H.; Irshad, F.; Yan, Y.; Li, C. Loss of function of the RRMF domain in *OsROS1a* causes sterility in rice (*Oryza sativa* L.). *Int. J. Mol. Sci.* **2022**, *23*, 11349. [CrossRef] [PubMed]
11. Zhang, L.; Si, Q.; Yang, K.; Zhang, W.; Okita, T.W.; Tian, L. mRNA localization to the endoplasmic reticulum in plant endosperm cells. *Int. J. Mol. Sci.* **2022**, *23*, 13511. [CrossRef] [PubMed]
12. Zhu, L.; Tian, Y.; Ling, J.; Gong, X.; Sun, J.; Tong, L. Effects of storage temperature on indica-japonica hybrid rice metabolites, analyzed using liquid chromatography and mass spectrometry. *Int. J. Mol. Sci.* **2022**, *23*, 7421. [CrossRef] [PubMed]
13. Cao, Y.; Zhang, J.; Wang, R.; Sun, H.; Yan, Y. Molecular characterization and SNP-based molecular marker development of two novel high molecular weight glutenin Genes from *Triticum spelta* L. *Int. J. Mol. Sci.* **2022**, *23*, 11104. [CrossRef] [PubMed]
14. Su, H.; Tan, C.; Liu, Y.; Chen, X.; Li, X.; Jones, A.; Zhu, Y.; Song, Y. Physiology and molecular breeding in sustaining wheat grain setting and quality under spring cold stress. *Int. J. Mol. Sci.* **2022**, *23*, 14099. [CrossRef] [PubMed]
15. Zhou, H.; Xia, D.; He, Y. Rice grain quality- traditional traits for high quality rice and health-plus substances. *Mol. Breed.* **2020**, *40*, 1. [CrossRef]

Disclaimer/Publisher's Note: The statements, opinions and data contained in all publications are solely those of the individual author(s) and contributor(s) and not of MDPI and/or the editor(s). MDPI and/or the editor(s) disclaim responsibility for any injury to people or property resulting from any ideas, methods, instructions or products referred to in the content.

International Journal of *Molecular Sciences*

MDPI

Article

Allelic Diversification of the *Wx* and *ALK* Loci in Indica Restorer Lines and Their Utilisation in Hybrid Rice Breeding in China over the Last 50 Years

Li-Xu Pan [1,†], Zhi-Zhong Sun [2,†], Chang-Quan Zhang [1,3], Bu Li [1], Qing-Qing Yang [1], Fei Chen [1], Xiao-Lei Fan [1,3], Dong-Sheng Zhao [1,3], Qi-Ming Lv [2], Ding-Yang Yuan [2,*] and Qiao-Quan Liu [1,3,*]

1 Jiangsu Key Laboratory of Crop Genomics and Molecular Breeding, State Key Laboratory of Hybrid Rice, Jiangsu Key Laboratory of Crop Genetics and Physiology, Yangzhou University, Yangzhou 225009, China; panlixu288@163.com (L.-X.P.); cqzhang@yzu.edu.cn (C.-Q.Z.); libu1026@126.com (B.L.); y19895320793@163.com (Q.-Q.Y.); feich12345@outlook.com (F.C.); xlfan@yzu.edu.cn (X.-L.F.); dszhao@yzu.edu.cn (D.-S.Z.)

2 State Key Laboratory of Hybrid Rice, Hunan Hybrid Rice Research Center, Changsha 410125, China; szznihaoa@163.com (Z.-Z.S.); lvqiming00@163.com (Q.-M.L.)

3 Key Laboratory of Plant Functional Genomics of the Ministry of Education, Jiangsu Co-Innovation Center for Modern Production Technology of Grain Crops, College of Agriculture, Yangzhou University, Yangzhou 225009, China

* Correspondence: yuandingyang@hhrrc.ac.cn (D.-Y.Y.); qqliu@yzu.edu.cn (Q.-Q.L.); Tel.: +86-136-3742-6104 (D.-Y.Y.); +86-130-5631-6473 (Q.-Q.L.)

† These authors contributed equally to this work.

check for updates

Citation: Pan, L.-X.; Sun, Z.-Z.; Zhang, C.-Q.; Li, B.; Yang, Q.-Q.; Chen, F.; Fan, X.-L.; Zhao, D.-S.; Lv, Q.-M.; Yuan, D.-Y.; et al. Allelic Diversification of the *Wx* and *ALK* Loci in Indica Restorer Lines and Their Utilisation in Hybrid Rice Breeding in China over the Last 50 Years. *Int. J. Mol. Sci.* **2022**, *23*, 5941. https://doi.org/10.3390/ijms23115941

Academic Editors: Jinsong Bao and Jianhong Xu

Received: 30 March 2022
Accepted: 23 May 2022
Published: 25 May 2022

Publisher's Note: MDPI stays neutral with regard to jurisdictional claims in published maps and institutional affiliations.

Copyright: © 2022 by the authors. Licensee MDPI, Basel, Switzerland. This article is an open access article distributed under the terms and conditions of the Creative Commons Attribution (CC BY) license (https://creativecommons.org/licenses/by/4.0/).

Abstract: Hybrid rice technology has been used for more than 50 years, and eating and cooking quality (ECQ) has been a major focus throughout this period. *Waxy* (*Wx*) and alkaline denaturation (*ALK*) genes have received attention owing to their pivotal roles in determining rice characteristics. However, despite significant effort, the ECQ of restorer lines (RLs) has changed very little. By contrast, obvious changes have been seen in inbred rice varieties (IRVs), and the ECQ of IRVs is influenced by *Wx*, which reduces the proportion of Wx^a and increases the proportion of Wx^b, leading to a decrease in amylose content (AC) and an increase in ECQ. Meanwhile, *ALK* is not selected in the same way. We investigated *Wx* alleles and AC values of sterile lines of female parents with the main mating combinations in widely used areas. The results show that almost all sterile lines were Wx^a-type with a high AC, which may explain the low ECQ of hybrid rice. Analysis of hybrid rice varieties and RLs in the last 5 years revealed serious homogenisation among hybrid rice varieties.

Keywords: *Oryza sativa* L.; hybrid rice; ECQ; *Waxy*; *ALK*; *indica* restorer; rice breeding

1. Introduction

Increasing rice (*Oryza sativa* L.) yield has been the main breeding objective for several decades, and the yield potential of irrigated rice has already experienced two quantum leaps [1]. The first advance was semidwarf breeding [2,3], and the second was utilisation of hybrid rice technology [4]. In 1971, the team of Professor Yuan Longping discovered *O. rufipogon*, a wild male sterile plant, in Hainan Province, China, which revealed the production potential of hybrid rice, paving the way for the development of F_1 hybrid rice introduced in 1976 [4]. Through the introduction of hybrid rice, the development of heterotic markers led to the second breakthrough and consequent leap in production. The achievements of the 1960s and 1970s increased rice production by ~50% in 10 years in many countries, including China, Indonesia and Vietnam. Since then, hybridisation has been the core focus of rice breeding in China [5,6].

Although *xian/indica*, *XI* (*O. sativa* L. *subsp. xian/indica*)-*geng/japonica*, *GJ* (*O. sativa* L. *subsp. geng/japonica*) and *GJ-GJ* hybrids have been reported, the most common rice hybrids are *XI-XI* hybrids, or *XI* hybrids for short. There are two types of *XI* hybrid rice: three-line hybrids and two-line hybrids [7–11]. In the early stages, there were many materials collected from various countries, and then more and more materials began to be cultivated in China. Among them, South China (SC) took the lead in breeding a large number of varieties. Since the 1990s, a large number of materials have been bred in the upper reaches of the Yangtze River (UYR) and the middle and lower reaches of the Yangtze River (MLYR). Among the materials in the UYR, 3-RLs predominate, while 2-RLs are mainly from the MLYR. Hybrid rice ECQ is directly controlled by both restorer and sterile lines.

As people's living standards increase with economic development, rice ECQ receives attention. In hybrid rice breeding, a high and stable yield is the main goal, and rice ECQ indices are essential to this end [12,13]. Primary factors affecting rice ECQ are AC, gelatinisation temperature (GT), gel consistency (GC) and viscosity [14–16]. For a long time, it was generally believed that hybrid rice in China had a high yield but poor quality [17]. Higher AC and a chalky grain rate are the main reasons why the quality of hybrid rice is lower than that of inbred rice [18]. In order to improve the efficacy of breeding for rice quality, the Ministry of Agriculture and Rural Affairs of China promulgated the agricultural industry standards NY20-1986 (1986), 'quality edible rice', and NY/T-593-2002 (2002), 'cooking rice variety quality'.

These factors are mainly regulated by starch-synthesis-related genes. Starch is synthesised by the orchestrated functional interactions of four classes of enzymes; ADP-glucose pyrophosphorylase, starch branching enzyme (BE), starch synthase (SS) and starch debranching enzyme (DBE) [19–23]. The granule-bound starch synthase I enzyme (GBSSI) is required for amylose synthesis in rice, and the gene encoding GBSSI was named *Waxy* (*Wx*) [24]. Many alleles of *Wx* (such as Wx^{a}, Wx^{b}, Wx^{in}, Wx^{op}, Wx^{mp}, Wx^{lv}, *wx* and $Wx^{mw/la}$) lead to regional changes in rice AC and affect consumer preferences [24–30]. GT is mainly controlled by the alkaline denaturation (*ALK*) gene encoding modified starch synthase IIa (SSIIa) [31]. Several studies reported that at least three single-nucleotide polymorphisms (SNPs) of *ALK* are associated with the diversity of GT in rice. Ex8-733 bp (A/G) and Ex8-864/865 bp (G/T and C/T) of *ALK* in exon 8 (Ex8) are closely related to changes in GT [14,31–33]. These functional SNPs produce three haplotypes/alleles, including ALK^{a} (A-GC) and ALK^{b} (G-TT), which control low GT, and ALK^{c} (G-GC), which controls high GT [32,34]. Recently, a new *ALK* allele, ALK^{d} Ex1-294 (G/T), was reported [15].

For rice, seed development begins with double fertilisation, which leads to the development of the embryo and the endosperm. After double fertilisation, two sperm enter the embryo sac through the pollen tube. One sperm cell fuses with the egg cell to form a zygote, and the other sperm cell fuses with the central cell to form a triploid primary endosperm cell. Zygotes and primary endosperm cells develop into an embryo and transfer genetic material from parents to the next generation, and endosperm nourishes developing embryos/seedlings [35–37]. Both sterile lines and restorer lines will affect the ECQ of hybrid rice. Although ECQ evolutionary trends in hybrid rice breeding can be realised directly from the perspective of restorer lines, the process of improving the ECQ of hybrid rice can be achieved indirectly.

Various materials are employed for the cultivation of parents of hybrid rice, especially RLs, which have been mainly used to systematically explore the ECQ of hybrid rice. At present, there are few studies on the ECQ of hybrid rice. Previous studies have mainly focused on rice quality in the F_2 generation of hybrid rice [13,38,39]. However, the results are limited to specific combinations and cannot be widely used to guide the cultivation of hybrid rice. Therefore, based on the analysis of restorer line materials, we investigated improving RLs to indirectly predict the evolutionary process driving the ECQ of hybrid rice.

2. Results

2.1. Characteristics of ECQ among Indica RLs

RL materials were subjected to various analyses, including AAC, differential scanning calorimetry (DSC), Rapid Visco-Analyzer (RVA) and ECQ (Figures 1c–h and S1). AAC values fell into two main ranges: 10–18% and 22–30%. Similarly, T_p values fell into two main ranges: 65–75 °C and 75–85 °C. The results reveal a negative correlation between AAC and taste value (r = −0.515, $p < 0.01$; Table S2). T_p was not significantly correlated with AAC or taste value, but T_p was significantly correlated with BDV and SBV, positively in the case of BDV (r = 0.517, $p < 0.01$; Table S2) and negatively in the case of SBV (r = −0.511, $p < 0.01$; Table S2).

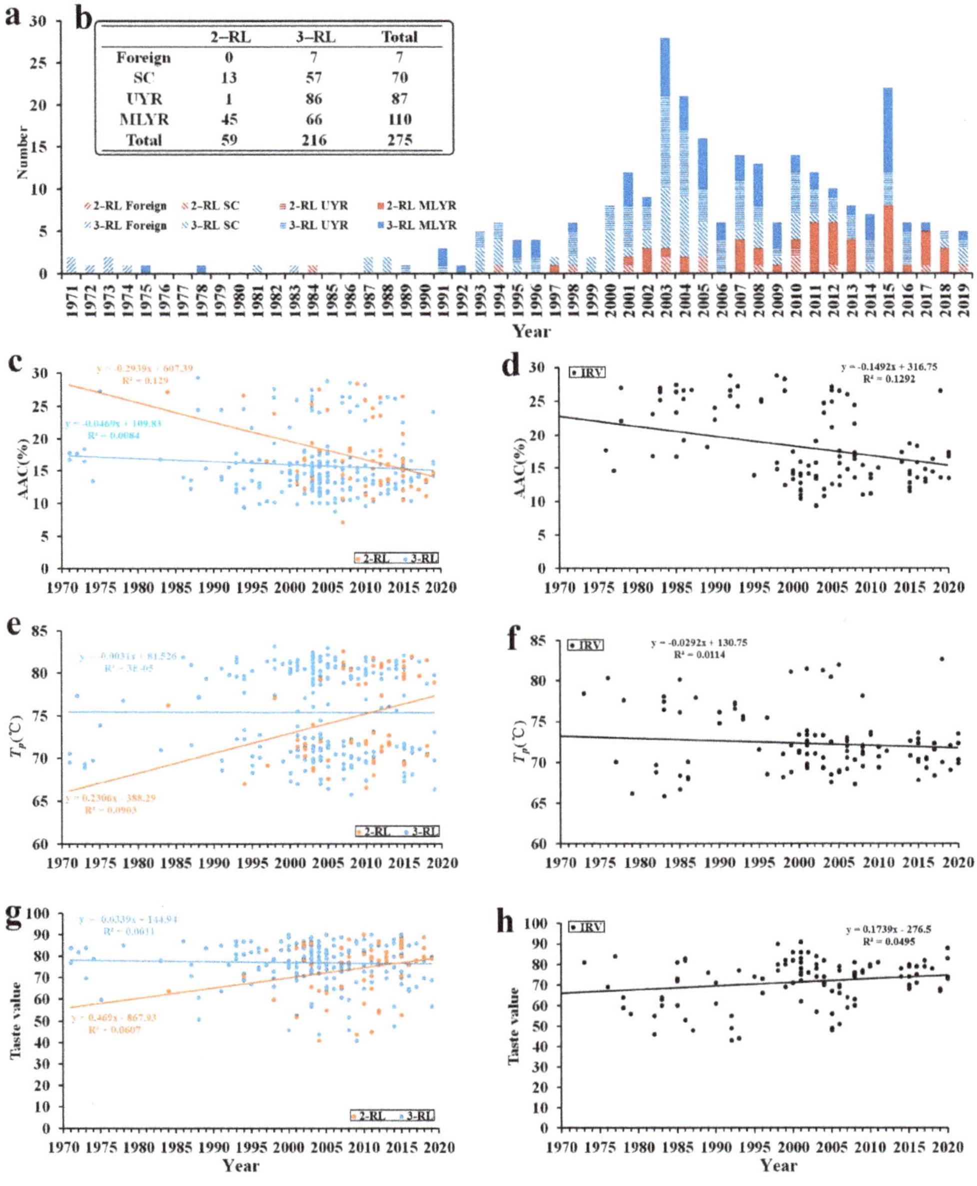

Figure 1. Phenotypic distribution statistics over time. (**a**,**b**) Distributions of materials in terms of breeding time, region and type. (**c**,**e**,**g**) 2-RLs and 3-RLs. (**d**,**f**,**h**) IRVs. (**c**,**d**) Apparent amylose content. (**e**,**f**) Peak temperature of gelatinisation. (**g**,**h**) Taste value.

2.2. ECQ Differentiation with Years and Places

In the past 50 years, none of the indices of RL materials have changed significantly. By contrast, for IRVs, AAC dropped from ~25% in 1970 to ~15% in 2000 (stage 1), and AAC has remained stable at ~15% since (stage 2). Taste value increased over time before the 21st century, and has remained stable since. T_p was high before 2000, and has decreased since. RVA and other indices followed similar trends. There was no significant change in RLs over time, but there was an obvious regular pattern for IRVs (Figures 1c–h and S1). Regarding rice ECQ, this gradually improved over time for IRVs, but remained stable for RLs, although it remained higher for RLs than for IRVs.

According to the classification of the area used for collecting materials, all materials were divided into four groups: Foreign, SC, UYR and MLYR. Since entries for the Foreign group were fewer and less distributed than in other areas after introduction, this group did not receive our focus. In general, T_p and ECQ values for UYR were higher than for other regions. Differences in hardness, stickiness and PKV were identical in the overall trend compared to ECQ (Figure S3). Regarding hardness and SBV, 2-RLs were significantly higher than 3-RLs, and PKV and BDV displayed the opposite phenomena. Except for these differences in indicators, there were no significant differences between 2-RLs and 3-RLs. All ECQ indices were significantly different between IRVs and RLs. In general, IRVs had higher AAC values and lower GT and ECQ values than RLs (Figure S2).

2.3. Allelic Diversification of the Wx Locus in Indica RLs

The data used in this experiment were obtained from the resequencing data of 1143 xian/indica rice lines reported by Lv et al., [11]. *Wx* and *ALK* segments were extracted from the data (physical location: Shuhui 498, R498). R498 genome assembly and annotation can be found at http://www.mbkbase.org/R498 (accessed 8 August 2021). Analysis of the *Wx* haplotype was performed using the chromosome 6 physical interval (1643113–1648065). A total of 33 SNPs were found to be in linkage disequilibrium with index SNPs (Figure 2b). Five haplotypes were identified from the above SNPs, and AAC values for materials corresponding to these haplotypes showed differences, but there were no significant differences between haplotypes 4 and 5 (Figure 2d). According to the analysis of known functional loci, there were no differences between haplotypes 4 and haplotype 5, which indicates that (508) in the intron and (215) in exon C-T changed CCC (proline) to CCT (proline); hence, there was no change in amino acids due to codon degeneracy.

Four *Wx* alleles were detected in the RLs. The predominant *Wx* gene was Wx^b (228), followed by Wx^a (44), and Wx^{lv} and Wx^{in} were also detected in smaller numbers (Figure 2a). The Hap1 material was Wx^{lv}-type, and Hap1 and Wx^{lv} were consistent in functional sites. The AAC of Hap1 was >25%, consistent with Wx^{lv}-type rice. Similarly, Hap2 material was Wx^a, whereas Hap3 material was Wx^{in}. It is worth noting that Hap4 and Hap5 were Wx^b, and these were consistent at functional sites. The AAC of Hap4 and Hap5 was ~15%, consistent with Wx^b-type rice (Figure 2d,e). Additionally, the *Wx* genes of IRV materials were also analysed (Figure S4).

Based on comprehensive analysis of breeding years, materials were divided into seven time periods: <1980, 1980–1989, 1990–1999, 2000–2004, 2005–2009, 2010–2014 and ≥2015 (Figure 3a,d,g). The AAC of Wx^a was ~25%, and for Wx^b, it was ~15%, and the higher the proportion of Wx^a, the greater the increase in AAC. According to the combined genotype and phenotype data, the reason for the decrease in AAC for IRVs before the 2000s was that the allelic proportion of the Wx^b type increased, and the proportion remained stable (~15%) after the 2000s (Figure S7a). The allelic proportion of the Wx^b type of 2-RLs also increased over time, but this was masked by 3-RLs. It should be noted that the early stage of 2-RLs was screened by test crossing with IRVs. The proportion of the Wx^b allele did not differ between areas (Figure S8c). Evidently, the Wx^b-type allele of IRVs was scarcer than that of RLs. Analysis of different types of materials showed that the Wx^a/Wx^b ratio of IRVs and 2-RLs decreased, while the ratio of 3-RLs did not change significantly (Figures 3a,d,g and S7a).

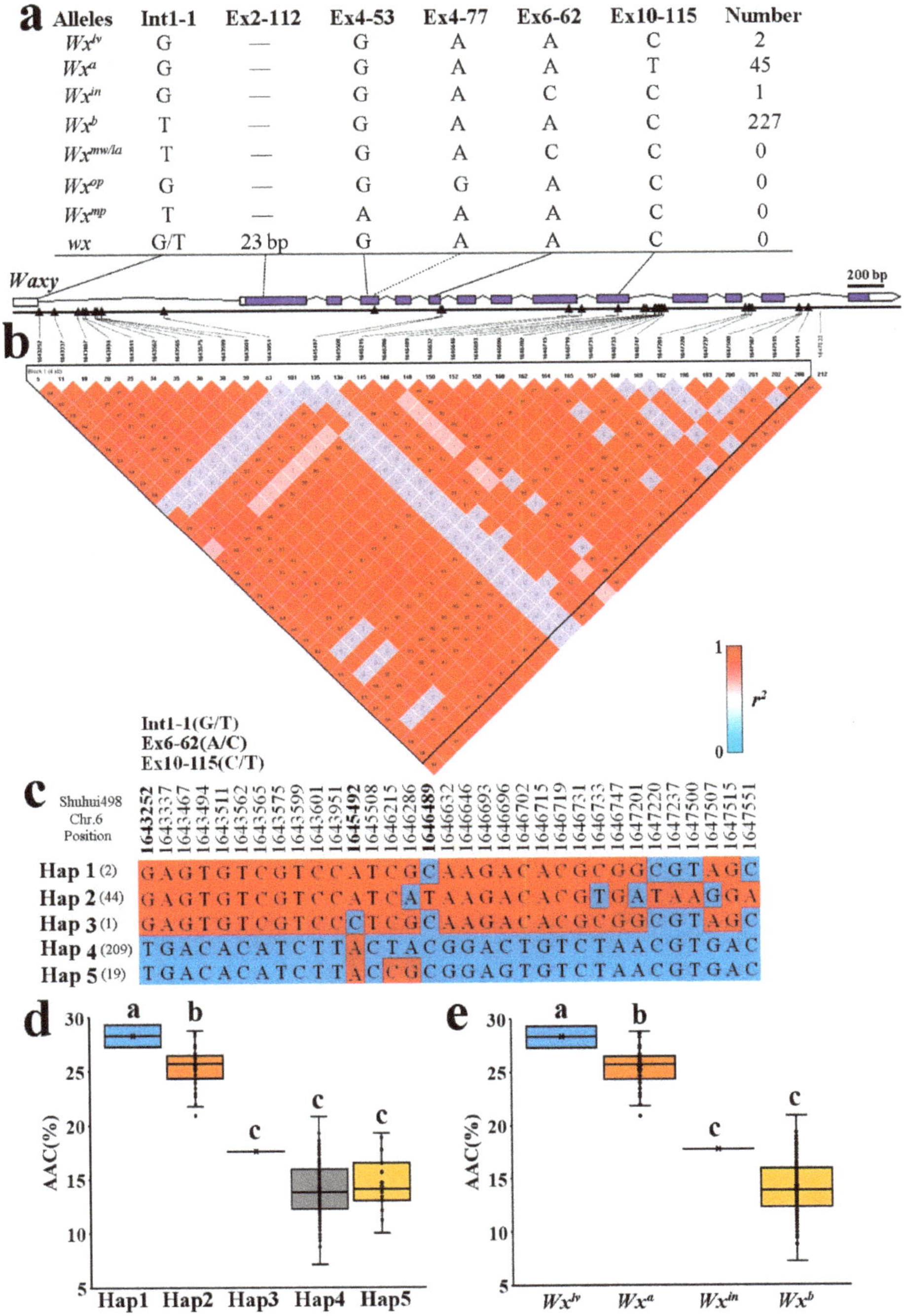

Figure 2. Analysis of different *Wx* alleles and haplotypes. (**a**) *Wx* allele types and numbers screened by KASP marker. (**b**) Linkage disequilibrium (LD) analysis. The r^2 value is reflected on the matrix diagram. (**c**) Results of haplotype analysis. (**d**) Apparent amylose content of different haplotypes. (**e**) Apparent amylose content of different allele types. Different letters indicate significant differences ($p < 0.05$).

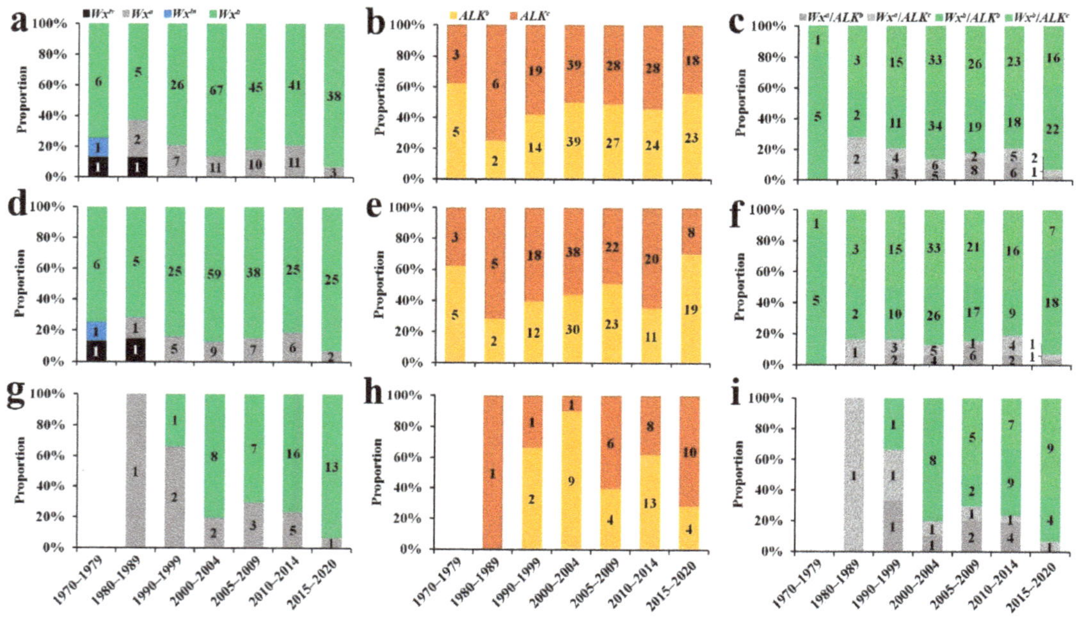

Figure 3. Frequency distribution of different rice materials based on different alleles. (**a**,**d**,**g**) Frequency distribution of different alleles of *Wx*. (**b**,**e**,**h**) Frequency distribution of different alleles of *ALK*. (**a**–**c**) Total restorer lines. (**d**–**f**) 3-RLs. (**g**–**i**) 2-RLs.

2.4. Allelic Diversification of the ALK Locus in Indica RLs

Analysis of *ALK* haplotype was performed using chromosome 6 physical interval 6811279–6816183. A total of 30 SNPs were found to be in linkage disequilibrium with index SNPs (Figure 4b). Five haplotypes were identified by the SNP above, and the T_p values of materials corresponding to these haplotypes showed differences, but it should be noted that the T_p of haplotype 1 and another four haplotypes were significantly different (Figure 4c,d). According to known functional loci analysis, there were differences between haplotypes 1 and the other four haplotypes, but there were no differences between haplotypes 2–5. In addition to 6811957, 6812194, 6815238, 6815273, 6815342 and 6815357, other SNP differences were found in the noncoding region, and except for 6815238 (A–G) encoding a serine to glycine change, all amino acids remained the same. According to known functional loci analysis, there were no differences between haplotypes 2 and 5; there were only differences between haplotype 1 and other haplotypes in functional loci of Ex8-864/865 (Figure 3d).

Additionally, the *ALK* genes of IRV materials were analysed (Figure S5). Two *ALK* alleles were detected in RLs (Figure 3a), most of which were ALK^b (134), followed by ALK^c (141). Haplotypes 2–5 were different in SNPs, but the GT was similar with no significant differences between haplotypes. Relative differences between haplotypes were further explored by cluster analysis of *ALK* genes with rice varieties as controls. A phylogenetic tree and genetic distance analyses showed that all varieties could be divided into three categories: Hap1 and 4, Hap2 and 3 and Hap5. The Hap1 allele clustered with the Hap4 allele, indicating a close genetic relationship between these two alleles, and they were distributed in *XI* and *GJ*. Hap2, Hap3 and Hap5 were distributed in *XI*, and Hap2 and Hap3 were closely genetically related (Figure S6 and Table S3).

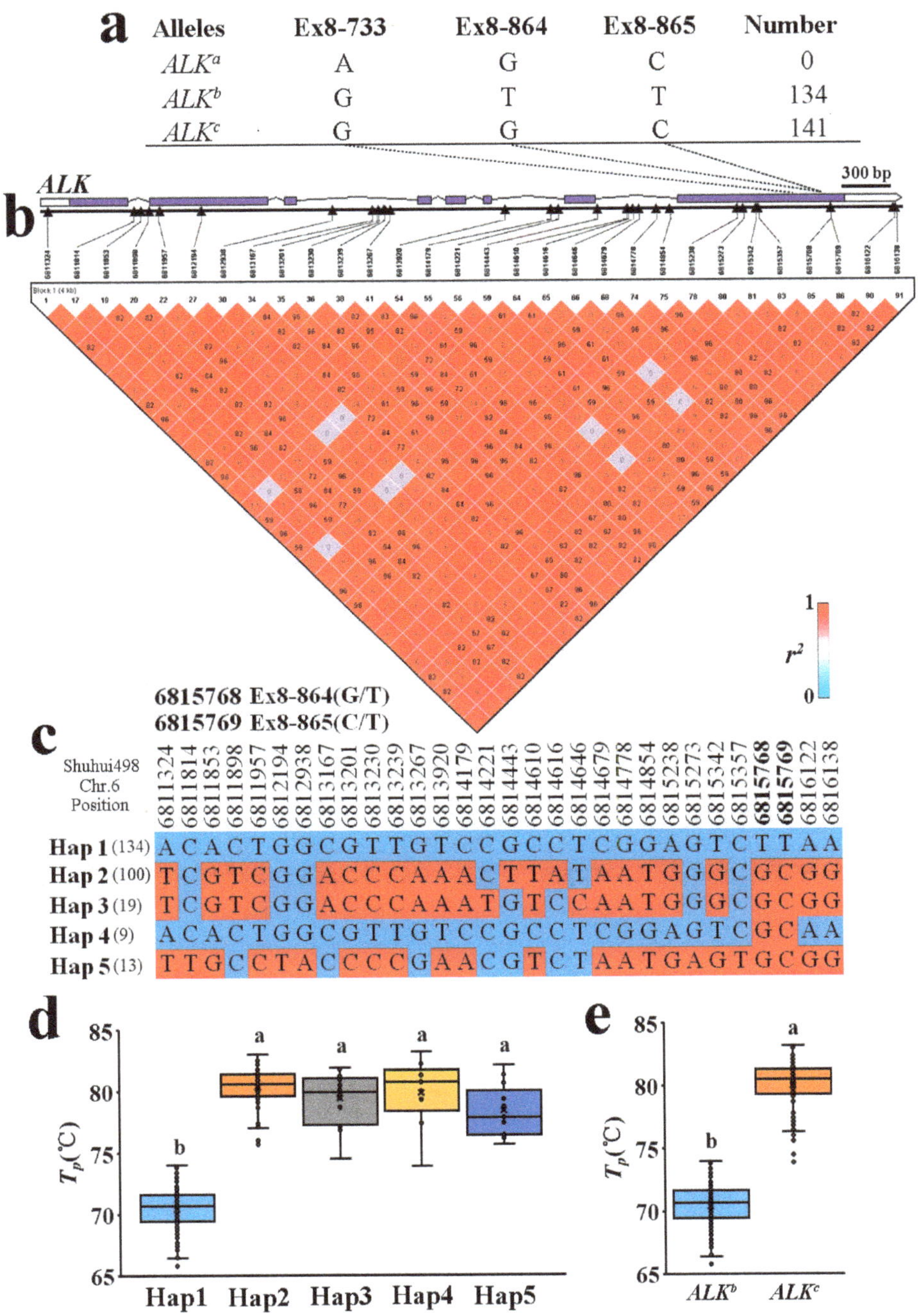

Figure 4. Analysis of different alleles and haplotypes of *ALK*. (**a**) Allele types and number of *ALK*s screened by KASP markers. (**b**) Linkage disequilibrium (LD) analysis. The r^2 value is shown on the matrix diagram. (**c**) Haplotype analysis. (**d**) Peak temperature of gelatinisation of different haplotypes. (**e**) Peak temperature of gelatinisation of different allele types. Different letters indicate significant differences ($p < 0.05$).

Previous studies showed that *ALK* had three major alleles: ALK^a, ALK^b and ALK^c. There were only two types (ALK^b and ALK^c) in the materials studied herein. Hap1 materials were ALK^b, and Hap1 and ALK^b were consistent at functional sites. The T_p of Hap1 was ~70 °C, consistent with the T_p of the ALK^b type. Hap2–5 materials were ALK^c, and they were consistent in functional sites. The T_p of Hap2–5 was ~80 °C, consistent with the T_p of the ALK^c type (Figure 3d,e).

The *ALK* gene was divided into ALK^b and ALK^c types, and ALK^c had a higher GT than ALK^b. As a result, the ALK^b-type rate of 2-RLs and 3-RLs did not change regularly over time; rather, fluctuations were irregular. The allelic proportion of the ALK^b-type rate of IRVs displayed two stages. The first stage (before the year 2000) had a lower proportion (~50%) than the second stage (after 2000), which was ~95% (Figures 3b,e,h and S7b). Regarding area distribution, the allelic proportion of the ALK^b type in the UYR was lower than in SC and the MLYR, which led to T_p in the UYR being significantly higher than that in SC and the MLYR (Figure S8f). The proportion of the ALK^b haplotype was higher in IRVs than RLs (Figures 3b,e,h and S8d,e). For the *ALK* gene, the ALK^b/ALK^c ratio of RLs did not change significantly, but for IRVs, the ratio displayed an increasing trend (Figure S7b).

2.5. Combining Wx and ALK Alleles and Their Utilisation in Indica RLs

Wx and *ALK* are both on the short arm of chromosome 6, with physical positions 164313−164806 and 6811279−6816183, respectively, in Nipponbare. Because *Wx* was mainly composed of Wx^a and Wx^b, the two genes are divided into four types: Wx^a/ALK^b, Wx^a/ALK^c, Wx^b/ALK^b and Wx^b/ALK^c. For different types of materials, we performed analysis of *ALK* and *Wx* gene combinations (Figure S9a), and there were no significant differences between ALK^b and ALK^c in IRVs, 2-RLs or 3-RLs under the Wx^a background. There were no significant differences between ALK^b and ALK^c in RLs under the Wx^b background, but the gene frequency of ALK^b was much higher than that of ALK^c in IRVs. This shows that there was selective pressure on *ALK* under the Wx^b background for IRVs. Previous results show that the Wx^b type has been widely used in rice breeding over the years (Figure S9b). Thus, it was speculated that different alleles of the *ALK* gene were not selected in the early breeding process, but began to be selected in the later stages. To explore this further, the distribution of breeding time under different *Wx* backgrounds was analysed. The results show that Wx^a materials were generally bred in the early stages, while Wx^b materials were generally bred more recently, consistent with the prediction.

2.6. Contributions of Elite Ls in the Development of Hybrid Rice

In this experiment, 275 RLs were selected, including 2-RLs and 3-RLs. Hybrid rice varieties bred using these RLs were collected and analysed. The results show that during this period, 80 RLs were used in hybrid rice breeding, and 1057 varieties were bred (https://www.ricedata.cn/variety/) (accessed 1 January 2022). Moreover, 24 RLs served as male plants in the breeding of >10 hybrid rice varieties (Figure 5a and Table S5). According to the statistical results of the number of hybrid rice varieties bred using RLs, almost all *Wx* genotypes of the elite RLs were Wx^b, while the genotypes ALK^b and ALK^c were almost equal among *ALK* genes (Figure 5a). This shows that from the perspective of RLs used in hybrid rice breeding, elite RLs mainly used Wx^b with a low AC, while Wx^a with a high AC has rarely been used. The two main alleles (ALK^b and ALK^c) of the *ALK* gene regulating GT are not present in RLs used in hybrid rice breeding.

In general, regarding RLs, the selection of functional alleles and the ECQ of the *Wx* gene of elite RLs used in hybrid rice breeding met the requirements of high ECQ. Notably, 132 hybrid rice varieties were bred with Huazhan as RLs, accounting for 12.49% of the total (Figure 5b). Huazhan was bred by the China National Rice Research Institute and Guangdong Academy of Agricultural Sciences. The breeding departments of these varieties were distributed all over China, and suitable planting areas were also widely distributed. Another extremely important RL in the history of hybrid rice breeding is Minghui63, used as a male plant to breed 35 hybrid rice varieties, considerably fewer varieties than Huazhan

(Figure 5a). The possible reason for this is that after 2014, the state successively launched the green channel for variety approval and the consortium test channel, and methods for testing rice varieties diversified, resulting in a sharp increase in the number of rice varieties included in regional tests, and it was ultimately approved. The above results confirm that in the past 5 years of hybrid rice breeding, a few materials have been heavily relied upon; in addition to Huazhan, Wushansimiao (28) and Chenghui727 (24) have been widely applied (Figure 5c). Further details are included in Table S5.

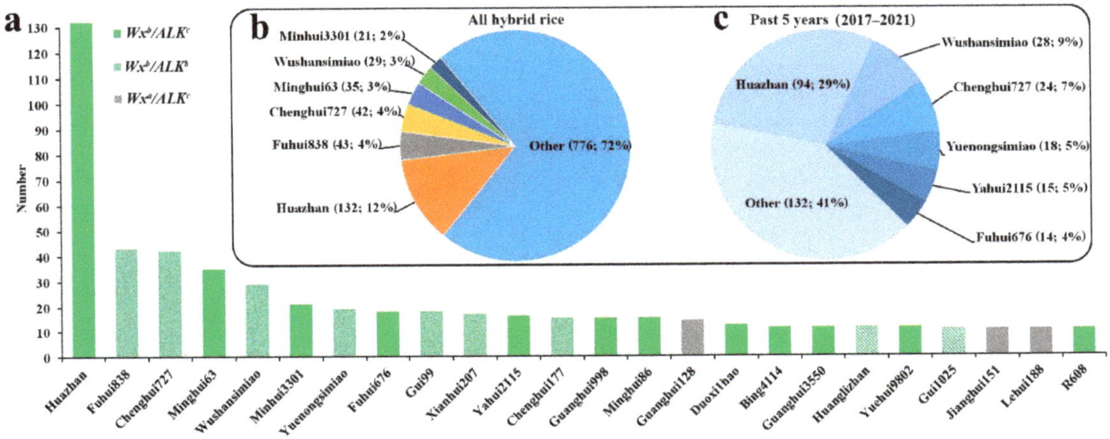

Figure 5. Hybrid rice varieties bred using RLs. (**a**) RLs of male plants used to breed > 10 hybrid rice varieties. (**b**) All hybrid varieties cultivated using the restorer line in this experiment. (**c**) Hybrid varieties cultivated in the past 5 years using the restorer line in this experiment.

3. Discussion

Herein, we assessed 275 rice samples from RLs of foreign, SC, UYR and MLYR origin. Earlier materials were mainly introduced from foreign sources, but a large number of materials were developed in SC, followed by more in the Yangtze River region. The 3-RLs were evenly distributed in different areas, and 2-RLs were mainly distributed in MLYR.

The phenotypes of selected materials were investigated, including AAC, DSC, RVA and taste value, and the results are consistent with the ranges reported in previous studies [14,15,29,38,39]. Correlation analysis of different indices revealed a significant correlation between AAC and taste value ($r = -0.515$, $p < 0.01$), with a significant negative correlation. Specifically, varieties with a higher AAC had higher hardness and a lower taste value [14]. There was no significant correlation between GT and ECQ, consistent with previous studies [15]. Significant correlations were identified between certain RVA profiles and texture characteristics. Therefore, RVA profiles are commonly used to evaluate milled rice ECQ [40,41]. Our experimental results reveal a significant positive correlation between eating value and BDV, and a significant negative correlation with SBV. These results are consistent with previous reports.

In the past 50 years, indices of RL materials have not changed regularly over time. Rather, RLs introduced from foreign sources were mainly Wx^b-type, and this appears to be reflected in ECQ. RLs introduced in the later stages were mainly used in recent breeding, resulting in low AAC. By contrast, obvious regular changes occurred in IRVs. Analysis of AAC of IRVs over this time interval, from the 1970s to the 2000s, showed a decreasing trend with a final decrease to ~15%. After the 2000s, AAC stabilised at 15% without any further decline. Reducing AAC could improve the ECQ of rice. There was no clear standard for ECQ in China before the first standard NY20-1986 (1986), 'quality edible rice', was issued in the 1980s. NY122-86, including ECQ indicators, limited the permissible range of AAC, GC and GT in high-quality rice. There is a significant negative correlation between ECQ and

AAC, since reducing AAC to improve rice ECQ has become a major goal of plant breeders. By contrast, GT has no significant correlation with ECQ, and the GT of different grades of rice is identical among the same rice types of rice, reflecting the fact that GT has not been considered in the breeding process. The Ministry of Agriculture and Rural Affairs of China promulgated new agricultural industry standards in the form of NY/T 593-2013, 'cooking rice variety quality'. This includes GT corresponding to alkali spreading value (ASV) > 6 or GT < 70 °C, which will encourage rice breeders to cultivate varieties with a lower GT.

Wx and *ALK* are the main genes regulating rice ECQ. *Wx* encodes the soluble starch synthase GBSS1, and many allele variations of *Wx* have been identified, including Wx^a, Wx^b, Wx^{in}, Wx^{op}, Wx^{mp} Wx^{lv}, $Wx^{mw/la}$ and *wx*, leading to regional changes in AC that affect consumer preferences [24–29,42,43]. The Wx^b allele type was found to be predominant in the experimental materials, followed by Wx^a, while Wx^{lv} and Wx^{in} were also present. Over the years, the Wx^a/Wx^b ratio of RL materials did not change regularly; rather, Wx^a/Wx^b of 2-RLs and IRVs decreased from the 1970s to the 2000s, then stabilised after then 2000s, consistent with changes in AAC. In accordance with the breeding regions, the Wx^a/Wx^b ratio of UYR was lower than those of SC and MLYR, consistent with AAC, indicating that *Wx* is the main gene regulated by AAC synthesis.

ALK/SSIIA encodes soluble starch synthase IIA (SSIIA) in rice, which plays a specific role in the synthesis of long B1 chains by elongating the short A and B1 chains of amylopectin in the endosperm [15,32,34,44]. SSIIa is the key enzyme controlling GT in rice [45]. Several studies have reported that at least four SNPs of *ALK* are associated with the four alleles in rice (ALK^a, ALK^b, ALK^c and ALK^d) [15,32,33,46,47]. Only two allele types, ALK^b and ALK^c, were detected in the experimental material tested herein. Previous results show that GT was higher for ALK^c and lower for ALK^b, consistent with our current results. ALK^a was generally present in *GJ*, and all rice accessions selected were *XI*, consistent with previous reports [34,48,49]. Over the years, the ALK^c/ALK^b ratio of RLs fluctuated irregularly, and based on area, UYR was significantly higher than others, consistent with the GT data. This indicates that *ALK* is the main gene regulating GT in rice. The ALK^c/ALK^b ratio in IRVs was lower than in 2-RLs and 3-RLs, consistent with T_p values reported in previous work.

The genetic material affecting rice ECQ in hybrid rice comes from RLs and sterile lines, and the two main genes (*Wx* and *ALK*) regulating rice ECQ are separated in F_2 plants, resulting in heterozygous rice eaten by consumers [50]. In theory, the endosperm of hybrid combinations includes four genotypes, $Wx^b/Wx^b/Wx^b$, $Wx^b/Wx^b/Wx^a$, $Wx^b/Wx^a/Wx^a$ and $Wx^a/Wx^a/Wx^a$, and the ratio of genetic separation of these four genotypes is 1:1:1:1 [51]. The AC phenotype in the heterozygous state containing Wx^a tends to be $Wx^a/Wx^a/Wx^a$. If the sterile line is the Wx^a type, even if the RL is Wx^b, 3/4 of the final edible hybrid rice will be high in AC, resulting in hybrid rice with a lower ECQ. *Wx* alleles and AC of sterile lines have been assessed alongside the main mating combinations II-32 A, Jin23 A, Zhenshan97 A, Zhong9 A, Bo A, Long tefu A, Y58A, Xieqing Zao A, Peiai 64S and Gang 46A, the widely used areas Zhenshan 97A, Wei 20A, II-32A, Jin 23A, Gang 46A, Xieqing Zao A, Bo A, Long Tefu A, Peiai 64S and Zhong 9A and corresponding maintainer line materials (https://www.ricedata.cn/variety/) (accessed 1 January 2022). Based on the statistics for sterile lines of females matching RLs in our experiment, the results are consistent with the above website (Figure S9). Almost all sterile lines were Wx^a with a high AC, which may be the main reason for the low ECQ of hybrid rice [17,24]. In addition, *ALK* will also be separated in the F_2 generation, leading to differences in the degree of gelatinisation of hybrid rice during cooking, which further reduces ECQ [35].

There are two types of *XI* hybrid rice, three-line hybrids and two-line hybrids, and the corresponding RLs are 2-RLs and 3-RLs [11]. Hybrid rice bred from RLs were analysed, and the results show that in the breeding process, we relied heavily on several or even a single backbone RL, and many varieties were bred over a short time period; hence, the assimilation of these varieties was rapid. Hybrid rices and their male parents (RLs) that had been employed in the last 5 years were analysed, and of all RLs, only 80 were bred as male parents. It is worth mentioning that many RLs have been eliminated by breeders. Hybrid rice bred as a parent from several varieties accounts for the majority of hybrids developed over the past 5 years. In particular, Huazhan has been used for 94 varieties, accounting for 29% of the total. In the process of hybrid rice breeding, the high-frequency utilisation of a few backbone RLs will decrease the diversity of rice quality traits. Wushansimiao and Chenghui727 have also been heavily relied upon, although not to the same extent as Huazhan. ECQ analysis of RL materials used with high frequency may reveal information on taste value, and almost all varieties tested herein were Wx^b haplotypes.

We need to consider improving the yield and broad-spectrum resistance of hybrid rice in relation to RLs [50]. In addition, we need to pay attention to rice ECQ, which can be enhanced by introducing major genes such as *Wx* alleles [51]. Rice with a low AC is less expansive, fluffy and soft, and hence popular among consumers [52]. Because of its adaptability to almost all climatic situations, many varieties of rice are present worldwide, with differences in ECQ, and a favoured type of rice for one cultural group may not be favoured by others [17]. The rice (*XJ*) variety available in northern and eastern parts of Asia is sticky and soft when cooked. Introducing alleles that regulate the synthesis of AAC and lower it compared with Wx^b, such as Wx^{op}, Wx^{mp} and $Wx^{mw/la}$ can improve ECQ [23,29,30]. Similarly, according to the different requirements for populations in different regions for high-quality rice, we should cultivate targeted rice varieties.

Previous studies and our current results show that *ALK* had no significant effect on taste value, and there was no strict requirement for the GT of high-quality rice, resulting in no selection pressure for *ALK*. However, previous studies showed that the taste value of ALK^c rice after cooling was significantly lower than that of ALK^b rice [51]. The gelatinisation temperature increases the cooking time of rice, increasing the need for fuel, which has a great impact on rice-related food processing. Ultimately, diversifying varieties should provide breeders with more raw materials (parents) for selecting and developing desirable traits.

In summary, we compared RLs and conventional materials employed in the rice breeding process over the past 50 years. The overall ECQ of RLs has remained high and not fluctuated significantly. The main reason for the improvement in ECQ in hybrid rice may be derived from sterile lines of female parents. RLs can use other alleles of *Wx*, such as Wx^{op}, Wx^{mp} and Wx^{mw}, to further reduce AC and improve ECQ, which may further improve hybrid rice ECQ. RLs can increase the utilisation of the ALK^b allele and improve the ECQ of hybrid rice varieties. New results should be compared with those of previous studies, and any findings and their implications should be discussed in the broadest possible context.

4. Materials and Methods

4.1. Plant Materials and Growth Conditions

A total of 387 indica rice accessions provided by the Hunan Hybrid Rice Research Center, Changsha, China, were used in this study (Figure 1a and Table S1). They included 112 varieties of IRVs, 59 2-RLs and 216 3-RLs, representing the parents of the majority of *XI* hybrid rice accessions that have been widely planted in China over the last 50 years, along with some varieties from foreign sources. All cultivation was performed in the same field in Changsha, China. The planting season was from May to November.

4.2. Polished Rice and Flour Preparation

Seeds were de-husked using a rice huller SY88-TH (SSANG YONG Motor, Seoul, Korea) and milled with a grain polisher (Kett, Tokyo, Japan). Intact rice was selected to measure the taste value, and another portion of polished rice samples was ground into flour in a mill (FOSS 1093 Cyclotec Sample Mill; Foss A/S, Hillerød, Denmark) and passed through a 100-mesh sieve. Flour was incubated in a 37 °C oven for 48 h, then in the natural environment for 2 days, and stored at 4 °C. Taste value data included taste value, hardness and stickiness.

4.3. Determination of Grain Physicochemical Properties

AAC of flour was measured by iodine colorimetry. [53] with modifications. GT of flour was determined by DSC (DSC200F3, NETZSCH, Bavaria, Germany) [16]. DSC data included enthalpy of gelatinisation (ΔH), onset temperature of gelatinisation (T_o), peak temperature of gelatinisation (T_p) and terminating temperature of gelatinisation (T_e). Pasting properties were determined by RVA (Newport Scientific, Warriewood, Australia) [54]. RVA data included hot-paste viscosity (HPV), peak paste viscosity (PKV), breakdown value (BDV), cool-paste viscosity (CPV) and setback value (SBV). All tests were performed in triplicate.

4.4. Measurement of Rice Taste Value

A STA1B rice taste meter (Sasaki, Tokyo, Japan) was used to determine the taste values of all materials. ECQ mainly reflects the taste value, hardness and stickiness of rice. All tests were performed in triplicate.

4.5. Sequencing and Genotype Annotation

DNA sequencing and SNP calling for all rice accessions were performed as described previously [11]. Tag SNPs were then identified from each clump using De Bakker's algorithm implemented in Haploview [55]. Linkage disequilibrium (LD) blocking these tag SNPs was then assessed using Gabriel's algorithm [56] and visualised using Haploview [42]. SNPs within a clump were selected based on $p < 0.01$ and an LD value of $r^2 > 0.5$ with index SNPs.

4.6. Detection of *Wx* and *ALK* Alleles

Genomic DNA was extracted from fresh leaves of all parental lines using a modified CTAB method. *Wx* and *ALK* allelic variations were analyzed by KASP genotyping [38]. The details of allele-specific primers are shown in Supplementary Table S4.

4.7. Statistical Analysis

DNA sequences of *ALK* genes of different materials were obtained, and homologous alignment was performed using the NCBI website (http://www.ncbi.nlm.nih.gov/) (accessed 1 January 2022). Nucleotide sequence similarity and multiple sequence alignment analyses were carried out using ClustalX and GeneDoc. A phylogenetic tree was constructed by the neighbour-joining (NJ) method with the Kimura 2-parameter (K2-P) model using MEGA version 5.0. Experiments were carried out in triplicate, and results are reported as mean values ± standard deviations (SD). One-way analysis of variance (ANOVA) and Tukey's multiple comparison test were used to determine significant differences among mean values using SPSS 16.0 statistical software (IBM, Armonk, NY, USA).

Supplementary Materials: The following supporting information can be downloaded at: https://www.mdpi.com/article/10.3390/ijms23115941/s1.

Author Contributions: L.-X.P. performed experiments, analysed data and wrote the manuscript; Z.-Z.S. and Q.-M.L. performed rice cultivation experiments; D.-Y.Y. and Q.-Q.L. designed experiments and edited the manuscript; B.L. and F.C. performed experiments; Q.-Q.Y. performed the KASP analysis; D.-S.Z., X.-L.F. and C.-Q.Z. reviewed the manuscript. All authors have read and agreed to the published version of the manuscript.

Funding: This work was supported by the National Science Foundation of China (31825019, and U19A2032), Hainan Yazhou Bay Seed Lab (B21HJ8105), the programs from Jiangsu Province Government (JBGS [2021]001, BZ2021017, and PAPD) and Postgraduate Research & Practice Innovation Program of Jiangsu Province (XKYCX18_078).

Institutional Review Board Statement: Not applicable.

Informed Consent Statement: Not applicable.

Data Availability Statement: Datasets supporting the conclusions of this article are included within the article (and its additional files).

Conflicts of Interest: The authors declare no conflict of interest.

Abbreviations

ECQ	Eating and cooking quality
RL	Restorer line
IRV	Inbred rice variety
AC	Amylose content
GT	Gelatinisation temperature
DSC	Differential scanning calorimetry
RVA	Rapid visco analyzer
T_p	Peak temperature of gelatinisation
SNP	Single-nucleotide polymorphism
KASP	Kompetitive allele-specific PCR
BDV	Breakdown viscosity
SBV	Setback viscosity
SC	South China
MLYR	Middle and lower reaches of the Yangtze River
UYR	Upper reaches of Yangtze River

References

1. Zeng, D.; Tian, Z.; Rao, Y.; Dong, G.; Yang, Y.; Huang, L.; Leng, Y.; Xu, J.; Sun, C.; Zhang, G.; et al. Rational design of high-yield and superior-quality rice. *Nat. Plants* **2017**, *3*, 17031. [CrossRef] [PubMed]
2. Angira, B.; Addison, C.K.; Cerioli, T.; Rebong, D.B.; Wang, D.R.; Pumplin, N.; Ham, J.H.; Oard, J.H.; Linscombe, S.D.; Famoso, A.N. Haplotype characterization of the *sd1* semidwarf gene in United States rice. *Plant Genome* **2019**, *12*, 190010. [CrossRef] [PubMed]
3. Ashikari, M.; Sakakibara, H.; Lin, S.; Yamamoto, T.; Takashi, T.; Nishimura, A.; Angeles, E.R.; Qian, Q.; Kitano, H.; Matsuoka, M. Cytokinin oxidase regulates rice grain production. *Science* **2005**, *309*, 741–745. [CrossRef] [PubMed]
4. Yuan, L.-P. Strategic hypotheses of breeding for hybrid rice. *Hybrid Rice* **1987**, *1*, 1–3.
5. Cheng, H.; Zhuang, Y.; Fan, Y.; Du, H.; Cao, Y. Progress in research and development on hybrid rice: A super-domesticate in China. *Ann. Bot.* **2007**, *100*, 959–966. [CrossRef]
6. Peng, S.; Khush, G.; Virk, P.; Tang, Q.; Zou, Y. Progress in ideotype breeding to increase rice yield potential. *Field Crop. Res.* **2008**, *108*, 32–38. [CrossRef]
7. Chen, L.-T.; Liu, Y.-G. Male sterility and fertility restoration in crops. *Annu. Rev. Plant Biol.* **2014**, *65*, 579–606. [CrossRef]
8. Huang, X.; Yang, S.; Gong, J.; Zhao, Q.; Feng, Q.; Zhan, Q.; Zhao, Y.; Li, W.; Cheng, B.; Xia, J.; et al. Genomic architecture of heterosis for yield traits in rice. *Nature* **2016**, *537*, 629–633. [CrossRef]
9. Chen, E.; Huang, X.; Tian, Z.; Wing, R.-A.; Han, B. The genomics of *Oryza* species provides insights into rice domestication and heterosis. *Annu. Rev. Plant Biol.* **2019**, *70*, 639–665. [CrossRef]
10. Xie, Y.; Shen, R.; Chen, L.; Liu, Y.-G. Molecular mechanisms of hybrid sterility in rice. *Sci. China Life Sci.* **2019**, *62*, 737–743. [CrossRef]
11. Lv, Q.; Li, W.; Sun, Z.; Ouyang, N.; Jing, X.; He, Q.; Wu, J.; Zheng, J.; Zheng, J.; Tang, S.; et al. Resequencing of 1,143 *indica* rice accessions reveals important genetic variations and different heterosis patterns. *Nat. Commun.* **2020**, *11*, 4778. [CrossRef] [PubMed]
12. Xiao, Y. Corrigendum: Correlations between parental lines and *indica* hybrid rice in terms of eating quality traits. *Front. Nutr.* **2021**, *8*, 663504.
13. Peng, Y.; Mao, B.; Zhang, C.; Shao, Y.; Wu, T.; Hu, L.; Hu, Y.; Tang, L.; Li, Y.; Tang, W.; et al. Influence of physicochemical properties and starch fine structure on the eating quality of hybrid rice with similar apparent amylose content. *Food Chem.* **2021**, *353*, 129461. [CrossRef] [PubMed]
14. Tian, Z.; Qian, Q.; Liu, Q.; Yan, M.; Liu, X.; Yan, C.; Liu, G.; Gao, Z.; Tang, S.; Zeng, D.; et al. Allelic diversities in rice starch biosynthesis lead to a diverse array of rice eating and cooking qualities. *Proc. Natl. Acad. Sci. USA* **2009**, *106*, 21760–21765. [CrossRef] [PubMed]
15. Zhang, C.; Yang, Y.; Chen, Z.; Chen, F.; Pan, L.; Lu, Y.; Li, Q.; Fan, X.; Sun, Z.; Liu, Q. Characteristics of grain physicochemical properties and the starch structure in rice carrying a mutated *ALK/SSIIa* gene. *J. Agric. Food Chem.* **2020**, *68*, 13950–13959. [CrossRef]

16. Zhang, C.; Chen, S.; Ren, X.; Lu, Y.; Liu, D.; Cai, X.; Li, Q.; Gao, J.; Liu, Q. Molecular structure and physicochemical properties of starches from rice with different amylose contents resulting from modification of OsGBSSI activity. *J. Agric. Food Chem.* **2017**, *65*, 2222–2232. [CrossRef]
17. Suwannaporn, P.; Linnemann, A.R. Rice-eating quality among consumers in different rice grain preference countries. *J. Sens. Stud.* **2008**, *23*, 1–13. [CrossRef]
18. Zhao, J. Comparison of Grain Yield and Quality between Hybrid Rice and Inbred Rice in China. *Hybrid Rice* **2008**, *23*, 1–4.
19. Ball, S.G.; Morell, M.K. From bacterial glycogen to starch: Understanding the biogenesis of the plant starch granule. *Annu. Rev. Plant Biol.* **2003**, *54*, 207–233. [CrossRef]
20. Nakamura, Y. Towards a better understanding of the metabolic system for amylopectin biosynthesis in plants: Rice endosperm as a model tissue. *Plant Cell Physiol.* **2002**, *43*, 718–725. [CrossRef]
21. Smith, A.M.; Denyer, K.; Martin, C. The synthesis of the starch granule. *Annu. Rev. Plant Phys.* **1997**, *48*, 67–87. [CrossRef] [PubMed]
22. Zeeman, S.; Kossmann, J.; Smith, A. Starch: Its metabolism, evolution, and biotechnological modification in plants. *Annu. Rev. Plant Biol.* **2010**, *61*, 209–234. [CrossRef] [PubMed]
23. Nakata, M.; Miyashita, T.; Kimura, R.; Nakata, Y.; Takagi, H.; Kuroda, M.; Yamaguchi, T.; Umemoto, T.; Yamakawa, H. MutMapPlus identified novel mutant alleles of a rice starch branching enzyme *IIb* gene for fine-tuning of cooked rice texture. *Plant Biotechnol. J.* **2018**, *16*, 111–123. [CrossRef]
24. Zhang, C.; Zhu, J.; Chen, S.; Fan, X.; Li, Q.; Lu, Y.; Wang, M.; Yu, H.; Yi, C.; Tang, S.; et al. *Wx(lv)*, the ancestral allele of rice *Waxy* gene. *Mol. Plant* **2019**, *12*, 1157–1166. [CrossRef] [PubMed]
25. Liu, L.; Ma, X.; Liu, S.; Zhu, C.; Jiang, L.; Wang, Y.; Shen, Y.; Ren, Y.; Dong, H.; Chen, L.; et al. Identification and characterization of a novel *Waxy* allele from a Yunnan rice landrace. *Plant Mol. Biol.* **2009**, *71*, 609–626. [CrossRef] [PubMed]
26. Mikami, I.; Aikawa, M.; Hirano, H.; Sano, Y. Altered tissue-specific expression at the *Wx* gene of the opaque mutants in rice. *Euphytica* **1999**, *105*, 91–97. [CrossRef]
27. Mikami, I.; Dung, L.; Hirano, H.; Sano, Y. Effects of the two most common *Wx* alleles on different genetic backgrounds in rice. *Plant Breed.* **2000**, *119*, 505–508. [CrossRef]
28. Mikami, I.; Uwatoko, N.; Ikeda, Y.; Yamaguchi, J.; Hirano, H.Y.; Suzuki, Y.; Sano, Y. Allelic diversification at the *wx* locus in landraces of Asian rice. *Theor. Appl. Genet.* **2008**, *116*, 979–989. [CrossRef]
29. Zhang, C.; Yang, Y.; Chen, S.; Liu, X.; Zhu, J.; Zhou, L.; Lu, Y.; Li, Q.; Fan, X.; Tang, S.; et al. A rare *Waxy* allele coordinately improves rice eating and cooking quality and grain transparency. *J. Integr. Plant Biol.* **2021**, *63*, 889–901. [CrossRef]
30. Huang, L.; Sreenivasulu, N.; Liu, Q. *Waxy* editing: Old meets new. *Trends Plant Sci.* **2020**, *25*, 963–966. [CrossRef]
31. Gao, Z.; Zeng, D.; Cheng, F.; Tian, Z.; Guo, L.; Su, Y.; Yan, M.; Jiang, H.; Dong, G.; Huang, Y.; et al. *ALK*, the key gene for gelatinization temperature, is a modifier gene for gel consistency in rice. *J. Integr. Plant Biol.* **2011**, *53*, 756–765. [PubMed]
32. Bao, J.-S.; Corke, H.; Sun, M. Nucleotide diversity in starch synthase IIa and validation of single nucleotide polymorphisms in relation to starch gelatinization temperature and other physicochemical properties in rice (*Oryza sativa* L.). *Theor. Appl. Genet.* **2006**, *113*, 1171–1183. [CrossRef] [PubMed]
33. Waters, D.; Henry, R.; Reinke, R.; Fitzgerald, M. Gelatinization temperature of rice explained by polymorphisms in starch synthase. *Plant Biotechnol. J.* **2006**, *4*, 115–122. [CrossRef] [PubMed]
34. Chen, Z.; Lu, Y.; Feng, L.; Hao, W.; Li, C.; Yang, Y.; Fan, X.; Li, Q.; Zhang, C.; Liu, Q. Genetic dissection and functional differentiation of ALK^a and ALK^b, two natural alleles of the *ALK/SSIIa* gene, responding to low gelatinization temperature in rice. *Rice* **2020**, *13*, 39. [CrossRef]
35. You, L.; Yu, L.; Liang, R.; Sun, R.; Hu, F.; Lu, X.; Zhao, J. Identification and analysis of genes involved in double fertilization in rice. *Int. J. Mol. Sci.* **2021**, *22*, 12850. [CrossRef]
36. Friedman, W.E. Comparative embryology of basal angiosperms. *Curr. Opin. Plant Biol.* **2001**, *4*, 14–20. [CrossRef]
37. Bleckmann, A.; Alter, S.; Dresselhaus, T. The beginning of a seed: Regulatory mechanisms of double fertilization. *Front. Plant Sci.* **2014**, *5*, 452. [CrossRef]
38. Shao, Y.; Peng, Y.; Mao, B.; Lv, Q.; Yuan, D.; Liu, X.; Zhao, B. Allelic variations of the *Wx* locus in cultivated rice and their use in the development of hybrid rice in China. *PLoS ONE* **2020**, *15*, e0232279. [CrossRef]
39. Zhang, L.; Zhang, C.; Yan, Y.; Hu, Z.; Wang, K.; Zhou, J.; Zhou, Y.; Cao, L.; Wu, S. Influence of starch fine structure and storage proteins on the eating quality of rice varieties with similar amylose contents. *J Sci Food Agric.* **2021**, *101*, 811–3818. [CrossRef]
40. Chen, H.; Chen, D.; He, L.-H.; Wang, T.; Lu, H.; Yang, F.; Deng, F.; Chen, Y.; Tao, Y.-F.; Li, M.; et al. Correlation of taste values with chemical compositions and Rapid Visco Analyser profiles of 36 *indica* rice (*Oryza sativa* L.) varieties. *Food Chem.* **2021**, *349*, 129176. [CrossRef]
41. Cai, X.-L.; Wang, Z.-Y.; Xing, Y.-Y.; Zhang, J.-L.; Hong, M.-M. Aberrant splicing of intron 1 leads to the heterogeneous 5′ UTR and decreased expression of *Waxy* gene in rice cultivars of intermediate amylose content. *Plant J.* **1998**, *14*, 459–465. [CrossRef] [PubMed]
42. Ball, S.; Guan, H.P.; James, M.; Myers, A.; Keeling, P.; Mouille, G.; Buléon, A.; Colonna, P.; Preiss, J. From glycogen to amylopectin: A model for the biogenesis of the plant starch granule. *Cell* **1996**, *86*, 349–352. [CrossRef]
43. Jiang, H.; Dian, W.; Liu, F.; Wu, P. Molecular cloning and expression analysis of three genes encoding starch synthase II in rice. *Planta* **2004**, *218*, 1062–1070. [CrossRef] [PubMed]

44. Nakamura, Y.; Francisco, P.B.; Hosaka, Y.; Sato, A.; Sawada, T.; Kubo, A.; Fujita, N. Essential amino acids of starch synthase IIa differentiate amylopectin structure and starch quality between *japonica* and *indica* rice varieties. *Plant Mol. Biol.* **2005**, *58*, 213–227. [CrossRef] [PubMed]
45. Umemoto, T.; Horibata, T.; Aoki, N.; Hiratsuka, M.; Yano, M.; Inouchi, N. Effects of variations in starch synthase on starch properties and eating quality of rice. *Plant Prod. Sci.* **2008**, *11*, 472–480. [CrossRef]
46. Li, C.; Cao, P.; Wu, P.; Yu, W.; Gilbert, R.G.; Li, E. Effects of endogenous proteins on rice digestion during small intestine (in vitro) digestion. *Food Chem.* **2020**, *344*, 128687. [CrossRef]
47. Butardo, V.M., Jr.; Luo, J.; Li, Z.; Gidley, M.J.; Bird, A.R.; Tetlow, I.J.; Fitzgerald, M.; Jobling, S.A.; Rahman, S. Functional genomic validation of the roles of soluble starch synthase IIa in *Japonica* rice endosperm. *Front. Genet.* **2020**, *11*, 289. [CrossRef]
48. Liu, J.; Li, M.; Zhang, Q.; Wei, X.; Huang, X. Exploring the molecular basis of heterosis for plant breeding. *J. Integr. Plant Biol.* **2020**, *62*, 287–298. [CrossRef]
49. Kim, Y.-J.; Zhang, D. Molecular control of male fertility for crop hybrid breeding. *Trends Plant Sci.* **2018**, *23*, 53–65. [CrossRef]
50. Wu, B.; Hu, W.; Xing, Y.Z. The history and prospect of rice genetic breeding in China. *Hereditas* **2018**, *40*, 841–857.
51. Zhang, C.; Feng, L.; Gu, M.; Liu, Q. Progress on inheritance and gene cloning for rice grain quality in Jiangsu province. *Hereditas* **2021**, *43*, 425–441.
52. Yao, S.; Zhang, Y.; Liu, Y.; Zhao, C.; Zhou, L.; Chen, T.; Zhao, Q.Y.; Pillay, B.; Wang, C. Effects of soluble starch synthase genes on eating and cooking quality in semi *Waxy* japonica rice with Wx^{mp}. *Food Prod. Process. Nutr.* **2020**, *2*, 22. [CrossRef]
53. Juliano, B.O.; Perez, C.M.; Blakeney, A.B.; Castillo, T.; Kongseree, N.; Laignelet, B.; Lapis, E.T.; Murty, V.V.S.; Paule, C.M.; Webb, B.D. International cooperative testing on the amylose content of milled rice. *Starch—Stärke* **1981**, *33*, 157–162. [CrossRef]
54. Tao, K.; Li, C.; Yu, W.; Gilbert, R.G.; Li, E. How amylose molecular fine structure of rice starch affects functional properties. *Carbohydr. Polym.* **2019**, *204*, 24–31. [CrossRef] [PubMed]
55. Barrett, J.C.; Fry, B.; Maller, J.; Daly, M.J. Haploview: Analysis and visualization of LD and haplotype maps. *Bioinformatics* **2005**, *21*, 263–265. [CrossRef] [PubMed]
56. Gabriel, S.B.; Schaffner, S.F.; Nguyen, H.; Moore, J.M.; Roy, J.; Blumenstiel, B.; Higgins, J.; DeFelice, M.; Lochner, A.; Faggart, M.; et al. The structure of haplotype blocks in the Human genome. *Science* **2002**, *296*, 2225. [CrossRef] [PubMed]

International Journal of *Molecular Sciences*

MDPI

Article

Development of Soft Rice Lines by Regulating Amylose Content via Editing the 5′UTR of the *Wx* Gene

Jinlian Yang, Xinying Guo, Xuan Wang, Yaoyu Fang, Fang Liu, Baoxiang Qin and Rongbai Li *

State Key Laboratory for Conservation and Utilization of Subtropical Agro-Bioresources, College of Agriculture, Guangxi University, Nanning 530004, China
* Correspondence: rongbaili@gxu.edu.cn

Abstract: The type of soft rice with low amylose content (AC) is more and more favored by consumers for its better eating and cooking quality, as people's quality of life continuously improves in China. The *Wx* gene regulates the AC of rice grains, thus affecting the degree of softness of the rice. Mei Meng B (MMB), Tian Kang B (TKB), and DR462 are three indica rice maintained lines with good morphological characters, but also with undesirably high AC. Therefore, CRISPR/Cas9 technology was used to edit the *Wx* gene of these lines to create a batch of soft rice breeding materials. New gene-edited lines MMB-10-2, TKB-21-12, and DR462-9-9, derived from the above parental lines, respectively, were selected in the T_2 generations, with an AC of 17.2%, 16.8%, and 17.8%, and gel consistency (GC) of 78.6 mm, 77.4 mm, and 79.6 mm, respectively. The rapid viscosity analysis (RVA) spectrum showed that the three edited lines had a better eating quality as compared to the corresponding wild type, and showing new characteristics, different from the high-quality soft rice popular in the market. There was no significant difference in the main agronomic traits in the three edited lines compared to the corresponding wild types. Moreover, the chalkiness of DR462-9-9 was reduced, resulting in an improved appearance of its polished rice. The present study created soft rice germplasms for breeding improved quality hybrid rice, without changing the excellent traits of their corresponding wild type varieties.

Keywords: rice; CRISPR/Cas9; *Wx* gene; eating and cooking quality (ECQ); amylose content

check for updates

Citation: Yang, J.; Guo, X.; Wang, X.; Fang, Y.; Liu, F.; Qin, B.; Li, R. Development of Soft Rice Lines by Regulating Amylose Content via Editing the 5′UTR of the *Wx* Gene. *Int. J. Mol. Sci.* **2022**, *23*, 10517. https://doi.org/10.3390/ijms231810517

Academic Editors: Jinsong Bao and Jianhong Xu

Received: 15 August 2022
Accepted: 6 September 2022
Published: 10 September 2022

Publisher's Note: MDPI stays neutral with regard to jurisdictional claims in published maps and institutional affiliations.

Copyright: © 2022 by the authors. Licensee MDPI, Basel, Switzerland. This article is an open access article distributed under the terms and conditions of the Creative Commons Attribution (CC BY) license (https://creativecommons.org/licenses/by/4.0/).

1. Introduction

Along with the improvement of our national living standards, consumers have higher requirements for the eating and cooking quality (ECQ) of rice. Rice with low AC has the advantages of palatability, good puffing properties, softness, elasticity, a low degree of retrogradation and hardening after cooling, and does not rot easily [1]. As the main component of the rice endosperm, AC is an important index for evaluating the ECQ of rice, which determines the taste and cooking properties of rice [2,3]. Using their AC, rice varieties can be divided into five categories: high (25~33%), medium (20~25%), low (12~20%), very low (5~12%), and waxy (0~5%) [4]. Countries and regions have different cultural favor for rice categories; therefore, it is necessary to develop multiple types of rice.

Since the 1980s, Japan has taken the lead in cultivating rice varieties with low amylose content [5]. In recent years, attention was gradually given to the development of cultivated rice varieties with low AC in rice breeding research in China, and a number of studies on the genetics, agronomy, processing quality and other aspects have been carried out in this regard [6–8]. A series of soft rice japonica varieties such as Yunjing 20, Yunjing 29, Yunjing 37, and Yunjing 41 bred in Yunnan; Huruan 1212 and songxiangjing1018 bred in Shanghai; Nanjing 46, Nanjing 9108, Fengjing 1606, and Wuxiangjing 113 bred in Jiangsu; and Jia 58 bred in Zhejiang have been popularized [9–11]. However, at present, the germplasm resources of rice with low AC that can be used as restorers and maintainers are still limited.

The waxy gene (*Wx*) encodes granule-bound starch synthase I (GBSSI), which is the key gene regulating amylose synthesis. The *Wx* gene directly affects AC in rice endosperm, thus affecting the degree of softness of rice. The CRISPR/Cas9 gene editing system has the advantages of high efficiency and easy operation and has been widely used in crop genetic improvement and breeding [12]. Previous studies have shown that the CRISPR/Cas9 system was used in the research on the *Wx* allele in rice and a series of germplasms were created for breeding [8,13–15]. Most researches focused on the editing of coding sequences to eliminate *Wx* expression, in order to generate glutinous rice. Teng (2021) used the CRISPR/Cas9 system to mediate the editing of the *Wx* gene in a Photothermosensitive Genic-Male-Sterile (PTGMS) line Y58S, which caused ultra-low AC mutations that produced a PTGMS glutinous rice strain with excellent waxiness [16]. Fu (2022) developed a rapid and highly efficient strategy through the CRISPR/Cas9 gene-editing system for generating *Wx* mutants from the background of five different rice varieties, which significantly reduced the AC and starch viscosity but did not affect the major agronomic traits [17]. Liu (2022) performed the targeted deletion of the first intron of the Wx^b allele via CRISPR/Cas9. The grain AC of mutant lines significantly increased from 13.0% to 24.0% [15].

In addition to creating glutinous rice, Huang (2020) generated novel Wx alleles by ed-iting the region of the Wxb promoter, in turn fine-tuning of Wx expression [3]. In this study, Meimeng B (MMB), Tiankang B (TKB), and DR462, which are high-quality indica rice parent materials commonly used in breeding, but with a hard rice quality, were used as test materials. They have high a yield performance but also high AC and low GC, resulting in poorer rice quality. It is of great significance to improve the quality of rice and increase the value of breeding and utilization by appropriately changing the AC and GC contents of these lines. However, the direct editing of the coding region of the *Wx* gene by using CRISPR/Cas9 often results in too low an AC content in edited lines, which directly results in glutinous rice lines. This does not meet the AC content requirement for soft rice. Therefore, in this study, by using CRISPR/Cas9 to edit the intron splice site (5′UISS) in the 5′UTR region of the *Wx* gene, the expression of the *Wx* gene was appropriately reduced, but the original protein coding sequence of the *Wx* gene was not changed. Thus, rice gene editing lines with lower AC and GC can be obtained, which can provide high-quality resources for the improvement of new rice varieties.

2. Results

2.1. Construction of CRISPR/Cas9-Wx Vector

According to the sequence of the parent *Wx* gene (*LOC_Os06g04200*), the CRISPR/Cas9 vector was constructed by targeting the intron splicing site (5′UISS) in the 5′UTR region of *Wx* (Figure 1A). As confirmed by sequencing and alignment, the sequences of target sites in MMB, TKB, and DR462 were consistent with the designed sequences (Figure 1B), indicating that MMB, TKB, and DR462 were suitable for editing.

A U6a-sgRNA expression cassette of the *Wx* gene was obtained through overlapping PCR, and the amplification product of the target construction fragment was 629 bp in length (Figure 2A). The constructed Cas9/sgRNA expression vector was transformed, cultured, and a single colony was selected. The primer PB-L/PB-R were used for PCR amplification and sequencing. The gel electrophoresis results showed that the *Wx* gene Cas9/sgRNA expression vector was present in 16 out of 20 colonies (Figure 2B), indicating that the CRISPR/Cas9-*Wx* expression vector was successfully constructed.

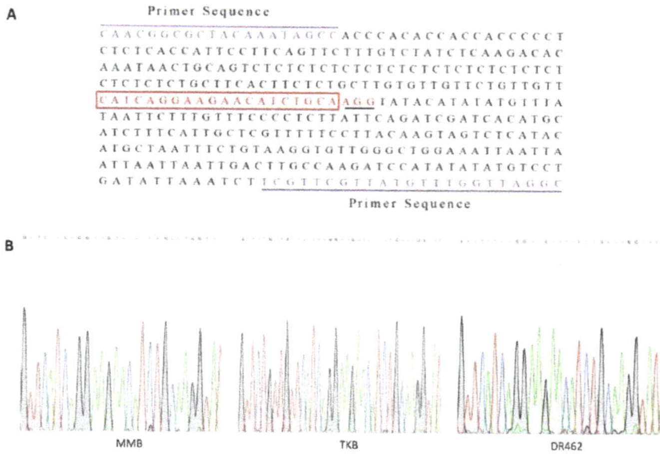

Figure 1. Location of sgRNA in the *Wx* gene and designed primers for amplification. (**A**): *Wx* gene partial sequence showing target positions and primer sequences; purple nucleotides: Upstream and downstream primers; Red Box: sgRNA sequences; black underline nucleotides: protospacer adjacent motif (PAM) region; (**B**): Sequencing peak map of sgRNA.

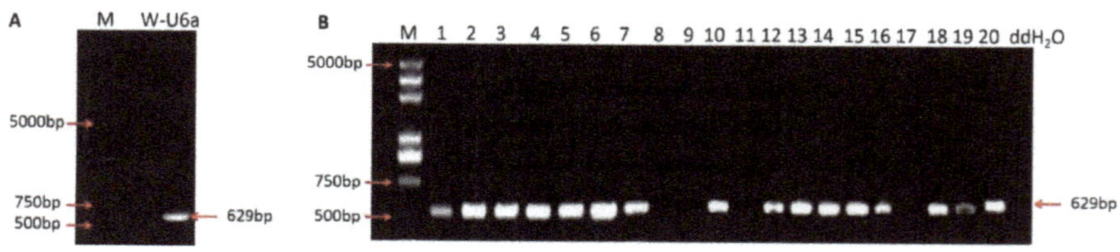

Figure 2. Construction of vector and verification of sgRNA. (**A**): Detection results of sgRNA expression cassette; (**B**) Verification of the size of *Wx* gene cas9/sgRNA expression vector segment; M: DL 5000 DNA marker; W-u6a: sgRNA-u6a fragment; ddH_2O is a negative control based on sterile water; 1–20 are the numbers of the positive clone bacterial solution picked at random; Lane 8, 9, 11, and 17 are empty plasmids; 5000 bp, 750 bp, 500 bp, 629 bp: fragment size.

2.2. Genetic Transformation of MMB, TKB, and DR462

The vector was introduced into MMB, TKB, and DR462 rice through *agrobacterium*-mediated transformation. Calli were induced from mature embryos of rice seeds after co-culture (Figure 3A). After two rounds of hygromycin resistance screening (Figure 3B–D), the well-growing calli were selected for plantlet differentiation and rooting (Figure 3E,F). As a result, a total of 30, 33, and 34 plants from MMB, TKB, and DR462, respectively, were obtained, which were regarded as the T_0 generation.

2.3. Mutation Analysis of T_0 Transformants

T_0 plants were confirmed with hygromycin resistance marker HPT-F/HPT-R. The amplification results of the target fragments of the T_0 plants are shown in Figure 4. There were 24, 26, and 26 positive transformed plants from MMB, TKB, and DR462, respectively, with the mutation rate of 80.0%, 78.8%, and 76.5%, respectively (Figure 4).

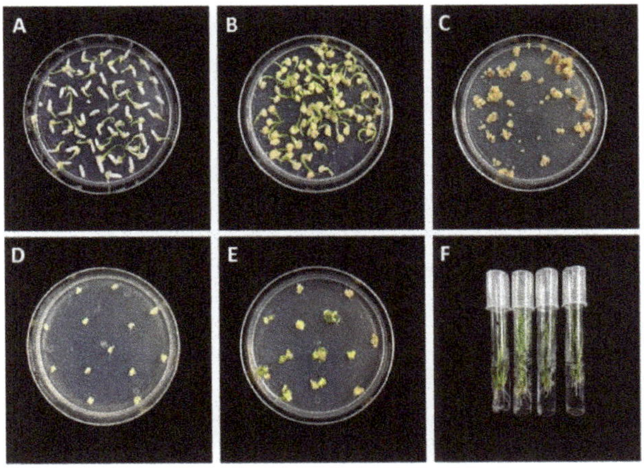

Figure 3. Process steps of genetic transformation. (**A**): Callus induction; (**B**): Callus screening; (**C**): The first screening of calli; (**D**): The second screening of calli; (**E**): Plantlet differentiation; (**F**): Rooting.

Figure 4. Amplification results of T_0 generation transformed plants of MMB (**A**), TKB (**B**), and DR462 (**C**). M: DL 5000 DNA Marker; ddH2O: a negative control based on sterile water; 5000 bp, 750 bp, 500 bp, 658 bp: fragment size.

The genotype analysis of target mutations in the T_0 generation of MMB, TKB, and DR462 showed that there were four types of mutations, including a homozygous mutant, bi-allelic mutant, heterozygous mutant, and wild-type mutant. Among them, the homozygous mutants accounted for the highest proportion. Respectively, there were 8, 9, and 9 with 33.3%, 34.6%, and 34.6% of the homozygous mutant rate in the T_0 generation of MMB, TKB, and DR462 (Table 1), indicating that this target has a high editing efficiency and is an ideal target for editing the *Wx* site. The analysis of target mutation types showed that most of the mutations were deletions only, with deletions ranging from 2 to 26 bases. Specifically, mutant MMB-10 from MMB showed a deletion of 11 bases, mutant TKB-21 from TKB showed a deletion of 26 bases, including an intron splice site (GT), while mutant DR462-9 from DR462 showed a deletion of two bases (Figure 5).

Table 1. Analysis of mutation types in the T_0 generation.

Parent	Homozygous		Bi-Allelic		Heterozygous		Wild		Total	
	No. of Plants	Rate (%)	No. of Plants	Rate (%)	No. of Plants	Rate (%)	No. of Plants	Rate (%)	No. of Plants	Rate (%)
MMB	8	33.3	7	29.2	4	16.7	5	20.8	24	100
TKB	9	34.6	7	26.9	5	19.2	6	23.1	26	100
DR462	9	34.6	8	30.8	5	19.2	4	15.4	26	100

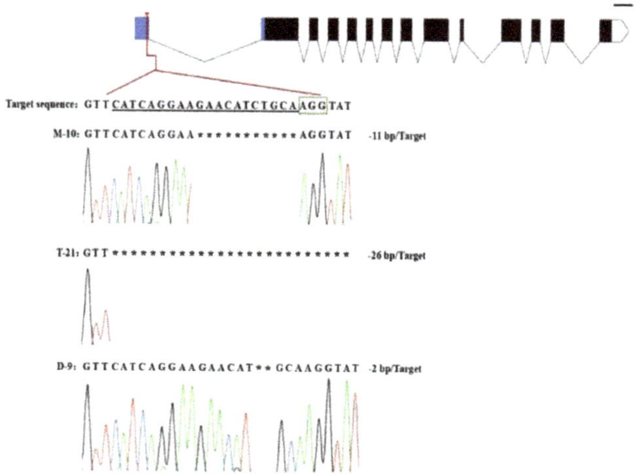

Figure 5. Target mutation types of three homozygous mutants in T_0 generation. M-10: Mutant MMB-10 from MMB; T-21: TKB-21 from TKB; D-9: DR462-9 from DR462: Green box: PAM: *: missing base.

2.4. Screening of T-DNA-Free Plants in T_1 Generations

The homozygous mutant plants MMB-10, TKB-21, and DR462-9 of the T_0 generation were selfed to develop the T_1 lines for T-DNA-free detection. The result showed that seven out of 20 T_1 plants from the line MMB-10 did not carry exogenous T-DNA fragments, all of which were homozygous by sequencing. Five out of 20 T_1 plants from TKB-21 did not carry exogenous T-DNA fragments and three of which were homozygous by sequencing. Four out of 20 T_1 plants from DR462-9 did not carry exogenous T-DNA fragments, all of which were homozygous by sequencing. Subsequently, these T-DNA-free homozygous mutants from MMB-10, TKB-21, and DR462-9 were grown and managed under greenhouse conditions (22 °C~25 °C) for harvesting the T_2 seeds of each line.

2.5. Quantitative Analysis of *Wx* Gene Expression in Mutant Lines in the T_2 Generation

The expression of the *Wx* Gene in partial homozygous mutant lines from the T_2 generation was detected by qRT-PCR, including five lines from MMB, three lines from TKB, and four lines from DR462. The results showed that the expression of *Wx* gene in mutant lines was significantly decreased as compared with the parents MMB, TKB, and DR462, indicating that the transcription efficiency of the *Wx* gene was inhibited (Figure 6).

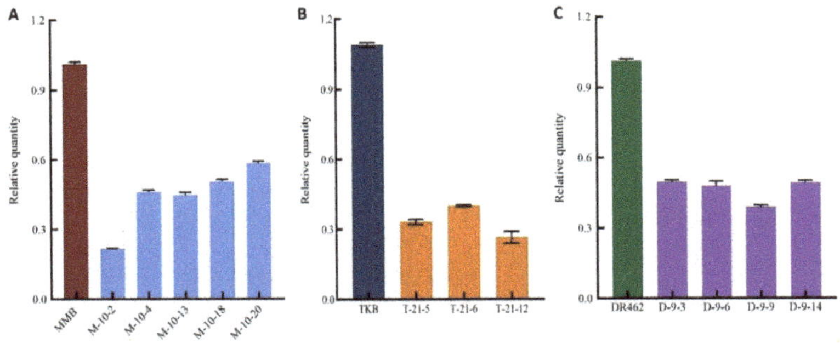

Figure 6. Relative expression level of *Wx* gene in mutant lines in T_2 generation and their respective wild type. (**A**): MMB; (**B**): TKB; (**C**): DR462; In the mutant lines, M, T, and R are short for MMB, TKB, and DR462.

2.6. Determination of AC in Mutant Lines in the T_2 Generation

As to the wild types, the ACs in MMB, TKB, and DR462 were 27.8%, 26.0%, and 25.4%, respectively. By comparison, the AC in the homozygous mutant lines decreased significantly (Figure 7). Among them, MMB-10-2, TKB-21-12, and DR462-9-9 had an AC of 17.2%, 16.8%, and 17.8%, reaching the first-grade high-quality standard concerning AC (13~18%) [18]. The results indicated that gene editing could effectively reduce the AC, so as to obtain lines with an ideal AC.

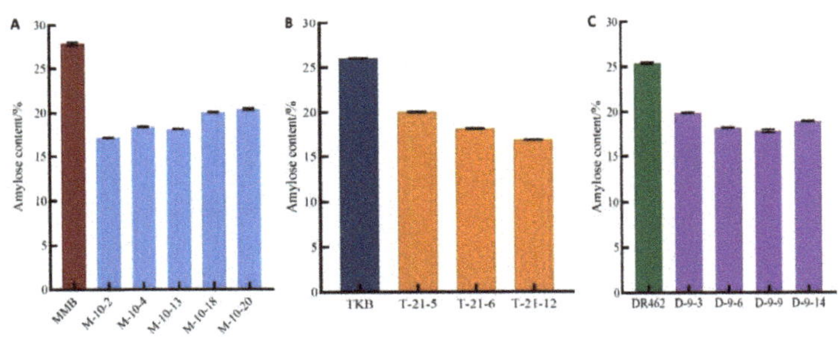

Figure 7. Amylose content in T_2 homozygous mutant lines and their respective wild types (**A**): MMB; (**B**): TKB; (**C**): DR462; In the mutant lines, M, T, and R are short for MMB, TKB, and DR462.

2.7. Determination of Gel Consistency (GC) in Mutant Lines in the T_2 Generation

Homozygous mutant lines MMB-10-2, TKB-21-12, and DR462-9-9 showed a GC of 78.6 mm, 77.4 mm, and 79.6 mm, while their parents MMB, TKB, and DR462 showed a GC of 60.2 mm, 61.2 mm, and 61.4 mm, respectively, in the analysis. The results indicated that the GC in the mutant lines was significantly increased compared with their respective wild type, and far exceeded the first-grade high-quality standard concerning GC (65 mm) [19] (Table 2).

Table 2. GC in partial T_2 mutant lines and wild type.

Wild Type	GC (mm)	Mutant Lines	GC (mm)
MMB	60.2 ± 0.2	MMB-10-2	78.6 ** ± 0.1
TKB	61.2 ± 0.2	TKB-21-6	77.4 ** ± 0.4
DR462	61.4 ± 0.6	DR462-9-9	79.6 ** ± 0.5

Note: ** $p < 0.01$ derived from one-way ANOVA with LSD test.

2.8. Analysis of RVA Profile in Mutant Lines in the T_2 Generation

The RVA analysis results of the edited lines showed that, in comparison with the parent MMB, the mutant line MMB-10-2 showed an increased breakdown viscosity (BDV) (994.0 cp from 355.0 cp), decreased setback viscosity (SBV) (1366.3 cp from 2353.6 cp), decreased consistency viscosity (CSV) (2360.3 cp from 2708.6 cp), and decreased viscosity index (Table 3; Figure 8A). In comparison with TKB, TKB-21-12 showed increased BDV(1533.6 cp from 641.3 cp), decreased SBV (729.6 cp from 1978.0 cp), decreased CSV (2263.3 cp from 2619.3 cp), and decreased viscosity index (Table 3; Figure 8B). In comparison with DR462, DR462-9-9 showed increased BDV (1293.3 cp from 508.7 cp), increased CSV(2738.0 cp from 1966.3 cp), and increased viscosity index, while maintaining an equivalent level in BDV (1444.6 cp from 1457.6 cp) (Table 3; Figure 8C).

Table 3. RVA profile characteristics of rice starch mutant lines in the T_2 generation and their respective wild types.

Lines	PKV	BDV	SBV	CSV	HPV	CPV
MMB	2729.6 ± 43.0	355.0 ± 6.5	2353.6 ± 21.5	2708.6 ± 26.8	2374.7 ± 40.8	5083.3 ± 37.2
MMB-10-2	2850.3 ± 39.0	994.0 ± 9.0	1366.3 ± 11.5	2360.3 ± 29.1	1856.3 ± 24.0	4216.7 ± 50.0
TKB	2832.3 ± 10.1	641.3 ± 22.5	1978.0 ± 14.7	2619.3 ± 36.1	2191.0 ± 28.1	4810.3 ± 39.7
TKB-21-12	3116.0 ± 6.03	1533.6 ± 15.1	729.6 ± 4.5	2263.3 ± 14.4	1539.0 ± 11.7	3802.0 ± 39.0
DR462	1539.6 ± 38.1	508.7 ± 18.0	1457.6 ± 19.8	1966.3 ± 24.5	1031.0 ± 25.0	2997.3 ± 49.2
DR462-9-9	2875.3 ± 34.6	1293.3 ± 21.0	1444.6 ± 3.5	2738.0 ± 21.1	1582.0 ± 14.1	4320.0 ± 35.0

Note: PKV: peak viscosity; BDV: breakdown viscosity; SBV: setback viscosity; CSV: consistency viscosity; HPV: hot paste viscosity; CPV: cool paste viscosity.

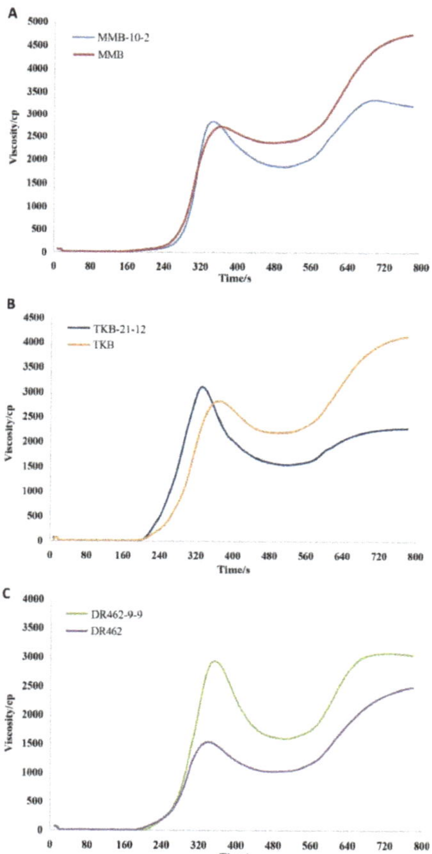

Figure 8. Comparison of RVA profiles between mutant lines in the T_2 generation and their respective wild types. (**A**): MMB; (**B**): TKB; (**C**): DR462.

Previous studies have demonstrated that BDV, SBV, and CSV are closely related to the hardness and viscosity of rice. Rice with higher BDV, lower SBV, and lower CSV appeared to be softer and more elastic [18]. The results indicated that the ECQ of the homozygous mutant lines was obviously improved as compared to the wild types.

2.9. Analysis of the Main Agronomic Traits in Mutant Lines in the T_2 Generation

To evaluate the selected mutant lines, we surveyed six main agronomic traits of MMB-10-2, TKB-21-12, and DR462-9-9 and their respective wild types. The results revealed that there was no significant difference between the mutant lines and their respective wild types, regarding traits including plant height, 1000-grain weight, panicle length, grain number per panicle, grain set rate, and effective panicle number (Table 4; Figure 9A,C,E). The results indicated that the mutations in the *Wx* genes do not change other agronomic traits. In addition, the grain appearance and chalkiness were observed and compared. The grain of MMB-10-2 and TKB-21-12 was not significantly different from the wild type in chalkiness, while they had a slight decrease in transparency and an increase in whiteness. The grain of DR462-9-9 had a decrease in chalkiness and chalky grain rate, and an increase in whiteness (Figure 9B,D,F).

Table 4. Agronomic traits of the mutant lines in the T_2 generation.

Lines	Plant Height (cm)	1000-Grain Weight (g)	Panicle Length (cm)	Grain Number per Panicle	Grain Set Rate (%)	Effective Spike Number
MMB	83.9 ± 1.3 a	19.4 ± 3.1 a	27.8 ± 0.1 a	186 ± 0.1 a	87.4 ± 2.0 a	8.5 ± 2.5 a
MMB-10-2	85.1 ± 0.9 a	19.1 ± 2.8 a	27.7 ± 0.3 a	196 ± 0.1 a	87.2 ± 2.5 a	8.7 ± 2.7 a
TKB	77.8 ± 1.2 b	27.4 ± 2.5 b	27.6 ± 0.3 b	190 ± 0.1 b	86.8 ± 1.8 b	8.2 ± 1.8 b
TKB-21-12	76.8 ± 2.1 b	27.0 ± 3.0 b	27.6 ± 0.3 b	186 ± 0.1 b	87.0 ± 2.0 b	8.3 ± 1.4 b
DR462	94.7 ± 1.6 c	24.1 ± 3.1 c	28.0 ± 0.2 c	188 ± 0.1 c	86.6 ± 1.9 c	7.9 ± 2.0 c
DR462-9-9	93.7 ± 1.4 c	23.8 ± 2.9 c	27.9 ± 0.2 c	180 ± 0.1 c	85.9 ± 1.8 c	7.7 ± 2.5 c

Note: a, b, c, $p = 0.05$ level, no significant difference in each line.

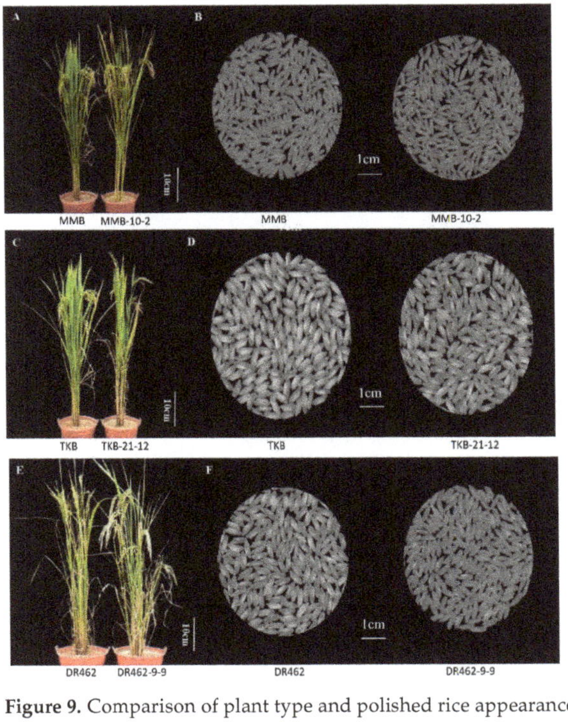

Figure 9. Comparison of plant type and polished rice appearance between mutant lines in the T_2

generation and their respective wild types. (**A**,**B**): Comparisons between MMB and MMB-10-2; (**C**,**D**): Comparisons between TKB and TKB-21-12; (**E**,**F**): Comparisons between DR462 and DR462-9-9.

3. Discussion

3.1. 5′UTR Region Regulates the Expression of Wx Gene

Previous studies have shown that mutation types such as base deletion and insertion in the non-coding region of *Wx* gene will affect its expression level, and then affect AC in the endosperm [20,21]. This study also revealed such a consistent conclusion. When the 5′UISS of the *Wx* gene was edited, the AC in the three elite indica rice lines was downregulated significantly (Figure 6). Different editing modes of the *Wx* gene have different effects on AC. For example, MMB-10 had a deletion of 11 bp, producing the progeny line MMB-10-2 with an AC of 17.4%; TKB-21 had a deletion of 26 bp, including a splice site (GT), creating the progeny line TKB-21-12 with an AC of 16.8%. In addition, mutations near 5′UISS also have a significant effect on the expression of *Wx*. For example, DR462-9 had a deletion of 2 bp near 5′UISS of the *Wx* gene, which led to the change of mRNA conformation and affected the correct splicing of introns, resulting in a decrease of AC, from 25.4% in the parent DR462 to 17.8% in the progeny line DR462-9-9. It could be confirmed that 5′UISS plays an important role in regulating the *Wx* gene.

There are relevant studies on the 5′UTR region of the *Wx* gene. Cai (1997, 2000) found an intron in the 5′UTR region of the *Wx* gene related to regulation in Indica Rice 232 and further found that the 16 base mutations in the first intron of the *Wx* gene can affect the function and efficiency of the its splice site, resulting in a decrease in the expression level of mature mRNA, thus reducing AC [22,23]. Cheng (2001) found that the natural mutation of G→T on the splice site at the 5′ end of the first intron of the *Wx* gene was the main reason for the decline of the *Wx* gene expression level and AC in rice varieties with medium and low AC [24]. In the study of highland barley starch synthesis, Li (2015) found that there are a large number of polymorphic sites in the 5′UTR region of the *Wx* gene, including large fragments of insertion and deletion. A deletion of about 400 bp in the 5′UTR region will reduce gene expression, resulting in the reduction of amylose content, and affecting the physicochemical and functional characteristics of starch [25].

3.2. Application of CRISPR/Cas9 System in Wx Gene Editing

Most studies using the CRISPR/Cas9 system to edit the *Wx* gene of rice directly turn the varieties into waxy ones [26]. Fan (2019) reduced the AC of variety Xiushui 134 from 19.78% to a glutinous rice level (AC < 2%) through gene editing [27]. Wang (2019) knocked out the *Wx* gene of rice line 209B and successfully obtained waxy maintainer *Wx* 209B with a lower AC. Then, the waxy male sterile line *Wx* 209A was developed using *Wx* 209B as male parent and 209A as female parent [28]. The above results show that editing the coding region of *Wx* Gene in rice will "kill" the gene, resulting in mutant lines with a very low AC content and glutinous rice with a high viscosity, which does not meet the requirement of consumers for soft rice quality. Therefore, accurately reducing AC and improving GC are the essence of creating germplasm resources for soft rice breeding. Zeng (2020) edited the 5′UISS locus in the *Wx*b 5′UTRs region of japonica rice material and obtained new germplasms with soft rice traits, and with a very low AC of 9.8~11.5% [29]. Similarly, this study also chose 5′UISS as the target in three indica rice lines with high AC content, and obtained homozygous mutant lines with an AC between 16.8% and 17.8%. The AC of the mutant lines obtained in this study is higher than that of Zeng (2020) and more suitable for consumers. The reasons for this may be (1) less base deletion, resulting in a lower splicing efficiency of mRNA; (2) nucleotide polymorphisms exist between Wx^a in indica rice and Wx^b in japonica rice, resulting in differences in the stability of gene mRNA splicing efficiency [30]. Even so, the three mutant lines in the T_2 generation obtained in this study, MMB-10-2, TKB-21-12, and DR462-9-9, had an AC of 17.2%, 16.8%, and 17.8%, reaching the first-grade high-quality standard concerning AC (13~18%).

3.3. Effects on Agronomic Traits of Plants When Editing the 5′UTR of Wx Gene

Gene editing may change the promoter sequence of the target gene, thereby affecting the agronomic traits of the plants [31,32]. Wang (2021) chose a target at the 5′UTR region of the *Wx* Gene of japonica rice variety Jiahua 1, and the mutant lines had main agronomic traits significantly different from that of the wild type, including a higher plant height, increased 1000-grain weight, shortened panicle length, and reduced seed setting rate [33]. In this study, the selected mutant lines in the T_2 generation had no significant differences in their main agronomic traits compared with the wild type and retained the excellent traits of the wild type. The whiteness of the endosperm in mutant lines increased, which may have been due to the changes in grain appearance caused by the decrease of AC and the increase of GC (Figure 9).

3.4. Advantages of CRISPR/Cas9 Technology in Improving ECQ of Rice

In traditional breeding, continuous backcross is usually used to introduce the *Wx* gene into rice varieties, to improve the ECQ of rice. However, this is time-consuming and it is difficult to break the linkage with undesirable traits. Subsequently, the RNAi, antisense RNA, and base/gene editing technologies were gradually developed to improve the quality of rice starch [34]. Zhao (2007) used RNAi technology to interfere with *GBSSII*, *Sbe3*, *SSI*, *SSII*, *Wx*, *PUL*, and *ISA* genes in rice, and created breeding intermediate materials with different AC, GC, and GT. The analysis of the ECG of transgenic plants showed that the AC of all mutants was decreased to various degrees, and most of them reached a significant level [35]. Terada (2000) used antisense RNA technology to introduced the antisense *Wx* gene into japonica and indica rice, and obtained the transgenic lines with ACs significantly reduced [36]. Mao (2013) used the CRISPR/Cas9 system to conduct gene editing research on Arabidopsis and rice genes [37]. Today, precise genome modifications with the CRISPR/Cas9 tool have revolutionized genome editing research [38]. It is a more efficient, accurate, easier, and cost-effective technique for achieving gene knock-out in a cell [39] and has made preliminary progress in the improvement of rice quality, especially in the research of *Wx* gene editing.

In this study, the CRISPR/Cas9 system was used to edit the 5′UISS of the *Wx* gene in indica rice lines MMB, TKB, and dr462. The sequencing of the target site in three materials showed consistency with the design target (Figure 1), and all of them performed efficient and accurate gene editing. This seems impossible for traditional breeding and other gene editing technologies. In addition, the mutant lines in the T_2 generation obtained in this study had decreased AC, increased GC, increased BDV, decreased SBV, and decreased CSV, without significant changes to their main agronomic traits. It is presented that the desirable traits can be stably inherited to the selfed offspring, indicating that CRISPR/Cas9 technology provides an effective strategy for improving the ECQ of rice.

4. Materials and Methods

4.1. Plant Materials and Growth Conditions

Three elite indica lines were selected, including maintainer Meimeng B (MMB), Tiankang B (TKB), and restorer DR462, which had high levels of AC. The seeds were supplied by the State Key Laboratory of Conservation and Utilization of Subtropical Agri-bioresources, Guangxi University. All rice lines were grown in a mesh room with isolation conditions in Nanning, China, during the normal rice-growing seasons, and treated according to the conventional planting management method.

4.2. Construction of CRISPR/Cas9 Vectors and Screening of Homozygous Mutants

The CRISPR/Cas9 targeting vector was constructed for editing the expression vector with reference to a previous study [16]. Vectors and promoters pYLCRISPR/Cas9Pubi-H and pYLsgRNA-LzU6a were provided by the State Key Laboratory of Conservation and Utilization of Subtropical Agri-bioresources, South China Agricultural University. Bacterial strain *E. coli* DH5α for plasmid propagation and preservation and agrobacterium EHA105 for genetic transformation were provided by the State Key Laboratory of Conservation and Utilization of Subtropical Agri-bioresources, Guangxi University.

According to the *Wx* (*LOC_Os06g04200*) genomic sequence provided by the China Rice Data Center [40]. The target sites in the *Wx* gene were designed via the CRISPR-GE [41] online toolkit and cloned into sgRNA, as previously described [16]. The constructs were respectively introduced into MMB, TKB, and DR462 by *Agrobacterium*-mediated transformation. The genomic DNA of transformed plants was extracted from young leaves using the CTAB method [42]. The target sites of the T_0, T_1, and T_2 generations were sequenced using the Sanger method [43]. The PCR products were detected using agarose gel electrophoresis. The homozygous *Wx* mutants without the T-DNA insertion were screened and selected for further experimental analysis. The primers used are listed in Table 5.

Table 5. Primer sequences used in this study.

Primer Name	Primer Sequence 5′-3′
Wx-text-F	TCCGCCACGGGTTCCAG
Wx-text-R	CTCCTACCTCAGCCACAACG
U-F	CTCCGTTTTACCTGTGGAATCG
gR-R	CGGAGGAAAATTCCATCCAC
Wx-U6a-1F	TGTGTGCTTACAGCCATGGCGTTTTAGAGCTAGAAAT
Wx-U6a-1R	GCCATGGCTGTAAGCACACACGGCAGCCAAGCCAGCA
Pps-GGL	TTCAGAGGTCTCTCTCGACTAGTATGGAATCGGCAGCAAAGG
Pgs-GGR	AGCGTGGGTCTCGACCGACGCGTATCCATCCACTCCAAGCTC
PB-R	GCGCGCGGTCTCTACCGACGCGTATCC
PB-L	GCGCGCgGTCTCGCTCGACTAGTATGG
HPT-F	ATTTGTGTACGCCCGACAGT
HPT-R	GTGCTTGACATTGGGGAGTT
CAS9-F	CTGACGCTAACCTCGACAAG
CAS9-R	CCGATCTAGTAACATAGATGACACC

Note: ACTAGT and ACGCGT are *Spe*I and *Mlu*I sites, respectively.

4.3. Quantitative Real-Time PCR (qRT-PCR) Expression Analysis

Total RNA was extracted from rice caryopses (with glumes removed) 10 days after flowering (DAF) using an RNAplant Plus Reagent kit (Tiangen, Beijing, China). First-strand cDNA was synthesized using a PrimeScript RT reagent kit (Takara, Tokyo, Japan), and qRT-PCR was performed on a CFX Connect Real-Time PCR Detection System (Bio-Rad, Hercules, CA, USA) using Cham Q SYBR qPCR master Mix (Vazyme, Nanjing, China). The Actin gene was used for normalization, and each experiment included three biological replicates.

4.4. Measuring Rice Grain Physicochemical Properties

AC, GC, and RVA were measured as described in the national agricultural industry standard (NY/T83-2017) [44]. In brief, AC was determined using a dual-wavelength spectrophotometric method, by drawing the reference wavelength analysis curve for AC and amylopectin (Figure 10A) and the standard curve for amylose (Figure 10B).

GC was determined by measuring the length of the rice gel after gelatinization and cooling. RVA spectrum was measured using a Techmaster RVA instrument (Pertentec-marster, Sweden, Nanning, China). The primary RVA parameters included PKV, BDV, SBV, CSV, HPV, and CPV. All tests were performed in triplicate.

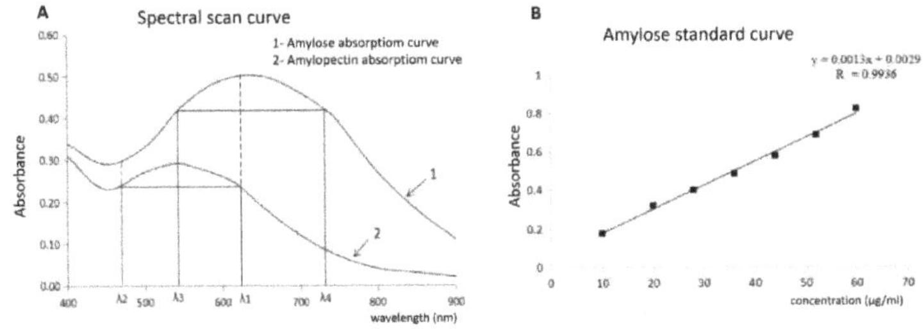

Figure 10. Drawing the standard curve of amylose. (**A**): Preparation analysis of amylose and amylopectin (**B**): Standard curve of amylose.

4.5. Agronomic Trait Investigation

To detect the agronomic traits of homozygous mutants in the T_2 generation, together with the corresponding wild type, 20 plants from each line were subjected to agronomic trait measurement at the maturation stage. The traits including plant height, thousand-grain weight, panicle length, grains per panicle, seed setting rate, and chalkiness degree. Field management followed normal agronomic practices.

4.6. Statistical Analysis

At least three replicates were performed for each experiment. Statistical and graphical analyses were performed using Excel 2016, GraphPad Prism 9, and Photoshop 7.0. All data were expressed as means $\pm$ standard deviations (means $\pm$ SD). One-way analysis of variance (ANOVA) was used to determine the level of significance (* and ** indicate significant differences at $p < 0.05$ and $p < 0.01$, respectively). Different lower-case letters indicate statistically significant differences at $p < 0.05$.

5. Conclusions

In this study, the elite indica rice varieties MMB, TKB, and DR462 were chosen as experimental materials, and the 5′UISS in the 5′UTR region of *Wx* gene was edited by CRISPR/cas9 gene editing technology. Screening was carried out to achieve homozygous gene-edited lines, and the relevant phenotypes of these lines were evaluated.

1. In the T_0 generation of transgenic plants, the genetic transformation efficiency was more than 80%. The sequencing analysis showed that, among the four mutation types, homozygous mutation occupied a relatively large proportion, which not only demonstrated the extremely high editing efficiency of the CRISPR/cas9 gene editing technology, but also implied a great potential for application in designing targets in the 5′UTR region;
2. The expression analysis showed that the expression level of *Wx* gene in the edited lines was significantly lower than that of their wild type parents, indicating that editing 5′UISS changed the splicing mode and efficiency of introns;
3. The ECQ analysis of MMB-10-2, TKB-21-12, and DR462-9-9 showed that, compared with the wild type parents, the AC of the editing lines was significantly reduced, reaching the first-grade high-quality standard concerning AC. The GC of the edited lines was significantly increased, far exceeding the first-grade high-quality standard concerning GC. The RVA spectrum of the edited lines showed that the grain of the edited lines became softer and more elastic, and the eating quality was greatly improved.
4. The main agronomic traits of MMB-10-2, TKB-21-12, and DR462-9-9 had no significant difference from their corresponding wild type, which implied that the trait changes caused by homozygous mutations could be stably inherited to self-bred offspring, reflecting the stability and reliability of CRISPR/cas9 gene editing technology in rice genetic improvement.

Author Contributions: Conceptualization, J.Y.; methodology, F.L., B.Q. and R.L.; validation, X.G. and X.W.; investigation, X.G. and Y.F.; resources, R.L.; writing—original draft preparation, J.Y.; reviews and revision, R.L.; funding acquisition, R.L. All authors have read and agreed to the published version of the manuscript.

Funding: This research was funded by Guangxi Zhuang Autonomous Region Science and Technology Department, grant numbers AB16380066 and AA17204070.

Conflicts of Interest: The authors declare no conflict of interest.

References

1. Zhong, Y.; Qu, J.; Li, Z.; Tian, Y.; Zhu, F.; Blennow, A.; Liu, X. Rice starch multi-level structure and functional relationships. *Carbohydr. Polym.* **2022**, *275*, 118777. [CrossRef] [PubMed]
2. Wani, A.A.; Singh, P.; Shah, M.A.; Schweiggert-Weisz, U.; Gul, K.; Wani, I.A. Rice Starch Diversity: Effects on structural, morphological, thermal, and physicochemical properties—A review. *Compr. Rev. Food Sci. Food Saf.* **2012**, *11*, 417–436. [CrossRef]
3. Huang, L.; Li, Q.-F.; Zhang, C.; Chu, R.; Gu, Z.; Tan, H.; Zhao, D.; Fan, X.; Liu, Q. Creating novel *Wx* alleles with fine-tuned amylose levels and improved grain quality in rice by promoter editing using CRISPR/Cas9 system. *Plant Biotechnol. J.* **2020**, *18*, 2164–2166. [CrossRef] [PubMed]
4. Juliano, B.O. Structure, chemistry, and function of the rice grain and its fractions. *Cereal Foods World* **1992**, *37*, 772–779. [CrossRef]
5. Zhu, D.W.; Zhang, H.C.; Guo, B.W.; Dai, Y.G.; Huo, Z.Y.; Xu, K.; Wei, H.Y.; Gao, H. Development and outlook of Chinese soft rice. *J. Yangzhou Univ. (Agric. Life Sci. Ed.)* **2015**, *36*, 47–52. (In Chinese) [CrossRef]
6. Yang, J.; Wang, J.; Fan, F. Development of AS-PCR marker based on a key mutation confirmed by resequencing of *Wx*-mp in milky princess and its application in japonica soft rice (*Oryza sativa* L.) breeding. *Plant Breed.* **2013**, *132*, 595–603. [CrossRef]
7. Zhu, D.-W.; Zhang, H.-C.; Guo, B.-W.; Xu, K.; Dai, Q.-G.; Wei, H.-Y.; Gao, H.; Hu, Y.-J.; Cui, P.-Y.; Huo, Z.-Y. Effects of nitrogen level on yield and quality of japonica soft super rice. *J. Integr. Agric.* **2017**, *16*, 1018–1027. [CrossRef]
8. Xu, Z.; Yu, M.; Yin, Y.; Zhu, C.; Ji, W.; Zhang, C.; Li, Q.; Zhang, H.; Tang, S.; Yu, H.; et al. Generation of selectable marker-free soft transgenic rice with transparent kernels by down regulation of SSII-2. *Crop J.* **2020**, *8*, 53–61. [CrossRef]
9. Lu, J.G.; Gao, R.C.; Li, P.; Xu, M.L.; Li, J.J. Breeding of low-amylose content late japonica rice and its taste quality analysis. *China Rice* **2014**, *20*, 41–45. (In Chinese) [CrossRef]
10. Lu, M.C. Rice variety Songxiangjing1018 and its direct seeding cultivation techniques. *China Seed Ind.* **2018**, *36*, 90–92. (In Chinese) [CrossRef]
11. Zhao, C.F.; Yue, H.L.; Huang, S.J.; Zhou, L.H.; Zhao, L.; Zhang, Y.D.; Chen, T.; Zhu, Z.; Zhao, Q.Y.; Yao, S.; et al. Eating quality and physicochemical properties in Nanjing rice varieties. *Sci. Agric. Sin.* **2019**, *52*, 909–920. (In Chinese) [CrossRef]
12. Wang, H.; La, R.M.; Qi, L.S. CRISPR/Cas9 in Genome Editing and Beyond. *Annu. Rev. Biochem.* **2016**, *85*, 227–264. [CrossRef] [PubMed]
13. Wang, X.; Han, Y.; Feng, X.; Gul, N.; Luo, L.; Liu, F.; Qin, B.X.; Liu, Y.G.; Li, R.B. Improvement of a traditional high-quality glutinous rice variety by Crispr-Cas9 gene editing system. *Mol. Plant Breed.* **2019**, *17*, 6332–6342. (In Chinese) [CrossRef]
14. Huang, L.C.; Gu, Z.W.; Tan, H.Y.; Zhao, W.; Xiao, Y.; Chu, R.; Fan, X.L.; Zhang, C.Q.; Li, Q.F.; Liu, Q.Q. Creating novel glutinous rice germplasms by editing *Wx* gene via CRISPR/Cas9 technology. *J. Plant Genet. Resour.* **2021**, *22*, 789–799. (In Chinese) [CrossRef]
15. Liu, X.; Ding, Q.; Wang, W.; Pan, Y.; Tan, C.; Qiu, Y.; Chen, Y.; Li, H.; Li, Y.; Ye, N.; et al. Targeted deletion of the first intron of the Wx^b allele via CRISPR/Cas9 significantly increases grain amylose content in rice. *Rice* **2022**, *15*, 1. [CrossRef] [PubMed]
16. Teng, K.; Wang, X.; Guo, X. Generation of a new glutinous photothermosensitive genic-male-sterile (PTGMS) line by CRISPR/Cas9-directed mutagenesis of *Wx* in rice (*Oryza sativa* L.). *Agriculture* **2021**, *11*, 1044. [CrossRef]
17. Fu, Y.; Luo, T.; Hua, Y.; Yan, X.; Liu, X.; Liu, Y.; Liu, Y.; Zhang, B.; Liu, R.; Zhu, Z.; et al. Assessment of the Characteristics of Waxy Rice Mutants Generated by CRISPR/Cas9. *Front. Plant Sci.* **2022**, *13*, 881964. [CrossRef]
18. Dou, L.L.; Xiao, L.; Li, J.Y. Improving tasting quality of the main parent line "93-11" of *Indica* hybrid rice. *J. Shanghai Norm. Univ. (Nat. Sci.)* **2014**, *43*, 245–250. (In Chinese) [CrossRef]
19. *NY/T593-2013;* Cooking Rice Variety Quality. Standards Press of China: Beijing, China, 2013.
20. Patron, N.J.; Smith, A.M.; Fahy, B.F.; Hylton, C.M.; Naldrett, M.J.; Rossnagel, B.G.; Denyer, K. The altered pattern of amylose accumulation in the endosperm of low-amylose barley cultivars is attributable to a single mutant allele of granule-bound starch synthase I with a deletion in the 5′-non-coding region. *Plant Physiol.* **2002**, *130*, 190–198. [CrossRef]
21. Yao, Y.; Thompson, D.B.; Guiltinan, M.J. Maize starch-branching enzyme isoforms and amylopectin structure. In the absence of starch-branching enzyme IIB, the further absence of starch-branching enzyme la leads to increased branching. *Plant Physiol.* **2004**, *136*, 3515–3523. [CrossRef]
22. Cai, X.; Wang, Z.; Zheng, F.; Hong, M. A regulation-related intron in 5′ untranslated region of rice *Waxy* gene. *Acta Phytophysiol. Sin.* **1997**, *23*, 257–261. (In Chinese)
23. Cai, X.; Wang, Z.; Xing, Y.; Zhang, J.; Hong, M. Alteration of RNA secondary structure of rice Waxy intron 1 caused by naturally ocurred mutations. *Acta Phytophysiol. Sin.* **2000**, *26*, 59–63. (In Chinese) [CrossRef]

24. Cheng, S.; Ge, H.; Wang, Z.; Hong, M. Analysis of influence of *Wx* intron 1 on gene expression in transgenic rice plant. *Acta Phytophysiol. Sin.* **2001**, *27*, 381–386. (In Chinese) [CrossRef]
25. Li, Q.; Pan, Z.F.; Deng, G.B. Effect of *Waxy* gene polymorphism on starch properties of highland barley in Qinghai Tibet plateau. *Conf. Chin. Genet. Soc.* **2015**, 64–65. [CrossRef]
26. Ma, X.; Zhang, Q.; Zhu, Q. A robust CRISPR/Cas9 system for convenient, high-efficiency multiplex genome editing in monocot and dicot plants. *Mol. Plant* **2015**, *8*, 1274–1284. [CrossRef]
27. Fan, M.Y.; Mei, F.T.; Zhu, Y.W.; Lin, Y.R.; Zhao, L.L.; Tian, G.L.; Wang, F. Greating new glutinous rice by CRISPR/Cas9-targeted mutagenesis in rice. *Fujian J. Agric. Sci.* **2019**, *34*, 503–508. (In Chinese) [CrossRef]
28. Wang, X.; Han, Y.; Feng, X.; Li, Y.-Z.; Qin, B.-X.; Luo, J.-J.; Wei, Z.; Qiu, Y.-F.; Liu, F.; Li, R.-B. Breeding of indica glutinous cytoplasmic male sterile line *Wx* 209A via CRISPR/Cas9 mediated genomic editing. *Czech J. Genet. Plant Breed.* **2019**, *55*, 93–100. [CrossRef]
29. Zeng, D.; Liu, T.; Ma, X. Quantitative regulation of *Waxy* expression by CRISPR/Cas9-based promoter and 5′UTR-intron editing improves grain quality in rice. *Plant Biotechnol. J.* **2020**, *18*, 2385–2387. [CrossRef]
30. Larkin, P.D.; Park, W.D. Association of *waxy* gene single nucleotide polymorphisms with starch characteristics in rice (*Oryza sativa* L.). *Mol. Breed.* **2003**, *12*, 335–339. [CrossRef]
31. Li, X.; Xie, Y.; Zhu, Q.; Liu, Y.-G. Targeted genome editing in genes and cis-regulatory regions improves qualitative and quantitative traits in crops. *Mol. Plant* **2017**, *10*, 1368–1370. [CrossRef]
32. Rodríguez-Leal, D.; Lemmon, Z.H.; Man, J.; Bartlett, M.E.; Lippman, Z.B. Engineering quantitative trait variation for crop improvement by genome editing. *Cell* **2017**, *171*, 470–480. [CrossRef] [PubMed]
33. Wang, W.W. *Creation of Waxy Allelic Mutants in Rice (Oryza sativa L.) by CRISPR/Cas9 System and Analysis of Starch Quality*; Shanghai Normal University: Shanghai, China, 2021. [CrossRef]
34. Li, Z.; Xiong, X.; Wang, F. Gene disruption through base editing-induced messenger RNA missplicing in plants. *New Phytol.* **2019**, *222*, 1139–1148. [CrossRef] [PubMed]
35. Zhao, H.J. *Research on Starch Quality Using Different Transgenic Constructs in the Endosperm of Rice (Oryza sativa L.)*; Yangzhou University: Yangzhou, China, 2007.
36. Terada, R.; Nakajima, M.; Isshiki, M. Antisense *waxy* genes with highly active promoters effectively suppress *Waxy* gene expression in transgenic rice. *Plant Cell Physiol.* **2000**, *41*, 881–888. [CrossRef]
37. Mao, Y.; Zhang, H.; Xu, N. Application of the CRISPR/Cas system for efficient genome engineering in plants. *Mol. Plant* **2013**, *6*, 2008–2011. [CrossRef] [PubMed]
38. Yarra, R.; Sahoo, L. Base editing in rice: Current progress, advances, limitations, and future perspectives. *Plant Cell Rep.* **2021**, *40*, 595–604. [CrossRef]
39. Gupta, D.; Bhattacharjee, O.; Mandal, D. CRISPR-Cas9 system: A new-fangled dawn in gene editing. *Life Sci.* **2019**, *232*, 116636. [CrossRef]
40. The China Rice Data Center. Available online: https://www.ricedata.cn/ (accessed on 20 March 2021).
41. The CRISPR-GE. Available online: http://skl.scau.edu.cn/ (accessed on 20 March 2021).
42. Murray, M.G.; Thompson, W.F. Rapid isolation of high molecular weight plant DNA. *Nucleic Acids Res* **1980**, *8*, 4321–4325. [CrossRef]
43. Sanger, F.; Coulson, A.R. A rapid method for determining sequences in DNA by primed synthesis with DNA polymerase. *J. Mol. Biol.* **1975**, *94*, 441–448. [CrossRef]
44. *NY/T83-2017*; Determination of Rice Quality. Standards Press of China: Beijing, China, 2017.

International Journal of *Molecular Sciences*

MDPI

Article

GLW7.1, a Strong Functional Allele of *Ghd7*, Enhances Grain Size in Rice

Rongjia Liu, Qinfei Feng, Pingbo Li, Guangming Lou, Guowei Chen, Haichao Jiang, Guanjun Gao, Qinglu Zhang, Jinghua Xiao, Xianghua Li, Lizhong Xiong and Yuqing He *

National Key Laboratory of Crop Genetic Improvement and National Centre of Plant Gene Research, Hubei Hongshan Laboratory, Huazhong Agricultural University, Wuhan 430070, China
* Correspondence: yqhe@mail.hzau.edu.cn

Abstract: Grain size is a key determinant of both grain weight and grain quality. Here, we report the map-based cloning of a novel quantitative trait locus (QTL), *GLW7.1* (*Grain Length, Width and Weight 7.1*), which encodes the CCT motif family protein, GHD7. The QTL is located in a 53 kb deletion fragment in the cultivar Jin23B, compared with the cultivar CR071. Scanning electron microscopy analysis and expression analysis revealed that *GLW7.1* promotes the transcription of several cell division and expansion genes, further resulting in a larger cell size and increased cell number, and finally enhancing the grain size as well as grain weight. *GLW7.1* could also increase endogenous GA content by up-regulating the expression of GA biosynthesis genes. Yeast two-hybrid assays and split firefly luciferase complementation assays revealed the interactions of GHD7 with seven grain-size-related proteins and the rice DELLA protein SLR1. Haplotype analysis and transcription activation assay revealed the effect of six amino acid substitutions on GHD7 activation activity. Additionally, the NIL with *GLW7.1* showed reduced chalkiness and improved cooking and eating quality. These findings provide a new insight into the role of *Ghd7* and confirm the great potential of the *GLW7.1* allele in simultaneously improving grain yield and quality.

Keywords: *GLW7.1*; GHD7; grain size; quality; rice

check for updates

Citation: Liu, R.; Feng, Q.; Li, P.; Lou, G.; Chen, G.; Jiang, H.; Gao, G.; Zhang, Q.; Xiao, J.; Li, X.; et al. *GLW7.1*, a Strong Functional Allele of *Ghd7*, Enhances Grain Size in Rice. *Int. J. Mol. Sci.* **2022**, *23*, 8715. https://doi.org/10.3390/ijms23158715

Academic Editors: Jinsong Bao and Jianhong Xu

Received: 9 July 2022
Accepted: 2 August 2022
Published: 5 August 2022

Publisher's Note: MDPI stays neutral with regard to jurisdictional claims in published maps and institutional affiliations.

Copyright: © 2022 by the authors. Licensee MDPI, Basel, Switzerland. This article is an open access article distributed under the terms and conditions of the Creative Commons Attribution (CC BY) license (https://creativecommons.org/licenses/by/4.0/).

1. Introduction

Rice (*Oryza sativa* L.) is the most important staple food crop in the world and feeds more than half of the world's population [1]. Therefore, to meet the food needs of a rapidly growing global population, increasing rice grain yield has been a major breeding goal. Rice yield is mainly determined by three major components: grain weight, number of grains per panicle and number of effective tillers per plant [2]. Among them, grain weight is largely determined by grain size, which includes grain length, width, and thickness [3]. In recent decades, many quantitative trait loci (QTLs)/genes regulating grain size have been isolated and shown to participate in multiple signaling pathways, including the G-protein signaling pathway, the ubiquitin–proteasome pathway, mitogen-activated protein kinase (MAPK) signaling pathway, phytohormone signaling and homeostasis, and transcriptional regulators [4].

G proteins are guanine nucleotide-binding trimeric proteins consisting of Gα, Gβ and Gγ subunits and regulate many biological processes. *GRAIN SIZE 3* (*GS3*), encoding an atypical Gγ protein, is the first identified major QTL negatively regulating grain size [5]. DENSE AND ERECT PANICLE 1 (DEP1), another atypical Gγ protein, positively regulates grain size by competitively binding to Gβ (RGB1) with GS3 [6]. Other G-proteins, including conventional Gγ proteins (RGG1 and RGG2), atypical Gγ protein (GGC2), Gα protein (RGA1) and Gβ protein (RGB1) could also control grain size [6–8]. In addition, OsMADS1, a MADS-domain transcription factor, negatively regulates grain length through directly interacting with GS3 and DEP1 [9].

Ubiquitination and deubiquitination are two opposite protein modifications which are involved in the regulation of rice grain size. GRAIN WIDTH 2 (GW2), a RING-type E3 ubiquitin ligase, negatively regulates grain width by ubiquitinating WG1 and targeting it for degradation via the 26S proteasome pathway [10]. CHANG LI GENG 1 (CLG1), another RING-type E3 ubiquitin ligase, positively regulates grain length by ubiquitinating GS3 and targeting it for degradation via the endosome pathway [11]. *WIDE AND THICK GRAIN 1* (*WTG1*), which encodes an otubain-like protease with deubiquitination activity, controls grain size and shape mainly by affecting cell expansion in the spikelet hull [12]. *LARGE GRAIN 1* (*LG1*), which encodes a constitutively expressed ubiquitin-specific protease15 (OsUBP15) with deubiquitination activity, positively regulates grain width and size [13].

The OsMKKK10-OsMKK4-OsMPK6 cascade has been revealed to positively regulate grain size by promoting cell proliferation in spikelet hulls [14–16]. GRAIN SIZE AND NUMBER 1 (GSN1)/OsMKP1, a MAPK phosphatase, negatively regulates grain size by directly interacting with and inhibiting the dephosphorylation of OsMPK6 [17]. In addition, the upstream gene *OsER1*, which encodes a receptor-like protein kinase, and the downstream transcription factor *OsWRKY53*, could both positively regulate grain size through the MAPK signaling cascade [18,19].

Phytohormones play various roles in plant growth and development, stress responses, and metabolism. Several genes controlling grain size have been reported to be involved in the brassinosteroid (BR) signaling pathway, such as *GRAIN WIDTH 5* (*GW5*) [20], *GRAIN LENGTH 2* (*GL2*) [21,22] and *GRAIN LENGTH 3.1* (*GL3.1*) [23]. Another set of genes regulate grain size through the auxin signaling pathway, such as *THOUSAND GRAIN WEIGHT 6* (*TGW6*) [24], *BIG GRAIN 1* (*BG1*) [25] and *THOUSAND GRAIN WEIGHT 3* (*TGW3*) [26]. Moreover, some gibberellic acid (GA)-signaling-pathway-related genes also regulate grain size, such as *GIBBERELLIN-DEFICIENT DWARF 1* (*GDD1*) [27], *SMALL AND ROUND SEED 3* (*SRS3*) [28] and *SMALL GRAIN AND DWARF 2* (*SGD2*) [29].

Many transcription factors participate in the regulation of grain size, including the SQUAMOSA promoter binding protein-like (SPL) family (*GRAIN LENGTH AND WIEIGHT 7* (*GLW7*)/*OsSPL13*, *GRAIN WIDTH 8* (*GW8*)/*OsSPL16*, *OsSPL18*) [30–32], the basic helix–loop–helix (bHLH) family (*Awn-1* (*An-1*), *OsbHLH079*, *OsbHLH107*) [33–35], APETALA2-type (AP2) transcription factors (*SMALL ORGAN SIZE1* (*SMOS1*), *SUPERNUMERARY BRACT* (*SNB*), *FRIZZY PANICLE* (*FZP*)) [36–38], and other transcription factors (*GRAIN SHAPE 9* (*GS9*), *SHORT GRAIN6* (*SG6*), *GRAIN LENGTH 4* (*GL4*)) [39–41].

Although many QTLs/genes regulating grain size have been identified, the understanding of grain size regulation is still fragmented. In this study, we report the mapping, cloning and initial characterization of a novel grain size QTL, *GLW7.1* (*Grain Length, Width and Weight 7.1*) in rice, which encodes the CCT (CONSTANS, CONSTANS-LIKE, and TIMING OF CHLOROPHYLL A/B BINDING1) motif family protein, GHD7. *Grain number, plant height, and heading date 7* (*Ghd7*) was first reported as a major regulator of heading date, and improved yield by increasing grain number [42]. Subsequent studies revealed that it participated in a variety of other developmental processes, such as stress responses, seed germination and nitrogen utilization [43–45]. Here, we performed scanning electron microscopic analysis, yeast two-hybrid assays, split firefly luciferase complementation (SFLC) assays and expression analysis to uncover the mechanism mediated by *GLW7.1* to regulate grain size. We also conducted haplotype analysis of *Ghd7* and transcription activation assay to uncover the reason underlying different effects between three allelic GHD7 proteins. Our results provide insights into the role of *Ghd7* in regulating grain size and the effect of different amino acid substitutions on transcriptional activation activity of GHD7 proteins, and we provide a promising *Ghd7* allele for breeding rice with high yield and superior quality.

2. Results

2.1. Identification of *GLW7.1*

To identify novel QTLs associated with grain size (Figure 1A), we selected two *indica* varieties, Jin23B (hereafter J23B) and CR071, that showed significant differences in grain size (Figure 1B,C) and constructed a set of 238 BC_3F_1 lines in the J23B background (Figure S1). Three QTLs were revealed by a subsequent QTL analysis, among which the QTL located between SSR markers RM501 and RM542 on chromosome 7 was the most significant contributor to grain length (Figure 1A, Table S1). To further evaluate the genetic effect of this QTL, we developed a near-isogenic line (NIL) in the genetic background of J23B (Figure S1). Genetic analysis of BC_4F_2 progenies derived from the NIL in 2015 showed that the dominant allele from CR071 could increase grain length, grain width and grain weight (Figure S2A–C). The similar genetic effect was further confirmed by BC_5F_3 progenies in 2017 (Figure S2D,F). Thus, we designated this QTL as *Grain Length, Width and Weight 7.1* (*GLW7.1*).

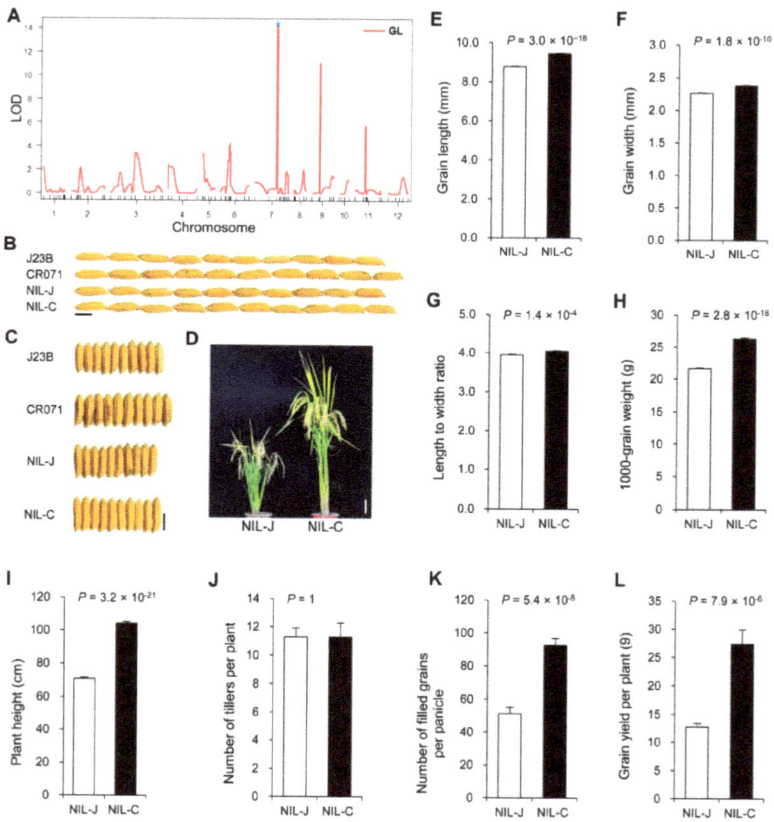

Figure 1. Field trial of *GLW7.1* NIL lines. (**A**) Primary mapping of QTLs for grain length using J23B/CR071 BC_3F_1 population (n = 238). The star symbol indicates *GLW7.1* locus. (**B**,**C**) Grain morphology. Scale bar: 5 mm. (**D**) The gross morphology of NIL plants. Scale bar: 10 cm. (**E**) Grain length. (**F**) Grain width. (**G**) Length to width ratio. (**H**) 1000-grain weight. (**I**) Plant height. (**J**) Number of tillers per plant. (**K**) Number of filled grains per panicle. (**L**) Grain yield per plant. All phenotypic data in (**E**–**L**) were measured from paddy-grown NIL plants grown under normal cultivation conditions. Data were represented as mean ± s.e.m. (n = 15). The Student's t-test was used to produce p values.

2.2. Characterization of GLW7.1

Two NIL plants carrying the homozygous J23B allele *glw7.1* and CR071 allele *GLW7.1* were developed and named as NIL-J and NIL-C, respectively. Compared to NIL-J, NIL-C displayed a higher value in grain length (increased by 8%) (Figure 1B,E) and grain width (increased by 5%) (Figure 1C,F), leading to an increase in length-to-width ratio by 2% (Figure 1G) and 1000-grain weight by 22% (Figure 1H). The plant height of NIL-C was about 33 cm higher than that of NIL-J (Figure 1D,I), but no difference was observed in tiller numbers per plant (Figure 1J). Meanwhile, NIL-C displayed more filled grains per panicle (increased by 80%) than NIL-J (Figure 1K). Thus, the increase in grain weight and grain number contributed to the increase in grain yield per plant by 114% in NIL-C, in comparison with NIL-J (Figure 1L).

The dominant *GLW7.1* locus with a yield-increasing effect has a good advantage in hybrid rice breeding. Considering that the simultaneous increase in rice grain length and width is usually accompanied by a decrease in rice quality [46,47], to further evaluate the prospects of *GLW7.1* in rice breeding, we then examined rice quality traits among NILs, including percentage of grains with chalkiness, amylose content, gel consistency, and taste value. Surprisingly, a significant reduction in grain chalkiness and a huge improvement in taste score were observed in NIL-C plants (Figure 2A,B,E), accompanied with a significant increase in amylose content and gel consistency (Figure 2C,D). These results demonstrate that the *GLW7.1* allele from CR071 is a pleiotropic gene conferring high yield and superior quality.

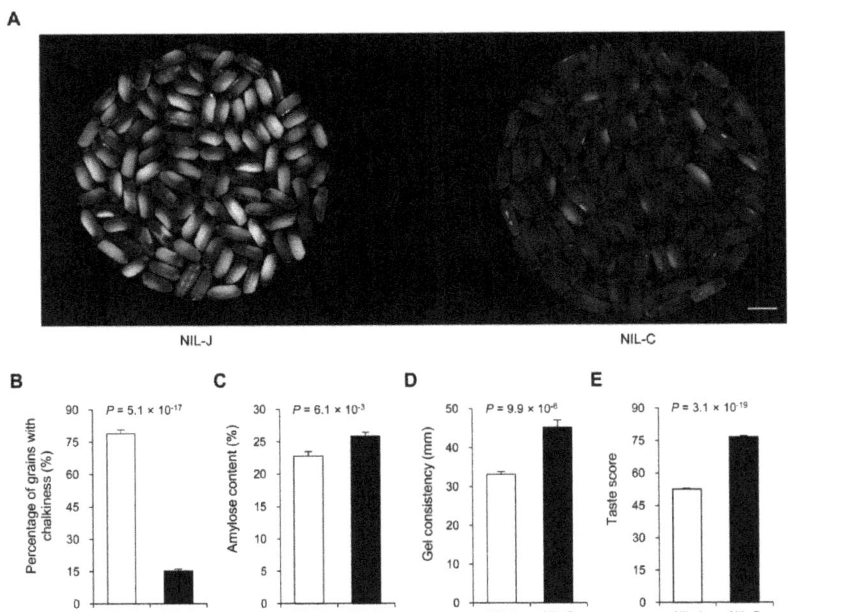

Figure 2. The effects of the *GLW7.1* allele on the physicochemical characteristics of milled rice. (**A**) Comparisons of chalkiness and endosperm transparency of milled rice between the *GLW7.1* NILs ($n = 100$). Scale bar: 1 cm. (**B**) Percentage of grains with chalkiness. (**C**) Amylose content. (**D**) Gel consistency. (**E**) Taste score. All phenotypic data in (**B–E**) were measured from paddy-grown NIL plants grown under normal cultivation conditions. Data are represented as mean ± s.e.m. ($n = 10$). The Student's *t*-test was used to produce *p* values.

The glume, including lemma and palea, determines the upper limit of grain size [39,46,48,49], and its size is determined by cell number and cell size. To uncover the cytological reason underlying the difference in grain size between NIL-J and NIL-C, we performed scanning electron microscopic analysis of the outer surfaces of lemmas (Figure 3A,B). Compared with NIL-J, the value of cell length, cell width and the number of longitudinal cells were significantly higher in NIL-C (Figure 3C–E), but the number of transverse cells showed no difference (Figure 3F). To further investigate how *GLW7.1* regulates cell number and cell size, we examined the expression levels of 43 genes involved in cell cycle and cell expansion using the young panicles (8–10 cm in length) of the two NILs. As expected, expression levels of 10 cell cycle related-genes (*CYCD1;1, E2F, MCM4, CDC20, CYCA2;3, CYCB1;1, CYClaZm, MAPK, CDKB* and *KN*) and 3 cell-expansion-related genes (*EXPA3, EXPA5* and *EXPB3*) were significantly up-regulated (fold-change > 1.5 and $p < 0.01$) in NIL-C (Figure 3G, Table S2). These results suggest that *GLW7.1* positively regulates grain size by promoting cell division and cell expansion to increase cell number and cell size of the glume during spikelet development.

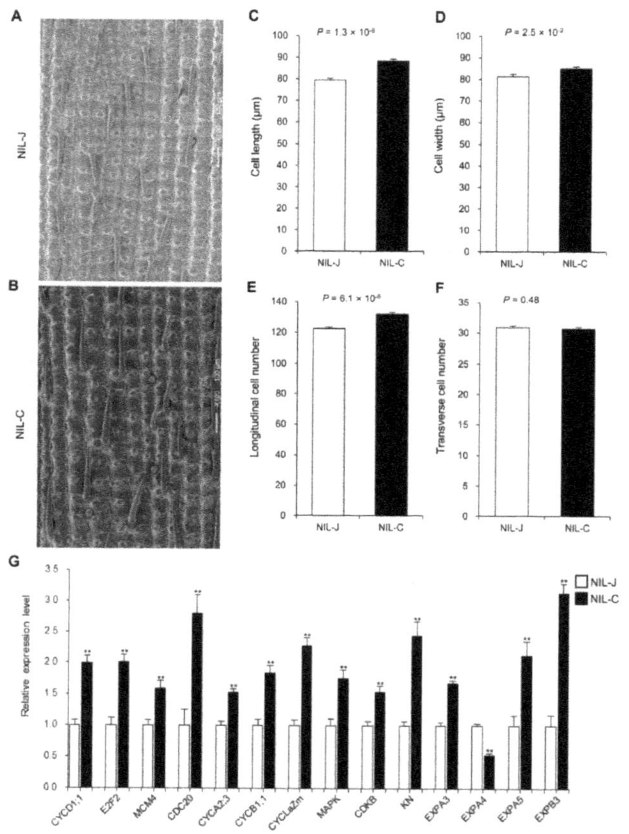

Figure 3. The effect of *GLW7.1* on cell number and cell size. (**A–F**) Scanning electron microscope analysis. Scale bar: 100 μm. (**A**) Outer epidermal cells of NIL-J. (**B**) Outer epidermal cells of NIL-C. (**C**) Average cell length. (**D**) Average cell width. (**E**) Total number of longitudinal cells. (**F**) Total number of transverse cells. Data are represented as mean ± s.e.m. ($n = 15$). The Student's *t*-test was used to produce *p* values. (**G**) Relative expression level of 10 cell cycle-related genes and 4 cell expansion genes in young panicles (8–10 cm in length) of NIL-J and NIL-C. *OsActin* (*LOC_Os03g50885*) was used as the control and the values of expression level in NIL-J were set to 1. Data are represented as mean ± s.e.m. ($n = 9$). The Student's *t*-test was used to produce *p* values (** indicates $p < 0.01$).

2.3. Fine Mapping of GLW7.1

To fine-map *GLW7.1*, we developed a random population consisting of 30,000 individuals from NIL-H lines (NIL plants with heterozygous allele *GLW7.1/glw7.1*) and screened recombinants in the target region using two newly developed markers (G7.1 and LG15). A total of 600 recombinants were identified, and further genotyping was conducted using 16 newly developed simple sequence repeats (SSR) and kompetitive allele-specific PCR (KASP) markers (Figure 4A, Table S1). The grain size of the 600 recombinants and 70 non-recombinants derived from the random population was investigated, and *GLW7.1* was mapped to the interval between LG18 and K5 by a subsequent QTL analysis (Figure S3A). Subsequently, we performed a progeny test by investigating the grain size of homozygous progenies derived from each recombinant, and three non-recombinant lines (NIL-J, NIL-C and NIL-H) were designated as controls. In the end, the *GLW7.1* locus was narrowed to the region between markers K17 and K19 (Figure 4B and Figure S3B).

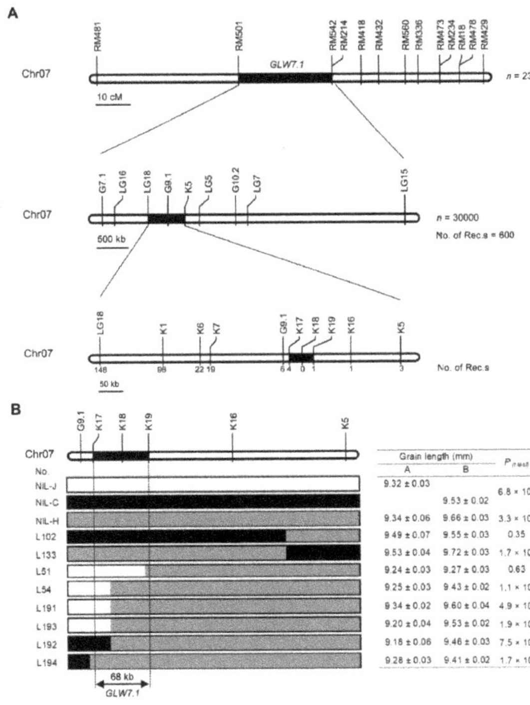

Grain length (mm) A	Grain length (mm) B	$P_{t\text{-}test}$	Progeny test
9.32 ± 0.03		6.8×10^{-7}	A
	9.53 ± 0.02		B
9.34 ± 0.06	9.66 ± 0.03	3.3×10^{-4}	H
9.49 ± 0.07	9.55 ± 0.03	0.35	B
9.53 ± 0.04	9.72 ± 0.03	1.7×10^{-2}	H
9.24 ± 0.03	9.27 ± 0.03	0.63	A
9.25 ± 0.03	9.43 ± 0.02	1.1×10^{-4}	H
9.34 ± 0.02	9.60 ± 0.04	4.9×10^{-6}	H
9.20 ± 0.04	9.53 ± 0.02	1.9×10^{-8}	H
9.18 ± 0.06	9.46 ± 0.03	7.5×10^{-4}	H
9.28 ± 0.03	9.41 ± 0.02	1.7×10^{-3}	H

Figure 4. Map-based cloning of *GLW7.1*. (**A**) Fine mapping of the *GLW7.1* using 30,000 BC_5F_2 segregants. Numbers below the line indicate the number of recombinants between *GLW7.1* and the marker shown. (**B**) Genotypes and phenotypes of the recombinants. Grain length (mean ± s.e.m.) of three near-isogenic lines (NIL), and recombinant BC_5F_3 lines (L102, L133, L51, L54, L191, L193, L192, L194). White bars represent chromosomal segments for J23B homozygote (progeny test named as A), black for CR071 homozygote (progeny test named as B), and grey for heterozygotes (progeny test named as H). Homozygous progenies from each line were harvested to compare phenotypic differences. The Student's *t*-test was used to produce *p* values.

By comparing the genomic sequences of Nipponbare (http://rice.uga.edu/, accessed on 18 March 2019), Zhenshan97 and Minghui63 (https://rice.hzau.edu.cn/rice_rs3/, accessed on 18 March 2019), we found a large fragment (~38 kb/55 kb) insertion between Zhenshan97 and Nipponbare/Minghui63 in the candidate region between markers K17 and K19 (Table S3). In order to fine-map the candidate gene, the whole genome of J23B and

CR071 were separately sequenced on Illumina and Nanopore (ONT) platforms to capture the target candidate segment sequences. Compared with CR071, J23B contained a 53 kb deletion in the candidate segment between markers K17 and K19 (68 kb in J23B and 121 kb in the CR071) (Table S3).

Three predicted open reading frames (ORFs) (*ORF1*, *ORF2* and *ORF4*) were located in the 68 kb target region of J23B, and four predicted ORFs (*ORF1*, *ORF2*, *ORF3* and *ORF4*) were located in the corresponding 121 kb region of CR071, excluding those ORFs encoding transposon and retrotransposon proteins (Figure 5A). *ORF1*, *LOC_Os07g15670*, encodes a putative peroxiredoxin. *ORF2*, *LOC_Os07g15680*, encodes a putative phospholipase D. *ORF3*, *LOC_Os07g15770*, encodes a CCT motif family protein, GHD7, which was reported to regulate heading date and yield potential in rice [42] (Figure 5B). *ORF4*, *LOC_Os07g15820*, encodes an expressed protein with unknown function.

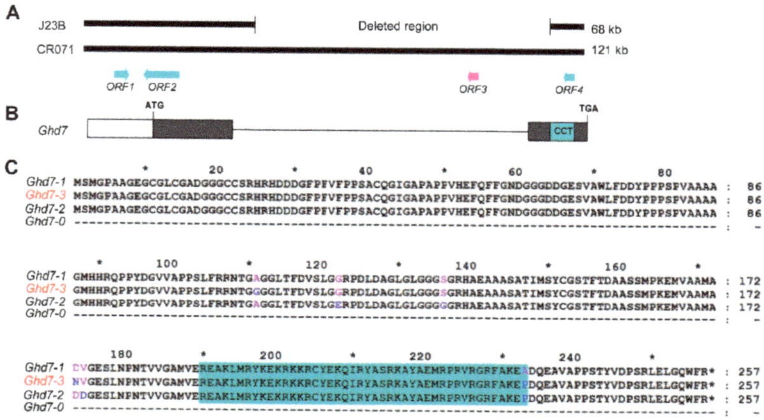

Figure 5. Candidate gene of *GLW7.1*. (**A**) A deletion of 53 kb was detected in J23B compared with CR071, which included the candidate gene of *GLW7.1* (the arrow symbol in magenta), while no variation was detected in the protein-coding regions of the other three ORFs (arrow symbols in cyan) and their corresponding promoter regions. (**B**) *ORF3* encodes the CCT motif family protein, GHD7. CCT domain is indicated in the cyan box. (**C**) The protein sequences of GHD7 for four *Ghd7* alleles. CR071 carried *Ghd7-3* allele of *Ghd7* compared with Minghui 63 (*Ghd7-1*), Nipponbare (*Ghd7-2*) and J23B (*Ghd7-0*). CCT domain (aa. 190–233) is indicated with cyan background. Polymorphic amino acids are indicated by different colors using the amino acids sequence of *Ghd7-1* allele as reference. Asterisks above amino acid sequences indicate positions of amino acids (10, 30, 50, 70, 90, 110, 130, 150, 170, 190, 210, 230 and 250) and asterisks within amino acid sequences indicate stop codons.

2.4. Positional Cloning of GLW7.1

To ascertain the candidate gene underlying *GLW7.1*, we compared the genomic sequences of the three ORFs (*ORF1*, *ORF2* and *ORF4*) from J23B and CR071, including promoter regions and protein-coding regions, and found no variation. Therefore, *ORF3* or *Ghd7*, which is located in the 53 kb deleted region of J23B, was likely to be the candidate of *GLW7.1*. We subsequently compared the genomic sequences of *Ghd7* from Minghui63, CR071 and Nipponbare. Different from the reported *Ghd7-1* allele of Minghui63, the allele of CR071 was termed as *Ghd7-3* because of three amino acid substitutions, and the allele of Nipponbare was termed as *Ghd7-2* because of four amino acid substitutions. The allele of J23B and Zhenshan97 was termed as *Ghd7-0* because of the loss of the complete gene region [42] (Figure 5C).

To determine whether the *Ghd7-3* allele underlies the QTL *GLW7.1*, we conducted a knockout experiment by editing the allele using the CRISPR-Cas9 system in the background of NIL-C. The sequence (c. 512 TGGCCAATGTTGGGGAGAGC) in the second exon was designed as the sgRNA target site to produce mutations neighboring the CCT domain coding region (Figure 6A). Three mutated alleles were obtained, named A1, A4 and A8. The allele A4 showed minor amino acids change (AN $\rightarrow$ D) and still retained the CCT domain. By contrast, both 1 bp insertion in allele A1 and 20 bp deletion in allele A8 resulted in frameshift mutations, which caused the loss of the CCT domain (Figure 6B). As expected, the alleles A1 and A8 produced smaller grains than the allele A4 in the NIL-C background (Figure 6C–F). We also conducted a complementation experiment by expressing the cDNA of the *Ghd7-3* allele driven by its native promoter in the NIL-J background. The grains produced from the complemented lines Com1, Com2 and Com3 were larger than those from negative transgenic plants NIL-J-Neg (Figure 6C–F). Additionally, the other two alleles of *Ghd7* (*Ghd7-1* and *Ghd7-2*) could also increase grain size in the Zhenshan97 background (Figure S4). Together, these data indicate that *Ghd7-3* is the functional gene underlying *GLW7.1*.

2.5. Haplotype Analysis of Ghd7

In order to investigate natural variations in *Ghd7*, we analyzed the sequencing data of 533 core germplasms in the *Ghd7* region [50]. Based on the nonsynonymous polymorphisms in the coding region that lead to amino acid substitutions or protein premature truncation (Table S4), nine haplotypes of *Ghd7* can be identified (Figure 7A), in agreement with the types reported [51]. Of these, four major haplotypes contained 498 accessions, while the remaining five rare haplotypes contained only 27 accessions. The four major haplotypes of *Ghd7* have been reported to be strong function, weak function and loss-of-function, respectively [42]. Three major haplotypes mainly existed in *indica*: Hap1 (*Ghd7-1*) represented by Minghui63 and 9311 was the type with strong function, Hap2 (*Ghd7-3*) represented by Teqing, and CR071 was another type with strong function, and the Hap9 (*Ghd7-0*) represented by Zhenshan97 and J23B was the type with loss of function. In addition, Hap4 (*Ghd7-2*) represented by Nipponbare and Zhonghua11, was the major haplotype with weak function in *japonica*.

Given that the middle region of CCT domain proteins was previously reported to have transactivation activity [52], we performed transcription activation assay in rice protoplasts prepared from leaf sheath of Nipponbare seedlings to verify the GHD7-mediated activation among three haplotypes. The three allelic GHD7 proteins (Hap1, Hap2 and Hap4 were respectively derived from Minghui63, CR071 and Nanyangzhan) were fused to the GAL4 DNA-binding domain (GAL4DBD) to generate effectors, and the firefly luciferase gene (*LUC*) was used as the reporter (Figure 7B) [53]. Compared with the GAL4 negative control, the three allelic GHD7 proteins showed dramatically different activation activity (Figure 7B and Figure S5B). The protein GHD7-Hap4 exhibited the strongest activation activity, compared with the weaker activity exhibited by GHD7-hap1 and the weakest activity by GHD7-Hap2. The difference in activation activity of the three allelic proteins may be due to amino acid substitutions.

To test this hypothesis, we added another four allelic GHD7 proteins (Hap3, artificial HapN1 combining exon1 region of Hap1 with exon2 region of Hap2, artificial HapN2 combining exon1 region of Hap1 with exon2 region of Hap4, and artificial HapN3 combining exon1 region of Hap2 with exon2 region of Hap4) to generate effectors (Figure 7B). A comparison of GHD7-Hap1 and GHD7-Hap3 revealed that the A233P substitution seems to have no effect on activation activity (Figure 7B and Figure S5C,D), consistent with its position in the CCT DNA-binding domain. By comparing GHD7-Hap2 and GHD7-HapN1, GHD7-HapN2 and GHD7-HapN3, we found that the A111G substitution seems to also have no effect on activation activity (Figure 7B and Figure S5C,D). A comparison of GHD7-Hap3, GHD7-Hap4 and GHD7-HapN2 revealed that both the G122E/S136G and V174D substitutions could strengthen the activation activity (Figure 7B and Figure S5C,D). By contrast, the D173N substitution could weaken the activation activity, by comparing GHD7-Hap3 and GHD7-HapN1 (Figure 7B and Figure S5C,D). These results suggest that the different genetic effects of the three major haplotypes may be due to the transcription activation activity difference among GHD7 proteins, which was promoted when G122E/S136G and V174D substitutions were contained and weakened when D173N substitution were contained.

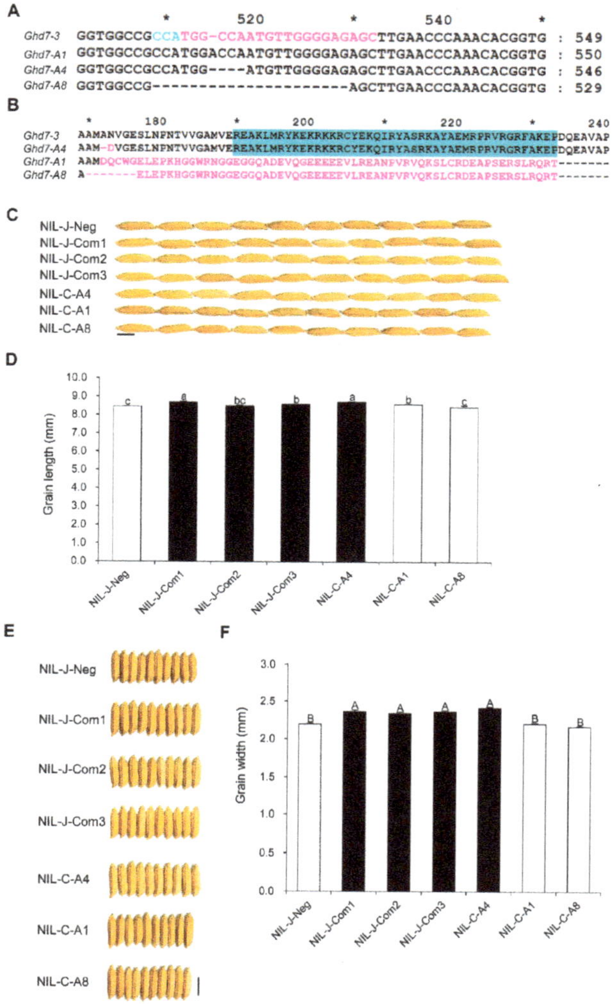

Figure 6. Transgenic validation of *GLW7.1*. (**A**,**B**) Gene editing targeting the upstream of CCT domain was conducted using CRISPR-Cas9 system. Compared with the amino acid change mutation (AN → D) caused by mutated allele A4, the mutated allele A1 with 1 bp insertion and A8 with 20 bp deletion resulted in frameshift mutations, which caused the loss of the CCT domain in the GHD7 protein. (**A**) The sgRNA target site is shown in magenta, the PAM sequence is shown in cyan, and asterisks above nucleotide sequences indicate positions of nucleotides (510, 530 and 550). (**B**) CCT domain is indicated in cyan background, polymorphic amino acids are indicated in magenta, and asterisks above amino acid sequences indicate positions of amino acids (170, 190, 210 and 230). (**C**,**E**) Grains morphology of transgenic lines. Scale bar: 0.5 cm. (**D**) Grain length of transgenic lines. (**F**) Grain width of transgenic lines. All phenotypic data were measured from paddy-grown transgenic lines grown under normal cultivation conditions. Data are represented as mean ± s.e.m. (n = 27 for NIL-J-Neg, 15 for NIL-J-Com1, 22 for NIL-J-Com2, 20 for NIL-J-Com3, 10 for NIL-C-A4, 14 for NIL-C-A1 and 4 for NIL-C-A8) and Duncan's multiple range tests were used to conduct statistical analysis (a, b and c indicate $p < 0.05$; A and B indicate $p < 0.01$).

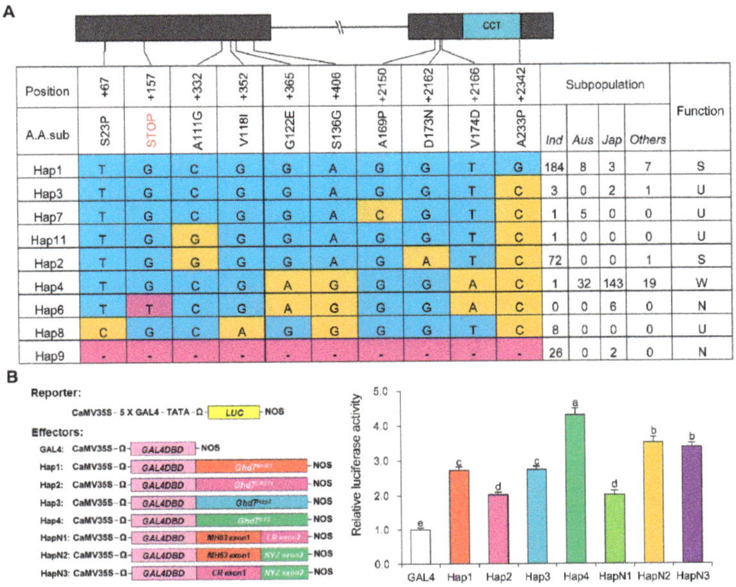

Figure 7. Haplotype analysis of *Ghd7*. (**A**) The nonsynonymous polymorphisms in the *Ghd7* CDS region that cause changes in the amino acid sequence of 525 cultivars. CCT domain is indicated in the cyan box. Polymorphic nucleotides that cause amino acid substitutions are indicated in yellow using the amino acids sequence of Hap1 as reference, and the nucleotides that cause frame-shift mutation or absence of the gene region are indicated in magenta. S, U, W and N represent strong functional, unknown functional, weak functional and nonfunctional alleles, respectively. (**B**) The transactivation activity of different allelic GHD7 proteins. Six allelic GHD7 were fused to the GAL4 DNA-binding domain (GAL4DBD). The relative activity of firefly luciferase (LUC) under control of the 5×GAL4-binding element was measured. Renilla luciferase (REN) activity was used as internal control. Data are represented as mean ± s.e.m. (n = 10) and Duncan's multiple range tests were used to conduct statistical analysis (a, b, c, d and e indicate $p < 0.05$).

2.6. GLW7.1 Determines Grain Size via Grain-Size Genes

To uncover the molecular pathway by which *GLW7.1* regulates grain size, we conducted a yeast two-hybrid (Y2H) screen using the C-terminal of GHD7 (aa. 208–257) as bait and a normalized prey library derived from young panicles of Zhenshan97. A total of 325 candidate positive clones were detected on synthetic growth medium without leucine, tryptophan, histidine and adenine. Of them, seven grain-size proteins (OsFBK12, FZP, OsNAC024, OsNAC025, OsNF-YC12, RICE STARCH REGULATOR 1 (RSR1) and SNB) [37,38,54–57] and the rice DELLA protein SLENDER RICE 1 (SLR1) [58] were selected for examination by reconstructing the prey vectors with the full-length protein sequences. The interactions of GHD7 with these proteins were then confirmed by X-α-gal filter lift assays (Figure 8A). Furthermore, the interactions were also demonstrated using split firefly luciferase complementation (SFLC) assays in tobacco leaf epidermal cells (Figure 8B). These results imply that the GHD7 protein may regulate the grain size and weight through interactions with the above grain-size proteins and DELLA protein.

To explore the downstream genes of *GLW7.1* in regulating grain size, we detected the expression levels of 63 grain-size genes in NIL-J and NIL-C by qRT-PCR analysis using the young panicles (8–10 cm in length). The *GLW7.1* locus significantly up-regulated the expression of nine positive grain-size genes (fold-change > 1.5 and $p < 0.01$), including genes encoding OsBZR1 (BES1/BZR1 homolog protein) [59], OsMAPK6 [15], GLW7 [31], OsbHLH107 [35], IDEAL PLANT ARCHITECTURE 1 (IPA1) [60], SRS3 (a kinesin motor domain protein) [61], SMALL AND ROUND SEED 5 (SRS5) (alpha-tubulin protein) [62] and OsWRKY53 [19] (Figure 9A). Furthermore, we observed an extremely significant down-regulation of *OsMADS1* (fold-change = 56.7 and $p = 6.1 \times 10^{-7}$), which negatively regulates grain length [9] (Figure 9A). Furthermore, transcription activation assay in rice protoplasts prepared from leaf sheath of Zhenshan97 seedlings shown that the LUC activity driven by the *OsMAPK6* promoter was significantly induced by GHD7 (Figure 9B), which indicates that GHD7 could directly activate the expression of *OsMAPK6*. Overall, these results suggest that *GLW7.1* positively regulates grain size through a series of grain-size genes.

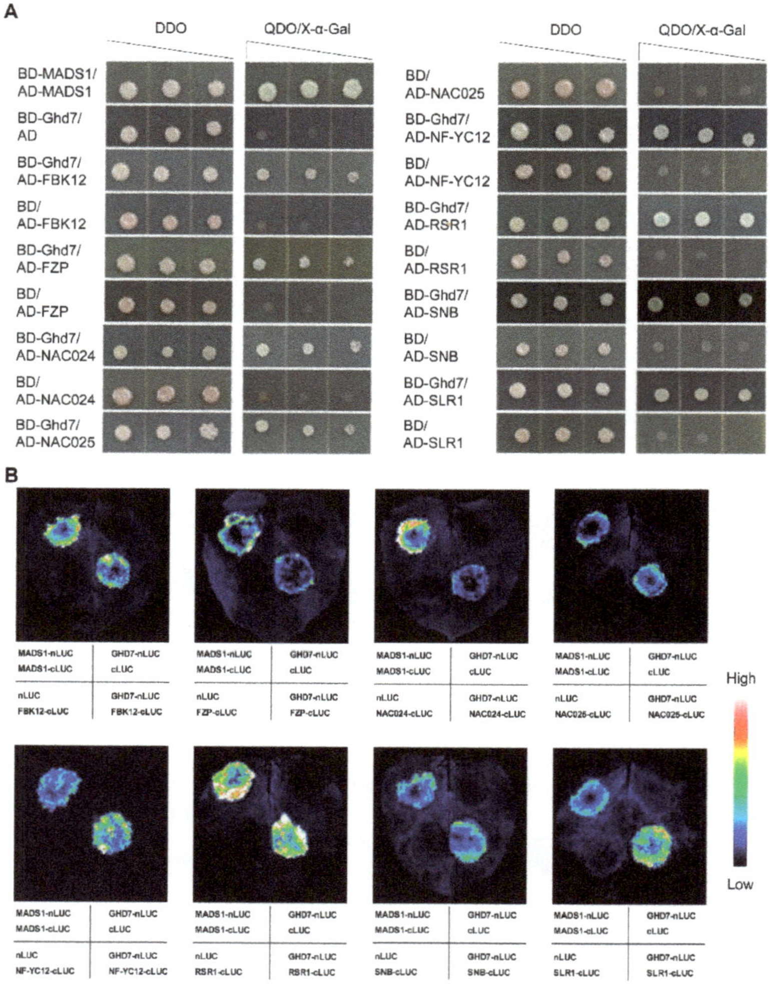

Figure 8. Eight candidate proteins that interact with GHD7. (**A**) Yeast two-hybrid assays. Serial dilutions of 10^3–10 transformed yeast cells were spotted on the control medium DDO (SD/-Trp/-Leu) and selective medium QDO (SD/-Trp/-Leu/-His/-Ade). The protein self-dimerization of OsMADS1 was used as positive controls. Co-transformed empty vectors pGADT7 (AD) and pGBKT7 (BD) were used as negative controls. (**B**) Split firefly luciferase complementation (SFLC) assays. nLUC-tagged GHD7 was co-transformed into tobacco leaves along with the cLUC-targeted candidate proteins. The protein self-dimerization of OsMADS1 was used as positive controls. Co-transformed empty vectors nLUC and cLUC were used as negative controls.

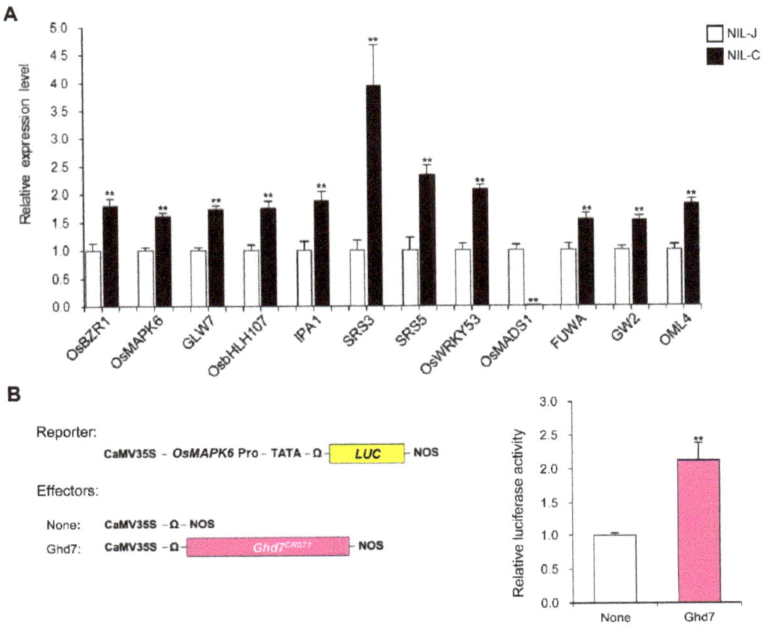

Figure 9. *GLW7.1* regulates the expression of grain-size genes. (**A**) Relative expression level of 12 grain-size-related genes in young panicles (8–10 cm) of NIL-J and NIL-C. *OsActin* (*LOC_Os03g50885*) was used as the control and the values of expression level in NIL-J were set to 1. Data were represented as mean ± s.e.m. ($n = 9$). The Student's t-test was used to produce p values (** indicates $p < 0.01$). (**B**) GHD7 induces the LUC activity driven by *OsMAPK6* promoter. Hap2 allelic GHD7 were overexpressed as effectors. The relative activity of firefly luciferase (LUC) under control of the promoter region of *OsMAPK6* was measured. Renilla luciferase (rLUC) activity was used as internal control. Data are represented as mean ± s.e.m. ($n \geq 5$). The Student's t-test was used to produce p values (** indicates $p < 0.01$).

2.7. GLW7.1 Positively Regulates GA Biosynthesis

Mutants with defection in GA biosynthesis, such as *gdd1* and *sgd2*, usually show reduced height and small seeds [27,29]. Using quantitative RT-PCR, we detected the up-regulated expression of four GA biosynthetic genes (*KS1, KO2, KAO* and *GA3ox2*), GA catabolic gene (*GA2ox3*) and the GA signaling pathway gene *SLR1* in the young panicles of NIL-C, compared to NIL-J (Figure 10A). To examine whether *Ghd7* was involved in GA biosynthesis, we analyzed the response of NIL-J and NIL-C to exogenous GA_3 and paclobutrazol (PAC, a GA biosynthesis inhibitor) treatment. The length of the second leaf sheath of NIL-J was significantly shorter than NIL-C and could be restored to the NIL-C level by exogenous GA_3 treatment (Figure 10B,C). In addition, the growth of both NIL lines was simultaneously inhibited by exogenous PAC treatment, and their second leaf sheaths were almost the same in length (1.51 cm in NIL-J and 1.41 cm in NIL-C) (Figure 10B,C). Moreover, the second leaf sheath of NIL-J was almost as long as NIL-C with different GA_3 concentrations treatment (Figure S6). Collectively, these findings suggest that *Ghd7* may be involved in GA biosynthesis rather than GA response.

To further confirm the role of *Ghd7* in GA biosynthesis, we measured endogenous GA_1 levels in 2-week-old seedlings of NIL-J and NIL-C. The GA_1 level in NIL-J was approximately 84.6% (0.69 ng/g) of that in NIL-C (0.81 ng/g) (Figure 10D). The effect of exogenous GA_3 treatment on *Ghd7* expression was also investigated by quantitative RT-PCR. The diurnal expression pattern of *Ghd7* was consistent with the previous report [42], and the expression level of *Ghd7* was significantly inhibited at 6 h after the exogenous

GA_3 treatment (Figure 10E). Interestingly, the expression levels of two GA biosynthetic genes, *GA20ox2* and *GA3ox2*, were also reported to be inhibited after the exogenous GA3 treatment [27]. Together, these results suggest that *Ghd7* positively regulates endogenous GA biosynthesis.

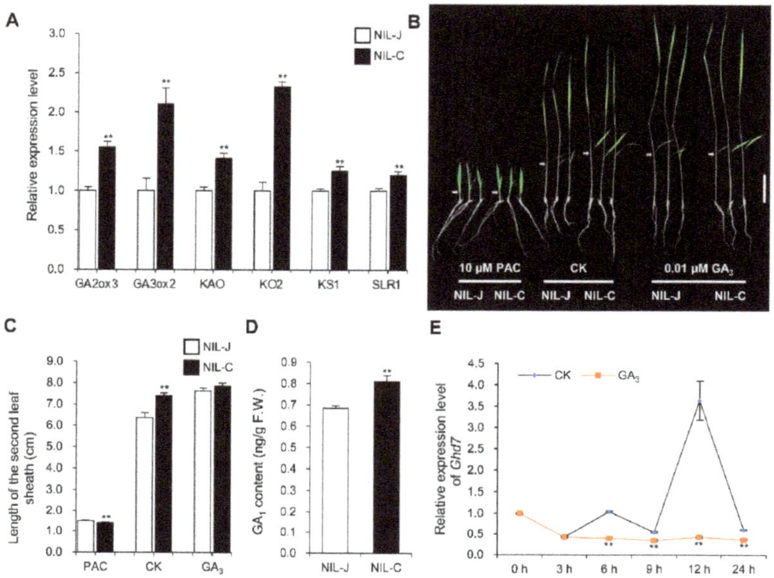

Figure 10. GLW7.1 participates in the biosynthesis of GA. (**A**) Relative expression level of 6 GA related genes in young panicles (8–10 cm) of NIL-J and NIL-C. *OsActin* (*LOC_Os03g50885*) was used as the control and the values of expression level in NIL-J were set to 1. Data were represented as mean ± s.e.m. (n = 9). Student's t-test was used to produce p values (** indicates $p < 0.01$). (**B**) Seedling phenotype of NIL-J was rescued by 0.01 μM GA_3. The germinated seeds were grown in the nutrient solution that contained 0.01 μM GA_3 or 10 μM paclobutrazol (PAC) and incubated at 28 °C under 13 h light/11 h dark conditions. CK, nutrient solution without any exogenous hormones. After 10 days, the seedlings were photographed. Scale bar: 5 cm. Arrow symbols indicate the second leaf sheaths. (**C**) The length of the second leaf sheaths were measured after treatment. Data are represented as mean ± s.e.m. ($n \geq 15$). Student's t-test was used to produce p values (** indicates $p < 0.01$). (**D**) Levels of endogenous GA1 in NIL-J and NIL-C 2-week-old seedlings grown in nutrient solution without treatment. Data are represented as mean ± s.e.m. (n = 4). Student's t-test was used to produce p values (** indicates $p < 0.01$). (**E**) Expression pattern of *Ghd7* under GA_3 treatment. The Zhonghua11 germinated seeds were grown in the nutrient solution. Two weeks later, half of the seedlings were moved to the nutrient solution that contained 50 μM GA_3, while the other half were moved to the nutrient solution without treatment as control. Relative expression level of *Ghd7* were detected. *OsActin* was used as the control and the values of expression level at 0 h were set to 1. Data were represented as mean ± s.e.m. ($n \geq 3$). Student's t-test was used to produce p values (** indicates $p < 0.01$).

3. Discussion

3.1. GLW7.1 Simultaneously Improves Grain Yield and Quality

In this study, we performed the map-based cloning of *GLW7.1*, a novel QTL regulating grain size, and confirmed that *Ghd7*, encoding a CCT motif family protein, underlies the QTL. *Ghd7* was previously reported as a major regulator of heading date and improved yield by increasing grain number [42]. Our results demonstrate that *GLW7.1*, or *Ghd7-3*, not only increases grain number, but also increases grain weight, manifested as increases in both grain length and width (Figure 1). On the other hand, *GLW7.1* has effects on reducing chalkiness and improving cooking and eating quality (Figure 2). Thus, *GLW7.1* or *Ghd7-3* is a positive regulator of both rice yield and quality. In contrast, many genes regulating grain size have negative effects on rice quality. For example, *GS2* and *GW2* increase not only grain size and weight, but also chalkiness simultaneously [46,47]. In addition, only 73 out of 533 accessions carry the *GLW7.1* allele (Figure 7A). Therefore, *GLW7.1* is a promising allele for simultaneously improving grain yield and quality during rice breeding.

3.2. Natural Variations Alter the Transcriptional Activity of GHD7

In this study, we performed haplotype analysis of *Ghd7* using a germplasm population consisting of 533 accessions and identified nine haplotypes based on nonsynonymous SNPs in the coding region (Figure 7A). Among those, the three reported functional alleles with different effects on heading date were included, which were Hap1 (*Ghd7-1*), Hap2 (*Ghd7-3*) and Hap4 (*Ghd7-2*) [42]. In order to uncover the reason underlying different effects of the three allelic GHD7 proteins, we focused on the six amino acid substitutions among them. The subsequent transcription activation activity assay showed that the G122E/S136G and V174D substitutions strengthen the activation activity, the D173N substitution weakens it, and the A111G and A233P substitutions have no effect on it (Figure 7B and Figure S5). It has been demonstrated that acidic amino acids in the activation domain are essential for the transcriptional activation of transcription factors. For example, when all the acidic amino acid residues in the activation domain of transcription factor OCT4 were replaced by alanine, its transcriptional activation activity decreased dramatically [63]. In our study, Hap4 allelic GHD7 contains two amino acid substitutions that increase acidic amino acids, G122E and V174D; Hap2 allelic GHD7 contains an amino acid substitution that decreases the acidic amino acid, D173N. Coincidentally, these acidic amino acid substitutions in the activation domain also cause similar changes in GHD7 transcriptional activation.

Furthermore, we noticed the transcriptional repression activity of GHD7 on *ARE1* [44], which seems to contradict our results. The study investigated the transcriptional repression of GHD7 on a specific downstream gene, *ARE1*, but did not investigate the transcriptional activation or repression activity of GHD7 itself. Actually, there was one study investigating the transcriptional repression activity of GHD7 [45]. Weng et al. found that the transcriptional activation activity of GAL4-VP16-GHD7 fusion protein was significantly weaker than that of GAL4-VP16, and they draw the conclusion that GHD7 had intrinsic transcriptional repression activity. We thought the conclusion was not rigorous because GHD7 fused to the C-terminal of VP16 may weaken the transcriptional activation activity dependent on VP16 C-terminal, and the transcriptional repression activity may not be caused by GHD7, but by C-terminal fusion. In our study, we investigated the transcriptional activation activity of GHD7 itself (Figure 7B and Figure S5) and observed its transcriptional activation of *OsMAPK6* (Figure 9B). However, it was reported that the ABI4 protein has different transcriptional activity for different downstream target genes. ABI4 could bind to promoters of *GA2ox7* and *NCED6* to activate the expression of these two genes [64], and could also bind to promoters of *CYP707A1* and *CYP707A2* to inhibit the expression of these two genes [65]. We hypothesized that GHD7 may function in a similar way on target genes.

In conclusion, natural variations in GHD7 proteins affect its transcriptional activity, which is likely to influence transcription of the downstream genes, and finally result in different effects on heading date and other traits. In addition, there are another three amino acid substitutions with unknown effects on its activation activity (Figure 7A), which may endue the remaining four haplotypes with different functions. Therefore, it would be of great importance to reveal the effect of each amino acid substitution and further construct different combinations of these natural variations, which may provide new ideas for future applications of *Ghd7* in rice breeding.

3.3. The Pathway of GLW7.1 Controlling Grain Size

In this study, for the first time, we showed that *Ghd7* is a major gene regulating grain size, which has long been recognized as a major regulator of heading date in rice [42,51]. Recently, the interactions of GHD7 with the CCAAT-box-binding transcription factors, OsNF-YB11 and OsNF-YC2, have been elucidated [66]. Similar to the interactions between GHD7 and OsNF-YB11/OsNF-YC2, a physical interaction between GHD7 with OsNF-YC12, another CCAAT-box-binding transcription factor, was detected in yeast cells and tobacco leaf epidermal cells (Figure 8). For *OsNF-YC12*, the mutants showed a decrease in grain length and width, and the overexpression showed an increase, implying a positive regulation in grain size [56]. We also found the interactions between GHD7 with three AP2 domain containing proteins, FZP, RSR1 and SNB (Figure 8) [37,38,57]. Of the three, FZP positively and SNB negatively regulate grain length and width by simultaneously affecting cell proliferation and expansion in spikelet hulls. Additionally, OsFBK12, an F-box protein containing a Kelch repeat motif, which positively regulates seed size by increasing cell size but decreasing cell number [54], and two NAC-type TFs, OsNAC024 and OsNAC025, which bind to the promoters of three grain size/weight regulating genes (*GW2*, *GW5* and *DWARF 11* (*D11*)) to modulate grain size [55], were also observed in the yeast two-hybrid assays and SFLC assays (Figure 8).

Expression analysis revealed that *GLW7.1* up-regulated the transcription of eight genes having positive effects on grain size, and down-regulated that of *OsMADS1*, a negative regulator of grain size, in NIL-C panicles (Figure 9A). Interesting, the expression of *OsMADS1* was significantly reduced in the mutants of *SNB* and *FZP*, which encode proteins interacting with GHD7, implying *OsMADS1* may be downstream of these two genes [67,68]. In addition, cytological observations showed that *GLW7.1* enhances grain size by promoting cell proliferation and expansion (Figure 3A–F). Among the up-regulated positive regulation genes of grain size, *OsBZR1* can affect grain length and grain width simultaneously, which are located at the downstream of BR signaling pathway [59]. The OsSPL13 protein encoded by *GLW7* could bind to the promoter of *SRS5* and activate its expression, and the two genes could promote glume cell expansion and jointly regulate grain length [31,62]. *OsMAPK6* and *OsWRKY53*, both of which belong to the MAPK signaling cascade, could promote cell proliferation and thus increase grain length and width in rice [15,19]. *OsMADS1*, a significantly down-regulated negative regulation gene of grain length, also regulates rice grain length by affecting cell proliferation [9]. However, the transcription activation activity assay showed that GHD7 could directly activate the expression of *OsMAPK6* (Figure 9B), which indicates that GHD7 may directly bind to the OsMAPK6 promoter to regulate its expression and thus control rice grain size.

We also investigated the expression of several genes that regulate cell cycle and cell expansion and found significantly higher expression levels of 10 cell-cycle-related genes and 3 cell expansion related-genes in NIL-C panicles (Figure 3G). However, we noted the up-regulation of three negative grain-size genes and the down-regulation of one expansion gene, which may indicate some unclear regulatory mechanisms. Therefore, based on the results above, we proposed that GHD7 interacts with several grain-size-related proteins, up-regulates the positive regulation genes of grain size, such as *OsMAPK6*, and down-regulates the negative, *OsMADS1*, thereby promoting the transcription of downstream cell division and expansion genes, and finally enhancing the grain size as well as grain weight (Figure S7).

3.4. GHD7 Participates in GA Pathway

In the GA signaling pathway, GA binds to the GID1 receptor, leading to the formation of a GID1-GA-DELLA complex, which further stimulates the interaction of DELLA with the SCF^{GID2} complex. Once recruited to SCF^{GID2} complex, DELLA is polyubiquitylated and then subsequently degraded through the 26S proteasome pathway [69]. Mutants with defection in the GA signaling pathway usually show reduced height and small seeds. For example, *GDD1* encodes a kinesin-like protein that directly regulates the expression of the

KO2 gene [27], and *SGD2* encodes an HD-ZIP II transcription factor that positively regulates the expression of GA biosynthesis genes [29], while their mutants reduce endogenous GA levels, leading to a decrease in cell size, thus resulting in a severely dwarfed, small-grain phenotype. Moreover, overexpressing RGG2, which mediates internal GA biosynthesis and participates in the GA signaling pathway, also causes dwarfism and small grains [7]. Moreover, the grain-size gene that was upregulated most in NIL-C, SRS3 (Figure 9A), the mutant of which showed reduced height and short seeds, was also reported to participate in regulating the expression of genes in the GA biosynthesis pathway [28]. In this study, we also observed that *Ghd7* could positively regulate endogenous GA biosynthesis (Figure 10).

In *Arabidopsis*, DELLA proteins physically interact with the CCT domain of CONSTANS (CO) and integrate gibberellic acid and photoperiod signaling to regulate flowering under long days [70]. Here, we identified the interactions of GHD7 with the rice DELLA protein, SLR1 (Figure 8), which has been reported to interact with transcription factors, such as GRF4-GIF1, NACs and OsPIL14, and inhibit their transcriptional activation of downstream genes [71–73]. Comprehensively considering the pleiotropic effect of *Ghd7* on plant height and grain size (Figure 1, Figure 6 and Figure S8), we hypothesized that GHD7 participates in the GA biosynthesis to increase grain size (Figure S7) and is regulated by the GID1-GA-DELLA module as feedback of the pathway.

4. Materials and Methods

4.1. Plant Materials and Growth Conditions

Two *indica* cultivars Jin23B (J23B) and CR071 were used to construct the QTL mapping population, with details described in Figure S1. The NIL populations used for the genetic analysis of *GLW7.1* in grain size in two years were, respectively, isolated from BC_4F_1 and BC_5F_2 plants with heterozygous allele *GLW7.1/glw7.1*. The progeny test was conducted in the BC_5F_4 generation. The NILs (NIL-J and NIL-C) involved in a series of subsequent experiments were isolated from the BC_5F_n ($n \geq 6$) plant with heterozygous allele *GLW7.1/glw7.1*. The NIL-J-Com lines were obtained by complementation test. The mutants (NIL-C-A lines) were obtained by CRISPR/Cas9-based genome editing. The T_1 generation materials were used for the analysis. The *Ghd7-2* allele lines (NIL-NYZ and NIL-ZS) were kindly provided by Guangming Lou. The Ghd7-1 allele lines (ZS-Com and ZS-Neg) were kindly provided by Lei Wang. The rice plants were grown at the experimental field of Huazhong Agricultural University in Wuhan during the summer with a density of 16 cm × 26 cm under normal field management.

4.2. Trait Measurement

Fully filled grains from each plant were used for measuring grain length, grain width, grain number, grain yield and 1000-grain weight by the yield traits scorer (YTS) platform [74] after 2014. (The phenotype of grain length for primary QTL mapping was measured by an electronic digital display caliper in 2012.) The plant height was measured from the main culm. The number of tillers per plant was counted as all fertile panicles in one plant. The percentage of grains with chalkiness, amylose content and gel consistency were measured according to the NY/T 593-2013 standard published by the Ministry of Agriculture, China (http://www.zbgb.org/27/StandardDetail1476335.htm, accessed on 3 October 2019). The taste score of milled rice was evaluated using a taste analyzer kit (Satake, RLTA10B-KC, Hiroshima, Japan) [75].

4.3. Linkage Analysis and QTL Mapping

A primary mapping population of 238 BC_3F_1 individuals was generated from a cross between J23B and CR071 (Figure S1). The plants were then genotyped by 157 polymorphic SSR markers covering the whole genome. The grain length was measured and the linkage analysis was carried out by composite interval mapping module of the software WinQTLCart 2.5 [76]. The R/qtl package [77] was employed to plot the linkage map using the output of WinQTLCart.

For fine mapping of the *GLW7.1*, we developed a BC_5F_3 population consisting of 30,000 individuals from NIL plants with heterozygous allele *GLW7.1/glw7.1*. *GLW7.1* was mapped to the interval between LG18 and K5 by a subsequent linkage analysis, and then narrowed to the region between markers K17 and K19 by progeny test. Relevant primer sequences were listed in Table S1.

4.4. Scanning Electron Microscopy

Lemmas of spikelets at the heading stage were collected and fixed in FAA solution (50% ethanol, 5% glacial acetic acid and 3.7% formaldehyde) for more than 16 h. The fixed samples were dehydrated in a graded ethanol series and then critical point dried, followed by being coated with gold. The samples were then observed with a scanning electron microscope (JEOL, JSM-6390LV, Tokyo, Japan) at an accelerating voltage of 10 kV and a spot size of 30 nm. The morphology of lemma cells was scanned at a magnification of 100× to measure cell length and cell width, and at 50× with three pictures that are combined to cover the entire lemma to measure the cell number. The cell size of the lemmas was measured from pictures using ImageJ software (NIH), and the cell number was counted manually.

4.5. RNA Extraction and Expression Analysis

Total RNA was extracted from young panicles (8–10 cm in length) and 2-week-old seedlings using TRIzol reagent (Invitrogen, 15596026, Shanghai, China). DNase I (Invitrogen, 18068015, Shanghai, China) pre-treated RNA was reverse-transcribed using the M-MLV Reverse Transcriptase kit (Promega, M170A, Madison, WI, USA) following the manufacturer's instructions. The qRT–PCR was then conducted in a total volume of 10 μL, which consisted of 5 μL of cDNA (10 ng/μL), 0.25 μL of each primer (10 μM), and 4.5 μL of 2× SYBR Green PCR Master Mix (Roche, 4913914001, Mannheim, Germany), using ABI Real-Time PCR systems (Q6 and ViiA7) according to the manufacturer's instructions. The *OsActin* gene (*LOC_Os03g50885*) was used as the internal control. The relative gene expression levels were calculated by the $2^{-\Delta\Delta Ct}$ method. Each measurement was performed with three biological samples and three replicates for each sample. Relevant primer sequences were listed in Table S2.

4.6. De Novo Assembly of Two Genomes and Sequences Comparison

In order to fine map the candidate genes, the whole genomes of J23B and CR071 were separately sequenced on Illumina and Nanopore (ONT) platforms to capture the target candidate segment sequences. For CR071, 50.3 Gb of ONT data (~135× genome coverage) and 3.8 Gb of Illumina data (~10× genome coverage) were used. For J23B, 10.2 Gb of ONT data (~28× genome coverage) and 5.2 Gb of Illumina data (~14× genome coverage) were used. The Nanopore reads were assembled using Canu [78] for CR071 and using wtdbg2 [79] for J23B. The contigs generated with Canu and wtdbg2 were polished with three rounds of Racon [80] based on Nanopore reads, followed by one round of Pilon [81] based on Illumina short reads. The assembled genomes were used in the subsequent analysis. We then captured the target candidate segment sequences in five genomes using the primer sequences of the two SNP markers and found a 53 kb deletion in J23B compared with CR071 (Table S3). The sequence comparison of J23B and CR071 was conducted by aligning the ONT reads and Illumina short reads of J23B and CR071 to the three annotated genomes (Zhenshan97, Minghui63 and Nipponbare) using minimap2 [82] and SAMtools [83] to detect the ORF underlying *GLW7.1*.

4.7. Haplotype Analysis

The variations in *Ghd7* in 533 accessions were queried from RiceVarMap v2.0 (http://ricevarmap.ncpgr.cn/, accessed on 5 July 2021) with variation IDs (vg0709154754, vg0709154664, vg0709154489, vg0709154469, vg0709154456, vg0709154415, vg0709152671, vg0709152659, vg0709152655, vg0709152479) in the coding region. We then identified nine haplotypes based on the diversity. Relevant data are listed in Table S4.

4.8. Vector Construction and Transformation

For preparing the complementation construct (Com), a 5 kb fragment, which consisted of 2.2 kb promoter and 2.8 kb genomic DNA of *Ghd7*, was amplified from CR071 and then cloned into the plant binary vector pCAMBIA1301. For preparing the CRISPR/Cas9 knockout construct, the sequence (c.512 TGGCCAATGTTGGGGAGAGC) in the second exon was designed as the sgRNA target site. The reverse complement sequence of the target site was inserted into the intermediate vector pER8-Cas9-U6 and then cloned into vector pCXUN-Cas9 [84]. The complementation construct was introduced into NIL-J and the knockout construct was introduced into NIL-C, respectively, by *Agrobacterium tumefaciens* (*EHA105*)-mediated transformation. The transgenic lines were further confirmed by PCR detection and direct sequencing. Relevant primer sequences were listed in Table S5.

4.9. Transcription Activation Assay

The coding sequences of exon1 and exon2 from different allelic *Ghd7* were amplified and combined to generate seven allelic *Ghd7*. These different allelic *Ghd7* were then fused with GAL4 DNA binding domain to generate effectors [53]. The firefly *LUC* gene was used as the reporter to analyze the transcriptional activity. Two allelic *OsMADS1* from Zhenshan97 and Nangyangzhan were fused with GAL4 DNA binding domain to generate the positive controls GALZ and GALN. To explore the downstream target genes, the Hap2 allelic *Ghd7* was fused into the 'None' effector vector, while the promoter of candidates (*OsBZR1*, *OsMADS1*, *OsMAPK6*, *OsSPL13* and *OsWRKY53*) was cloned into the '190LUC' reporter vector. The renilla *LUC* gene was used as an internal transformation control. The rice protoplasts prepared from the leaf sheath of Nipponbare and Zhenshan97 seedlings were transfected with different combinations of vectors by PEG-mediated transformation [85]. The firefly luciferase activity was detected after at least 12h using the Dual-Luciferase reporter kit (Promega, E1960, Madison, WI, USA), according to the manufacturer's protocol. Relevant primer sequences were listed in Table S5.

4.10. Yeast Two-Hybrid Assays

The prey library was derived from young panicles (5–15 cm in length) of Zhenshan97. The coding sequence of the C-terminal of GHD7 (aa. 208–257) was amplified and then cloned into the bait vector pGBKT7 (Clontech, 630443, Mountain View, CA, USA) for yeast two-hybrid screening. Full-length cDNAs of *OsFBK12*, *FZP*, *OsNAC024*, *OsNAC025*, *OsNF-YC12*, *RSR1*, *SNB* and *SLR1* were amplified and then cloned into the prey vector pGADT7 (Clontech, 630442, Mountain View, CA, USA), respectively, for subsequent yeast two-hybrid assays. All procedures were conducted according to the manufacturer's protocol. Relevant primer sequences are listed in Table S5.

4.11. SFLC Assays

Full-length cDNAs of *Ghd7*, *OsFBK12*, *FZP*, *OsNAC024*, *OsNAC025*, *OsNF-YC12*, *RSR1*, *SNB* and *SLR1* were amplified and then cloned into the nLUC vector (pCAMBIA1300-35S-HA-Nluc-RBS) or cLUC vector (pCAMBIA1300-35S-Cluc-RBS) [9], respectively, for subsequent split firefly luciferase complementation (SFLC) assays. Vectors for testing the protein–protein interactions (such as GHD7-nLUC and FBK12-cLUC), together with the p19 silencing vector, were co-transfected into tobacco (*N. benthamiana*) leaves via *Agrobacterium tumefaciens* (*EHA105*) infiltration. After at least 48 h, injected leaves were sprayed with 5 mM luciferin (Promega, E1605, Madison, WI, USA). The LUC signal was captured using a cooling CCD imaging apparatus (Tanon, Tanon-5200, Shanghai, China). Each assay was repeated at least three times. Relevant primer sequences were listed in Table S5.

4.12. Exogenous GA_3 and PBZ Treatment of Seedlings

The germinated seeds of NIL-J and NIL-C were grown in a nutrient solution that contained various concentrations of GA_3 (Sangon Biotech, A600738, Wuhan, China) or 10 μM Paclobutrazol (Sangon Biotech, A630332, Wuhan, China) and incubated at 28 °C under 13 h light/11 h dark conditions. After 10 days, the length of the second leaf sheaths was measured.

4.13. Measurement of GA_1

The shoots of 2-week-old seedlings were sampled, frozen in liquid nitrogen, and ground to fine powder. Tissues weighing 0.1 g were extracted with 1 mL 0.01 M PBS solution at 4 °C for 12 h. After centrifugation (12,000 rpm, 4 °C, 15 min), the supernatant was collected for GA_1 measurement. Endogenous GA_1 levels were detected by enzyme-linked immunosorbent assay (ELISA) following the manufacturer's instructions (Jingmei Biotechnology, JM-110038P2, Yancheng, China).

4.14. Statistical Analysis

ANOVA analysis or Student's *t*-test analysis were conducted using SPSS 22 (SPSS Inc., Chicago, IL, USA).

Supplementary Materials: The supporting information can be downloaded at: https://www.mdpi.com/article/10.3390/ijms23158715/s1.

Author Contributions: R.L. conducted most of the experiments, including fine mapping, gene cloning, genetic transformation, expression analysis, scanning electron microscopic analysis, Y2H analysis, transcription activation assay, SFLC assay and GA treatment. Q.F. participated in the Y2H analysis, transcription activation assay and SFLC assay. P.L. participated in revising the manuscript. G.L. conducted parts of the phenotyping. G.C. and H.J. participated in the development of the NILs. G.G., Q.Z., J.X., X.L. and L.X. participated in field management and logistics. Y.H. designed and supervised the study. Y.H. and R.L. analyzed the data and wrote the manuscript. All authors have read and agreed to the published version of the manuscript.

Funding: This work was supported by the National Natural Science Foundation of China (U21A20211, 91935303), the Ministry of Science and Technology (2021YFF1000200, 2020YFD0900302), Hubei Science and Technology (2021ABA011) and China Agriculture Research System (CARS-01-03).

Institutional Review Board Statement: Not applicable.

Informed Consent Statement: Not applicable.

Data Availability Statement: The whole-genome sequencing data in this paper can be found in the NCBI database under the following accession numbers: The whole-genome resequencing of J23B and CR071 (PRJNA791417), Illumina reads of J23B (SRR17299467), Nanopore reads of J23B (SRR17299468), Illumina reads of CR071 (SRR17299469), Nanopore reads of J23B (SRR17299470). Gene sequence in this paper can be found in the Rice Genome Annotation Project Database (http://rice.uga.edu/): *Ghd7* (*LOC_Os07g15770*).

Acknowledgments: We thank Lei Wang for providing genetic materials containing *Ghd7-1*.

Conflicts of Interest: The authors declare no conflict of interest.

References

1. International Rice Genome Sequencing Project; Sasaki, T. The map-based sequence of the rice genome. *Nature* **2005**, *436*, 793–800. [CrossRef] [PubMed]
2. Xing, Y.; Zhang, Q. Genetic and molecular bases of rice yield. *Annu. Rev. Plant Biol.* **2010**, *61*, 421–442. [CrossRef] [PubMed]
3. Zuo, J.; Li, J. Molecular genetic dissection of quantitative trait loci regulating rice grain size. *Annu. Rev. Genet.* **2014**, *48*, 99–118. [CrossRef] [PubMed]
4. Li, G.M.; Tang, J.Y.; Zheng, J.K.; Chu, C.C. Exploration of rice yield potential: Decoding agronomic and physiological traits. *Crop J.* **2021**, *9*, 577–589. [CrossRef]
5. Mao, H.; Sun, S.; Yao, J.; Wang, C.; Yu, S.; Xu, C.; Li, X.; Zhang, Q. Linking differential domain functions of the GS3 protein to natural variation of grain size in rice. *Proc. Natl. Acad. Sci. USA* **2010**, *107*, 19579–19584. [CrossRef]

6. Sun, S.; Wang, L.; Mao, H.; Shao, L.; Li, X.; Xiao, J.; Ouyang, Y.; Zhang, Q. A G-protein pathway determines grain size in rice. *Nat. Commun.* **2018**, *9*, 851. [CrossRef]
7. Miao, J.; Yang, Z.; Zhang, D.; Wang, Y.; Xu, M.; Zhou, L.; Wang, J.; Wu, S.; Yao, Y.; Du, X.; et al. Mutation of RGG2, which encodes a type B heterotrimeric G protein γ subunit, increases grain size and yield production in rice. *Plant Biotechnol. J.* **2019**, *17*, 650–664. [CrossRef]
8. Tao, Y.; Miao, J.; Wang, J.; Li, W.; Xu, Y.; Wang, F.; Jiang, Y.; Chen, Z.; Fan, F.; Xu, M.; et al. *RGG1*, involved in the cytokinin regulatory pathway, controls grain size in rice. *Rice* **2020**, *13*, 76. [CrossRef]
9. Liu, Q.; Han, R.; Wu, K.; Zhang, J.; Ye, Y.; Wang, S.; Chen, J.; Pan, Y.; Li, Q.; Xu, X.; et al. G-protein $\beta\gamma$ subunits determine grain size through interaction with MADS-domain transcription factors in rice. *Nat. Commun.* **2018**, *9*, 852. [CrossRef]
10. Hao, J.; Wang, D.; Wu, Y.; Huang, K.; Duan, P.; Li, N.; Xu, R.; Zeng, D.; Dong, G.; Zhang, B.; et al. The GW2-WG1-OsbZIP47 pathway controls grain size and weight in rice. *Mol. Plant* **2021**, *14*, 1266–1280. [CrossRef]
11. Yang, W.; Wu, K.; Wang, B.; Liu, H.; Guo, S.; Guo, X.; Luo, W.; Sun, S.; Ouyang, Y.; Fu, X.; et al. The RING E3 ligase CLG1 targets GS3 for degradation via the endosome pathway to determine grain size in rice. *Mol. Plant* **2021**, *14*, 1699–1713. [CrossRef] [PubMed]
12. Huang, K.; Wang, D.; Duan, P.; Zhang, B.; Xu, R.; Li, N.; Li, Y. *WIDE AND THICK GRAIN 1*, which encodes an otubain-like protease with deubiquitination activity, influences grain size and shape in rice. *Plant J.* **2017**, *91*, 849–860. [CrossRef]
13. Shi, C.; Ren, Y.; Liu, L.; Wang, F.; Zhang, H.; Tian, P.; Pan, T.; Wang, Y.; Jing, R.; Liu, T.; et al. Ubiquitin specific protease 15 has an important role in regulating grain width and size in rice. *Plant Physiol.* **2019**, *180*, 381–391. [CrossRef]
14. Duan, P.; Rao, Y.; Zeng, D.; Yang, Y.; Xu, R.; Zhang, B.; Dong, G.; Qian, Q.; Li, Y. *SMALL GRAIN 1*, which encodes a mitogen-activated protein kinase kinase 4, influences grain size in rice. *Plant J.* **2014**, *77*, 547–557. [CrossRef] [PubMed]
15. Liu, S.; Hua, L.; Dong, S.; Chen, H.; Zhu, X.; Jiang, J.; Zhang, F.; Li, Y.; Fang, X.; Chen, F. OsMAPK6, a mitogen-activated protein kinase, influences rice grain size and biomass production. *Plant J.* **2015**, *84*, 672–681. [CrossRef] [PubMed]
16. Xu, R.; Duan, P.; Yu, H.; Zhou, Z.; Zhang, B.; Wang, R.; Li, J.; Zhang, G.; Zhuang, S.; Lyu, J.; et al. Control of grain size and weight by the OsMKKK10-OsMKK4-OsMAPK6 signaling pathway in rice. *Mol. Plant* **2018**, *11*, 860–873. [CrossRef]
17. Guo, T.; Chen, K.; Dong, N.Q.; Shi, C.L.; Ye, W.W.; Gao, J.P.; Shan, J.X.; Lin, H.X. *GRAIN SIZE AND NUMBER1* negatively regulates the OsMKKK10-OsMKK4-OsMPK6 cascade to coordinate the trade-off between grain number per panicle and grain size in rice. *Plant Cell* **2018**, *30*, 871–888. [CrossRef]
18. Guo, T.; Lu, Z.Q.; Shan, J.X.; Ye, W.W.; Dong, N.Q.; Lin, H.X. *ERECTA1* acts upstream of the OsMKKK10-OsMKK4-OsMPK6 cascade to control spikelet number by regulating cytokinin metabolism in rice. *Plant Cell* **2020**, *32*, 2763–2779. [CrossRef]
19. Tian, X.; Li, X.; Zhou, W.; Ren, Y.; Wang, Z.; Liu, Z.; Tang, J.; Tong, H.; Fang, J.; Bu, Q. Transcription factor OsWRKY53 positively regulates brassinosteroid signaling and plant architecture. *Plant Physiol.* **2017**, *175*, 1337–1349. [CrossRef]
20. Liu, J.; Chen, J.; Zheng, X.; Wu, F.; Lin, Q.; Heng, Y.; Tian, P.; Cheng, Z.; Yu, X.; Zhou, K.; et al. *GW5* acts in the brassinosteroid signalling pathway to regulate grain width and weight in rice. *Nat. Plants* **2017**, *3*, 17043. [CrossRef]
21. Che, R.; Tong, H.; Shi, B.; Liu, Y.; Fang, S.; Liu, D.; Xiao, Y.; Hu, B.; Liu, L.; Wang, H.; et al. Control of grain size and rice yield by GL2-mediated brassinosteroid responses. *Nat. Plants* **2015**, *2*, 15195. [CrossRef] [PubMed]
22. Tang, Y.; Liu, H.; Guo, S.; Wang, B.; Li, Z.; Chong, K.; Xu, Y. OsmiR396d affects gibberellin and brassinosteroid signaling to regulate plant architecture in rice. *Plant Physiol.* **2018**, *176*, 946–959. [CrossRef] [PubMed]
23. Gao, X.; Zhang, J.Q.; Zhang, X.; Zhou, J.; Jiang, Z.; Huang, P.; Tang, Z.; Bao, Y.; Cheng, J.; Tang, H.; et al. Rice qGL3/OsPPKL1 Functions with the GSK3/SHAGGY-Like Kinase OsGSK3 to Modulate Brassinosteroid Signaling. *Plant Cell* **2019**, *31*, 1077–1093. [CrossRef] [PubMed]
24. Ishimaru, K.; Hirotsu, N.; Madoka, Y.; Murakami, N.; Hara, N.; Onodera, H.; Kashiwagi, T.; Ujiie, K.; Shimizu, B.; Onishi, A.; et al. Loss of function of the IAA-glucose hydrolase gene *TGW6* enhances rice grain weight and increases yield. *Nat. Genet.* **2013**, *45*, 707–711. [CrossRef] [PubMed]
25. Liu, L.; Tong, H.; Xiao, Y.; Che, R.; Xu, F.; Hu, B.; Liang, C.; Chu, J.; Li, J.; Chu, C. Activation of *Big Grain1* significantly improves grain size by regulating auxin transport in rice. *Proc. Natl. Acad. Sci. USA* **2015**, *112*, 11102–11107. [CrossRef]
26. Hu, Z.; Lu, S.J.; Wang, M.J.; He, H.; Sun, L.; Wang, H.; Liu, X.H.; Jiang, L.; Sun, J.L.; Xin, X.; et al. A novel QTL *qTGW3* encodes the GSK3/SHAGGY-like kinase OsGSK5/OsSK41 that interacts with OsARF4 to negatively regulate grain size and weight in rice. *Mol. Plant* **2018**, *11*, 736–749. [CrossRef] [PubMed]
27. Li, J.; Jiang, J.; Qian, Q.; Xu, Y.; Zhang, C.; Xiao, J.; Du, C.; Luo, W.; Zou, G.; Chen, M.; et al. Mutation of rice BC12/GDD1, which encodes a kinesin-like protein that binds to a ga biosynthesis gene promoter, leads to dwarfism with impaired cell elongation. *Plant Cell* **2011**, *23*, 628–640. [CrossRef]
28. Wu, T.; Shen, Y.; Zheng, M.; Yang, C.; Chen, Y.; Feng, Z.; Liu, X.; Liu, S.; Chen, Z.; Lei, C.; et al. Gene *SGL*, encoding a kinesin-like protein with transactivation activity, is involved in grain length and plant height in rice. *Plant Cell Rep.* **2014**, *33*, 235–244. [CrossRef]
29. Chen, W.; Cheng, Z.; Liu, L.; Wang, M.; You, X.; Wang, J.; Zhang, F.; Zhou, C.; Zhang, Z.; Zhang, H.; et al. *Small Grain and Dwarf 2*, encoding an HD-Zip II family transcription factor, regulates plant development by modulating gibberellin biosynthesis in rice. *Plant Sci.* **2019**, *288*, 110208. [CrossRef]
30. Wang, S.; Wu, K.; Yuan, Q.; Liu, X.; Liu, Z.; Lin, X.; Zeng, R.; Zhu, H.; Dong, G.; Qian, Q.; et al. Control of grain size, shape and quality by *OsSPL16* in rice. *Nat. Genet.* **2012**, *44*, 950–954. [CrossRef]

31. Si, L.; Chen, J.; Huang, X.; Gong, H.; Luo, J.; Hou, Q.; Zhou, T.; Lu, T.; Zhu, J.; Shangguan, Y.; et al. *OsSPL13* controls grain size in cultivated rice. *Nat. Genet.* **2016**, *48*, 447–456. [CrossRef] [PubMed]
32. Yuan, H.; Qin, P.; Hu, L.; Zhan, S.; Wang, S.; Gao, P.; Li, J.; Jin, M.; Xu, Z.; Gao, Q.; et al. *OsSPL18* controls grain weight and grain number in rice. *J. Genet. Genom.* **2019**, *46*, 41–51. [CrossRef] [PubMed]
33. Luo, J.; Liu, H.; Zhou, T.; Gu, B.; Huang, X.; Shangguan, Y.; Zhu, J.; Li, Y.; Zhao, Y.; Wang, Y.; et al. *An-1* encodes a basic helix-loop-helix protein that regulates awn development, grain size, and grain number in rice. *Plant Cell* **2013**, *25*, 3360–3376. [CrossRef] [PubMed]
34. Seo, H.; Kim, S.H.; Lee, B.D.; Lim, J.H.; Lee, S.J.; An, G.; Paek, N.C. The rice basic *Helix-Loop-Helix 79* (*OsbHLH079*) determines leaf angle and grain shape. *Int. J. Mol. Sci.* **2020**, *21*, 2090. [CrossRef]
35. Yang, X.; Ren, Y.; Cai, Y.; Niu, M.; Feng, Z.; Jing, R.; Mou, C.; Liu, X.; Xiao, L.; Zhang, X.; et al. Overexpression of *OsbHLH107*, a member of the basic helix-loop-helix transcription factor family, enhances grain size in rice (*Oryza sativa* L.). *Rice* **2018**, *11*, 41. [CrossRef]
36. Aya, K.; Hobo, T.; Sato-Izawa, K.; Ueguchi-Tanaka, M.; Kitano, H.; Matsuoka, M. A novel AP2-type transcription factor, SMALL ORGAN SIZE1, controls organ size downstream of an auxin signaling pathway. *Plant Cell Physiol.* **2014**, *55*, 897–912. [CrossRef]
37. Jiang, L.; Ma, X.; Zhao, S.; Tang, Y.; Liu, F.; Gu, P.; Fu, Y.; Zhu, Z.; Cai, H.; Sun, C.; et al. The APETALA2-Like transcription factor SUPERNUMERARY BRACT controls rice seed shattering and seed size. *Plant Cell* **2019**, *31*, 17–36. [CrossRef]
38. Ren, D.; Hu, J.; Xu, Q.; Cui, Y.; Zhang, Y.; Zhou, T.; Rao, Y.; Xue, D.; Zeng, D.; Zhang, G.; et al. *FZP* determines grain size and sterile lemma fate in rice. *J. Exp. Bot.* **2018**, *69*, 4853–4866. [CrossRef]
39. Zhao, D.S.; Li, Q.F.; Zhang, C.Q.; Zhang, C.; Yang, Q.Q.; Pan, L.X.; Ren, X.Y.; Lu, J.; Gu, M.H.; Liu, Q.Q. *GS9* acts as a transcriptional activator to regulate rice grain shape and appearance quality. *Nat. Commun.* **2018**, *9*, 1240. [CrossRef]
40. Wu, W.; Liu, X.; Wang, M.; Meyer, R.S.; Luo, X.; Ndjiondjop, M.N.; Tan, L.; Zhang, J.; Wu, J.; Cai, H.; et al. A single-nucleotide polymorphism causes smaller grain size and loss of seed shattering during African rice domestication. *Nat. Plants* **2017**, *3*, 17064. [CrossRef]
41. Zhou, S.R.; Xue, H.W. The rice PLATZ protein SHORT GRAIN6 determines grain size by regulating spikelet hull cell division. *J. Integr. Plant Biol.* **2020**, *62*, 847–864. [CrossRef] [PubMed]
42. Xue, W.; Xing, Y.; Weng, X.; Zhao, Y.; Tang, W.; Wang, L.; Zhou, H.; Yu, S.; Xu, C.; Li, X.; et al. Natural variation in *Ghd7* is an important regulator of heading date and yield potential in rice. *Nat. Genet.* **2008**, *40*, 761–767. [CrossRef] [PubMed]
43. Hu, Y.; Song, S.; Weng, X.; You, A.; Xing, Y. The heading-date gene Ghd7 inhibits seed germination by modulating the balance between abscisic acid and gibberellins. *Crop J.* **2021**, *9*, 297–304. [CrossRef]
44. Wang, Q.; Su, Q.; Nian, J.; Zhang, J.; Guo, M.; Dong, G.; Hu, J.; Wang, R.; Wei, C.; Li, G.; et al. The Ghd7 transcription factor represses *ARE1* expression to enhance nitrogen utilization and grain yield in rice. *Mol. Plant* **2021**, *14*, 1012–1023. [CrossRef]
45. Weng, X.; Wang, L.; Wang, J.; Hu, Y.; Du, H.; Xu, C.; Xing, Y.; Li, X.; Xiao, J.; Zhang, Q. *Grain number, plant height, and heading date7* is a central regulator of growth, development, and stress response. *Plant Physiol.* **2014**, *164*, 735–747. [CrossRef]
46. Hu, J.; Wang, Y.; Fang, Y.; Zeng, L.; Xu, J.; Yu, H.; Shi, Z.; Pan, J.; Zhang, D.; Kang, S.; et al. A rare allele of *GS2* enhances grain size and grain yield in rice. *Mol. Plant* **2015**, *8*, 1455–1465. [CrossRef]
47. Song, X.J.; Huang, W.; Shi, M.; Zhu, M.Z.; Lin, H.X. A QTL for rice grain width and weight encodes a previously unknown RING-type E3 ubiquitin ligase. *Nat. Genet.* **2007**, *39*, 623–630. [CrossRef]
48. Li, N.; Li, Y. Maternal control of seed size in plants. *J. Exp. Bot.* **2015**, *66*, 1087–1097. [CrossRef]
49. Shomura, A.; Izawa, T.; Ebana, K.; Ebitani, T.; Kanegae, H.; Konishi, S.; Yano, M. Deletion in a gene associated with grain size increased yields during rice domestication. *Nat. Genet.* **2008**, *40*, 1023–1028. [CrossRef]
50. Zhao, H.; Yao, W.; Ouyang, Y.; Yang, W.; Wang, G.; Lian, X.; Xing, Y.; Chen, L.; Xie, W. RiceVarMap: A comprehensive database of rice genomic variations. *Nucleic Acids Res.* **2015**, *43*, D1018–D1022. [CrossRef]
51. Zhang, J.; Zhou, X.; Yan, W.; Zhang, Z.; Lu, L.; Han, Z.; Zhao, H.; Liu, H.; Song, P.; Hu, Y.; et al. Combinations of the *Ghd7*, *Ghd8* and *Hd1* genes largely define the ecogeographical adaptation and yield potential of cultivated rice. *New Phytol.* **2015**, *208*, 1056–1066. [CrossRef] [PubMed]
52. Tiwari, S.B.; Shen, Y.; Chang, H.C.; Hou, Y.; Harris, A.; Ma, S.F.; McPartland, M.; Hymus, G.J.; Adam, L.; Marion, C.; et al. The flowering time regulator CONSTANS is recruited to the *FLOWERING LOCUS T* promoter via a unique *cis*-element. *New Phytol.* **2010**, *187*, 57–66. [CrossRef] [PubMed]
53. Hao, Y.J.; Wei, W.; Song, Q.X.; Chen, H.W.; Zhang, Y.Q.; Wang, F.; Zou, H.F.; Lei, G.; Tian, A.G.; Zhang, W.K.; et al. Soybean NAC transcription factors promote abiotic stress tolerance and lateral root formation in transgenic plants. *Plant J.* **2011**, *68*, 302–313. [CrossRef] [PubMed]
54. Chen, Y.; Xu, Y.; Luo, W.; Li, W.; Chen, N.; Zhang, D.; Chong, K. The F-box protein OsFBK12 targets OsSAMS1 for degradation and affects pleiotropic phenotypes, including leaf senescence, in rice. *Plant Physiol.* **2013**, *163*, 1673–1685. [CrossRef]
55. Dwivedi, N.; Maji, S.; Waseem, M.; Thakur, P.; Kumar, V.; Parida, S.K.; Thakur, J.K. The Mediator subunit OsMED15a is a transcriptional co-regulator of seed size/weight-modulating genes in rice. *Biochim. Biophys. Acta Gene Regul. Mech.* **2019**, *1862*, 194432. [CrossRef]
56. Bello, B.K.; Hou, Y.; Zhao, J.; Jiao, G.; Wu, Y.; Li, Z.; Wang, Y.; Tong, X.; Wang, W.; Yuan, W.; et al. NF-YB1-YC12-bHLH144 complex directly activates *Wx* to regulate grain quality in rice (*Oryza sativa* L.). *Plant Biotechnol. J.* **2019**, *17*, 1222–1235. [CrossRef]

57. Fu, F.F.; Xue, H.W. Coexpression analysis identifies Rice Starch Regulator1, a rice AP2/EREBP family transcription factor, as a novel rice starch biosynthesis regulator. *Plant Physiol.* **2010**, *154*, 927–938. [CrossRef]
58. Itoh, H.; Ueguchi-Tanaka, M.; Sato, Y.; Ashikari, M.; Matsuoka, M. The gibberellin signaling pathway is regulated by the appearance and disappearance of SLENDER RICE1 in nuclei. *Plant Cell* **2002**, *14*, 57–70. [CrossRef]
59. Zhu, X.; Liang, W.; Cui, X.; Chen, M.; Yin, C.; Luo, Z.; Zhu, J.; Lucas, W.J.; Wang, Z.; Zhang, D. Brassinosteroids promote development of rice pollen grains and seeds by triggering expression of Carbon Starved Anther, a MYB domain protein. *Plant J.* **2015**, *82*, 570–581. [CrossRef]
60. Miura, K.; Ikeda, M.; Matsubara, A.; Song, X.J.; Ito, M.; Asano, K.; Matsuoka, M.; Kitano, H.; Ashikari, M. *OsSPL14* promotes panicle branching and higher grain productivity in rice. *Nat. Genet.* **2010**, *42*, 545–549. [CrossRef]
61. Kitagawa, K.; Kurinami, S.; Oki, K.; Abe, Y.; Ando, T.; Kono, I.; Yano, M.; Kitano, H.; Iwasaki, Y. A novel kinesin 13 protein regulating rice seed length. *Plant Cell Physiol.* **2010**, *51*, 1315–1329. [CrossRef] [PubMed]
62. Segami, S.; Kono, I.; Ando, T.; Yano, M.; Kitano, H.; Miura, K.; Iwasaki, Y. *Small and round seed 5* gene encodes alpha-tubulin regulating seed cell elongation in rice. *Rice* **2012**, *5*, 4. [CrossRef] [PubMed]
63. Boija, A.; Klein, I.A.; Sabari, B.R.; Dall'Agnese, A.; Coffey, E.L.; Zamudio, A.V.; Li, C.H.; Shrinivas, K.; Manteiga, J.C.; Hannett, N.M.; et al. Transcription Factors Activate Genes through the Phase-Separation Capacity of Their Activation Domains. *Cell* **2018**, *175*, 1842–1855.e1816. [CrossRef] [PubMed]
64. Shu, K.; Chen, Q.; Wu, Y.; Liu, R.; Zhang, H.; Wang, P.; Li, Y.; Wang, S.; Tang, S.; Liu, C.; et al. ABI4 mediates antagonistic effects of abscisic acid and gibberellins at transcript and protein levels. *Plant J.* **2016**, *85*, 348–361. [CrossRef]
65. Shu, K.; Zhang, H.; Wang, S.; Chen, M.; Wu, Y.; Tang, S.; Liu, C.; Feng, Y.; Cao, X.; Xie, Q. ABI4 regulates primary seed dormancy by regulating the biogenesis of abscisic acid and gibberellins in arabidopsis. *PLoS Genet.* **2013**, *9*, e1003577. [CrossRef]
66. Shen, C.; Liu, H.; Guan, Z.; Yan, J.; Zheng, T.; Yan, W.; Wu, C.; Zhang, Q.; Yin, P.; Xing, Y. Structural insight into DNA recognition by CCT/NF-YB/YC complexes in plant photoperiodic flowering. *Plant Cell* **2020**, *32*, 3469–3484. [CrossRef]
67. Bai, X.; Huang, Y.; Mao, D.; Wen, M.; Zhang, L.; Xing, Y. Regulatory role of *FZP* in the determination of panicle branching and spikelet formation in rice. *Sci. Rep.* **2016**, *6*, 19022. [CrossRef]
68. Lee, D.Y.; An, G. Two AP2 family genes, *supernumerary bract* (*SNB*) and *Osindeterminate spikelet 1* (*OsIDS1*), synergistically control inflorescence architecture and floral meristem establishment in rice. *Plant J.* **2012**, *69*, 445–461. [CrossRef]
69. Daviere, J.M.; Achard, P. A pivotal role of dellas in regulating multiple hormone signals. *Mol. Plant* **2016**, *9*, 10–20. [CrossRef]
70. Xu, F.; Li, T.; Xu, P.B.; Li, L.; Du, S.S.; Lian, H.L.; Yang, H.Q. DELLA proteins physically interact with CONSTANS to regulate flowering under long days in *Arabidopsis*. *FEBS Lett.* **2016**, *590*, 541–549. [CrossRef]
71. Huang, D.; Wang, S.; Zhang, B.; Shang-Guan, K.; Shi, Y.; Zhang, D.; Liu, X.; Wu, K.; Xu, Z.; Fu, X.; et al. A gibberellin-mediated DELLA-NAC signaling cascade regulates cellulose synthesis in rice. *Plant Cell* **2015**, *27*, 1681–1696. [CrossRef] [PubMed]
72. Mo, W.; Tang, W.; Du, Y.; Jing, Y.; Bu, Q.; Lin, R. PHYTOCHROME-INTERACTING FACTOR-LIKE14 and SLENDER RICE1 interaction controls seedling growth under salt stress. *Plant Physiol.* **2020**, *184*, 506–517. [CrossRef] [PubMed]
73. Li, S.; Tian, Y.; Wu, K.; Ye, Y.; Yu, J.; Zhang, J.; Liu, Q.; Hu, M.; Li, H.; Tong, Y.; et al. Modulating plant growth-metabolism coordination for sustainable agriculture. *Nature* **2018**, *560*, 595–600. [CrossRef]
74. Yang, W.; Guo, Z.; Huang, C.; Duan, L.; Chen, G.; Jiang, N.; Fang, W.; Feng, H.; Xie, W.; Lian, X.; et al. Combining high-throughput phenotyping and genome-wide association studies to reveal natural genetic variation in rice. *Nat. Commun.* **2014**, *5*, 5087. [CrossRef]
75. Champagne, E.T.; Richard, O.A.; Bett, K.L.; Grimm, C.C.; Vinyard, B.T.; Webb, B.D.; McClung, A.M.; Barton, F.E.; Lyon, B.G.; Moldenhauer, K.; et al. Quality evaluation of US medium-grain rice using a Japanese taste analyzer. *Cereal Chem.* **1996**, *73*, 290–294.
76. Silva, L.D.C.E.; Wang, S.; Zeng, Z.-B. Composite Interval Mapping and Multiple Interval Mapping: Procedures and guidelines for using Windows QTL Cartographer. In *Quantitative Trait Loci (QTL): Methods and Protocols*; Rifkin, S.A., Ed.; Humana Press: Totowa, NJ, USA, 2012; pp. 75–119. [CrossRef]
77. Broman, K.W.; Wu, H.; Sen, S.; Churchill, G.A. R/qtl: QTL mapping in experimental crosses. *Bioinformatics* **2003**, *19*, 889–890. [CrossRef] [PubMed]
78. Koren, S.; Walenz, B.P.; Berlin, K.; Miller, J.R.; Bergman, N.H.; Phillippy, A.M. Canu: Scalable and accurate long-read assembly via adaptive k-mer weighting and repeat separation. *Genome Res.* **2017**, *27*, 722–736. [CrossRef]
79. Ruan, J.; Li, H. Fast and accurate long-read assembly with wtdbg2. *Nat. Methods* **2020**, *17*, 155–158. [CrossRef]
80. Vaser, R.; Sovic, I.; Nagarajan, N.; Sikic, M. Fast and accurate de novo genome assembly from long uncorrected reads. *Genome Res.* **2017**, *27*, 737–746. [CrossRef]
81. Walker, B.J.; Abeel, T.; Shea, T.; Priest, M.; Abouelliel, A.; Sakthikumar, S.; Cuomo, C.A.; Zeng, Q.; Wortman, J.; Young, S.K.; et al. Pilon: An integrated tool for comprehensive microbial variant detection and genome assembly improvement. *PLoS ONE* **2014**, *9*, e112963. [CrossRef]
82. Li, H. Minimap2: Pairwise alignment for nucleotide sequences. *Bioinformatics* **2018**, *34*, 3094–3100. [CrossRef] [PubMed]
83. Li, H.; Handsaker, B.; Wysoker, A.; Fennell, T.; Ruan, J.; Homer, N.; Marth, G.; Abecasis, G.; Durbin, R. The Sequence Alignment/Map format and SAMtools. *Bioinformatics* **2009**, *25*, 2078–2079. [CrossRef] [PubMed]

84. He, Y.; Zhang, T.; Yang, N.; Xu, M.; Yan, L.; Wang, L.; Wang, R.; Zhao, Y. Self-cleaving ribozymes enable the production of guide RNAs from unlimited choices of promoters for CRISPR/Cas9 mediated genome editing. *J. Genet. Genom.* **2017**, *44*, 469–472. [CrossRef]
85. Zhang, Y.; Su, J.; Duan, S.; Ao, Y.; Dai, J.; Liu, J.; Wang, P.; Li, Y.; Liu, B.; Feng, D.; et al. A highly efficient rice green tissue protoplast system for transient gene expression and studying light/chloroplast-related processes. *Plant Methods* **2011**, *7*, 30. [CrossRef] [PubMed]

International Journal of
Molecular Sciences

Article

Dynamic Change in Starch Biosynthetic Enzymes Complexes during Grain-Filling Stages in BEIIb Active and Deficient Rice

Yining Ying [1], Feifei Xu [1], Zhongwei Zhang [1], Piengtawan Tappiban [1] and Jinsong Bao [1,2,*]

1 Institute of Nuclear Agriculture Science, College of Agriculture and Biotechnology, Zhejiang University, Zijingang Campus, Hangzhou 310058, China
2 Hainan Institute of Zhejiang University, Hainan Yazhou Bay Seed Lab, Yazhou Bay Science and Technology City, Yazhou District, Sanya 572025, China
* Correspondence: jsbao@zju.edu.cn; Tel.: +86-571-86971932

Abstract: Starch is the predominant reserve in rice (*Oryza sativa* L.) endosperm, which is synthesized by the coordinated efforts of a series of starch biosynthetic-related enzymes in the form of a multiple enzyme complex. Whether the enzyme complex changes during seed development is not fully understood. Here, we investigated the dynamic change in multi-protein complexes in an *indica* rice variety IR36 (wild type, WT) and its BEIIb-deficient mutant (*be2b*) at different developmental stages. Gel permeation chromatography (GPC) and Western blotting analysis of soluble protein fractions revealed most of the enzymes except for SSIVb were eluted in smaller molecular weight fractions at the early developing stage and were transferred to higher molecular weight fractions at the later stage in both WT and *be2b*. Accordingly, protein interactions were enhanced during seed development as demonstrated by co-immunoprecipitation analysis, suggesting that the enzymes were recruited to form larger protein complexes during starch biosynthesis. The converse elution pattern from GPC of SSIVb may be attributed to its vital role in the initiation step of starch synthesis. The number of protein complexes was markedly decreased in *be2b* at all development stages. Although SSIVb could partially compensate for the role of BEIIb in protein complex formation, it was hard to form a larger protein complex containing over five proteins in *be2b*. In addition, other proteins such as PPDKA and PPDKB were possibly present in the multi-enzyme complexes by proteomic analyses of high molecular weight fractions separated from GPC. Two putative protein kinases were found to be potentially associated with starch biosynthetic enzymes. Collectively, our findings unraveled a dynamic change in the protein complex during seed development, and potential roles of BEIIb in starch biosynthesis via various protein complex formations, which enables a deeper understanding of the complex mechanism of starch biosynthesis in rice.

Keywords: rice; starch synthase; starch branching enzyme; protein–protein interaction; protein complex; grain filling

check for updates

Citation: Ying, Y.; Xu, F.; Zhang, Z.; Tappiban, P.; Bao, J. Dynamic Change in Starch Biosynthetic Enzymes Complexes during Grain-Filling Stages in BEIIb Active and Deficient Rice. *Int. J. Mol. Sci.* **2022**, *23*, 10714. https://doi.org/10.3390/ijms231810714

Academic Editor: Prem L. Bhalla

Received: 21 July 2022
Accepted: 7 September 2022
Published: 14 September 2022
Corrected: 11 May 2023

Publisher's Note: MDPI stays neutral with regard to jurisdictional claims in published maps and institutional affiliations.

Copyright: © 2022 by the authors. Licensee MDPI, Basel, Switzerland. This article is an open access article distributed under the terms and conditions of the Creative Commons Attribution (CC BY) license (https://creativecommons.org/licenses/by/4.0/).

1. Introduction

Rice (*Oryza sativa* L.) is one of the most important food crops worldwide. Starch, composed of amylose and amylopectin, is the most abundant component in the rice grain that provides nutrients to the developing embryo and seedling. The development of the rice endosperm starts after double fertilization, then the fertilized polar nuclei undergo mitotic cell proliferation and cellularization until 6 to 7 days after flowering (DAF), and it is at 5–20 DAF that genes encoding the enzymes of starch biosynthesis are vigorously expressed and accumulation of starch and other storage compounds occurs [1–3].

The starch biosynthesis pathway is orchestrated by multiple enzymes in rice endosperm, which mainly include ADP-glucose pyrophosphorylase (AGPase), granule-bound starch synthase (GBSS), soluble starch synthase (SSs), starch branching enzyme (BEs), starch debranching enzyme (DBEs), and phosphorylases (Phos). Seven AGPase subunits [4–7], eleven isoforms of SSs [8,9], three isoforms of BEs [10–12], four isoforms of DBEs [13–15], and two isoforms of Phos [16,17] were identified, and their functions in the process of starch synthesis have been elucidated by mutant and transgenic analyses. In addition, starch biosynthesis isozymes display as a complex form, rather than as monomers, to accomplish their activities in the starch biosynthesis pathway by associating with other enzymes. For example, starch plastidial phosphorylase (Pho1, also known as PHS1 and SP) form a complex with disproportionating enzyme (Dpe1) to use a broader range of sugars to synthesize malto-oligosaccharides (MOSs) and enhance the synthesis of long MOSs, which plays an important role in the initiation of starch synthesis [18]. Moreover, Pho1 may facilitate interactions between SSIIa and SSIIIa in amyloplasts; although, no direct association between SSIIa and SSIIIa was detected [19].

The direct evidence of physical interactions among starch biosynthetic enzymes was initially demonstrated in the developing endosperm of wheat [20]. Subsequently, further evidence for the identification of more phosphorylation-dependent complexes was accumulated in wheat [21], maize [22–27], barley [28], and rice [29–32]. Specifically, an approximately 230 kDa trimeric complex formed between SSI, SSIIa, and BEIIb is one of the best studied and well-characterized protein complexes, which is commonly found in cereals and contributes to the synthesis of short and intermediate amylopectin chains within the clusters [33]. A large protein complex (approximately 670 kDa) involving SSIII interacting with SSIIa, BEIIa, BEIIb, and several other proteins including pyruvate orthophosphate dikinase (PPDK) and AGPase has also been demonstrated in maize endosperm [23]. In addition, gel filtration analyses of soluble proteins from developing rice endosperm in a *japonica* background showed that rice starch biosynthetic enzymes, including SSI, SSIIa, SSIIIa, SSIVb, BEI, BEIIb, and PUL, form a larger protein complex (>700 kDa) than those found in other cereals [29].

BEs are essential for amylopectin biosynthesis because they are the only enzymes that introduce α-1,6 glycosidic bonds into α-polyglucans [34]. The three isoforms in rice (BEI, BEIIa, and BEIIb) play distinct roles, i.e., BEI transfers a variety of both short chains and intermediate chains (DP $\leq$ 40) while BEIIa and BEIIb preferentially form amylopectin short chains of DP 6-15 and DP 6-7, respectively [10,34]. Although BEI accounts for the largest relative activities (approximately 80%) [35], BEIIb (the major form of BEII in maize and rice) made the greatest contribution to amylopectin synthesis [10,12]. The inactivation of BEIIb resulted in the *amylose extender* (*ae*) mutant-containing starch molecules with longer amylopectin chains and fewer branches, increasing the amylose content, changing the crystalline pattern of starch from A-type to B- or C-type and producing an opaque seed phenotype in rice [10,12,36–38], whereas no such change has been so far reported in the *be1* and *be2a* mutants [11,39]. In addition, mutations in *BEIIb* gene altered the interactions of starch biosynthetic enzymes and changed the formation of multi-enzyme complexes in maize [24,25], barley [28], and rice [30]. These observations imply that BEIIb plays a specific role in the formation of multi-enzyme complexes in amylopectin synthesis. However, there is little information concerning the dynamic change in starch biosynthetic isozyme complexes at different developmental stages of rice endosperm. In addition, most studies on protein complex formation were focused on *japonica* cultivars [30–32], which have almost inactive SSIIa [40]. Therefore, whether and how novel protein complexes are present in *indica* rice with active SSIIa isozyme remains unknown.

In a previous study, we investigated the changes in starch fine structure and functional properties in *indica* rice variety IR36 (wild type, WT) and its BEIIb-deficient mutant with reduced BEIIb level (*be2b*) during three typical developmental stages from 5 to 15 DAF [41]. In the present study, we analyzed dynamic changes in total protein expression profiles and multi-enzyme complexes in the same materials by gel-filtration chromatography, co-immunoprecipitation, and Western blot analysis. In addition, proteomic analysis was further conducted to confirm the putative large complexes (>400 kDa) to reveal the presence of other proteins in the complexes and differentially expressed proteins between WT and *be2b* at different developmental stages. Finally, the essential role of protein phosphorylation and protein kinases in multi-protein complex formation was discussed. These results not only provide valuable insights into the roles of BEIIb in the formation of multi-enzyme complexes, but also provide a novel understanding of the complex network of starch biosynthetic enzymes involved in starch accumulation in developing endosperm.

2. Results

2.1. Accumulation of Starch Biosynthetic Related Enzymes (SSREs)

To investigate the effects of BEIIb deficiency on other starch biosynthetic-related enzyme (SSRE) accumulation in rice endosperm at different developmental stages, total proteins were isolated from the WT and *be2b* mutant [41], and 11 antibodies from GBSS, SS, BE, DBE, and Pho classes were applied to detect the expression of SSREs (Figure 1A). As shown in Figure 1B, quantification of Western blot bands revealed SSREs were differentially accumulated in WT and *be2b* at different grain filling stages. It is expected that the accumulation of BEIIb significantly decreased in *be2b*, and a significant reduction in SSI and SSIIa was also observed. However, the number of SSIVb, BEI, ISA1, PUL, and Pho1 was up-regulated in *be2b*. The total GBSSI level was lower in *be2b* than in WT at 5 and 10 DAF, but higher at 15 DAF, while SSIIIa and BEIIa were higher at 5 DAF, but subsequently decreased.

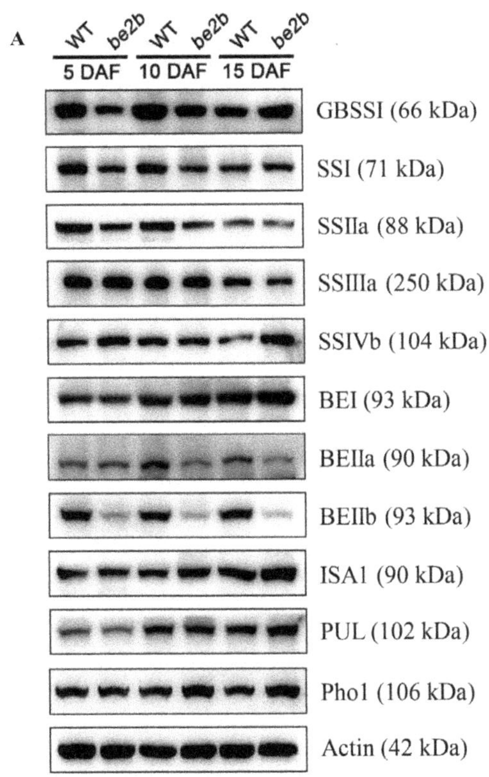

Figure 1. *Cont.*

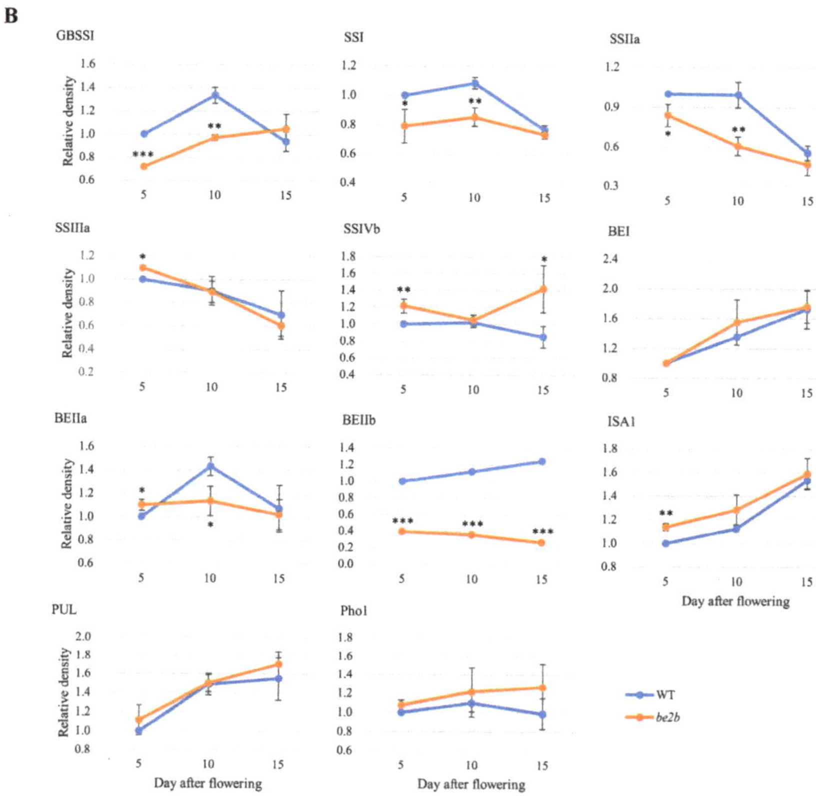

Figure 1. The expression of starch biosynthetic related enzymes (SSREs) during different grain-filling stages (5 DAF, 10 DAF, 15 DAF) in wild type (WT) and *be2b* mutant: (**A**) Total proteins were separated by SDS-PAGE. Each lane was loaded with 50 μg of proteins. Gels were analyzed by Western blotting using the indicated isozyme-specific antibodies. The anti-actin antibody is used as a loading control. Results shown are representative of two biological replicates. (**B**) Relative density of Western blotting results measured by Image J. Data are Mean ± SD from two biological replicates. The asterisks indicate statistical significance between WT and *be2b*, as determined by the Student's *t*-test (*, $p < 0.05$; **, $p < 0.01$; ***, $p < 0.001$).

The relative levels of SSREs vary with the dynamic changes in endosperm development in both WT and *be2b*; although, the protein levels fluctuated within a narrow range (less than twofold) as estimated by immunoblot (Figure 1B). In WT, BEI, BEIIb, ISA1, and PUL increased in the steady state level over development, while SSIIa and SSIIIa decreased, and the accumulated levels of GBSSI, SSI, SSIVb, BEIIa, and Pho1 peaked at the mid developmental stage. In the *be2b* mutant, the dynamic changes in some enzymes were strongly affected by a dramatic reduction in the BEIIb level. GBSSI and Pho1 displayed an upward trend from 5 to 15 DAF, and the relative level of SSIVb fell from 5 to 10 DAF, then increased. The changing pattern of other enzymes, SSI, SSIIa, SSIIIa, BEI, BEIIa, ISA1, and PUL, were identical to that of WT.

2.2. Molecular Weight Distribution of SSREs

Soluble proteins extracted from developing seeds of WT and *be2b* at different grain-filling stages were separated by gel-filtration chromatography. The eluted fractions were denatured and analyzed by Western blotting using enzyme-specific antibodies to examine possible changes in the aggregation state of the main SSREs involved in amylopectin biosynthesis. In addition, the phosphorylation status of BEs was also examined with site-specific phosphopeptide antibodies (BEI-Ser562, BEIIb-Ser685) identified previously [42]. Figure 2 shows that while some proteins could be detected at smaller than their expected monomeric size derived from partial fragments of isozymes, all starch biosynthetic proteins analyzed were eluted at higher molecular weights, suggesting the existence of protein multimerization in all samples. As expected, faint signals of the BEIIb band (ca. 87 kDa) were observed in the fractions of *be2b* compared with WT. In comparison with BEI and BEIIb (Figure 2B), phosphoproteins showed elution patterns with similar elution positions but narrower molecular weight ranges (Figure 2C). Although some differences in elution patterns of isozymes were noted at 5 and 10 DAF between WT and *be2b*, striking differences were observed at 15 DAF except SSIIIa. The majority of SSI, BEI, and BEIIb were eluted in fractions 2 to 5 (>400 kDa) in WT at 15 DAF, while in the *be2b* mutant, all of these enzymes were eluted mainly in fractions 6 to 9 (100–300 kDa). Similarly, SSIIa, BEIIa, ISA1, PUL, and Pho1 also eluted earlier from the column in WT than those in *be2b* extracts, indicating that these proteins are components of larger complexes in WT rather than in *be2b*. By contrast, SSIVb was eluted in a broad molecular weight range (fractions 2–9, >100 kDa) in *be2b* at 15 DAF, but in a low molecular weight form (fractions 7–9, 100–200 kDa) in WT.

Considering the dynamic changes in endosperm development, most of the enzymes were eluted in higher molecular weight fractions at later stages of endosperm development (10 and 15 DAF) than at the earlier stage (5 DAF), in both WT and *be2b*. Specifically, most of SSI, BEI, BEIIb, and PUL were eluted in low molecular weight fractions 7 to 9 (100–200 kDa) at 5 and 10 DAF in WT, whereas the amount of these enzymes eluted in fractions 2 to 6 (>300 kDa) was considerably elevated at 15 DAF. SSIIa was mainly detected in fractions 6 and 7 (200–300 kDa) at 5 and 10 DAF in WT, while only a small amount was found in these fractions, and most of them were eluted in fractions 4 and 5 (400–600 kDa) at 15 DAF. BEIIa was detected in fractions 7, fractions 6 and 7, and fractions 2 to 4 (>500 kDa) at 5, 10, and 15 DAF in WT, respectively. ISA1 and Pho1 were found in fractions 4 to 7 (200–600 kDa) at 5 and 10 DAF in WT, whereas they were mainly eluted in fractions 2 and 3 (>600 kDa) at 15 DAF. The elution patterns of SSIIIa in different samples were almost identical; although, the overall trend was that with the development of seed, the high molecular weight fractions were increased. Similar results were also obtained from the *be2b* mutant, indicating an apparent increase in molecular mass or aggregation state of these enzymes with the endosperm development. However, an opposite trend was found in the SSIVb protein; it was inclined to form a larger protein complex at earlier stages in WT, while in *be2b*, the elution patterns of SSIVb were similar among different developmental stages, suggesting SSIVb plays a different role in protein complex formation.

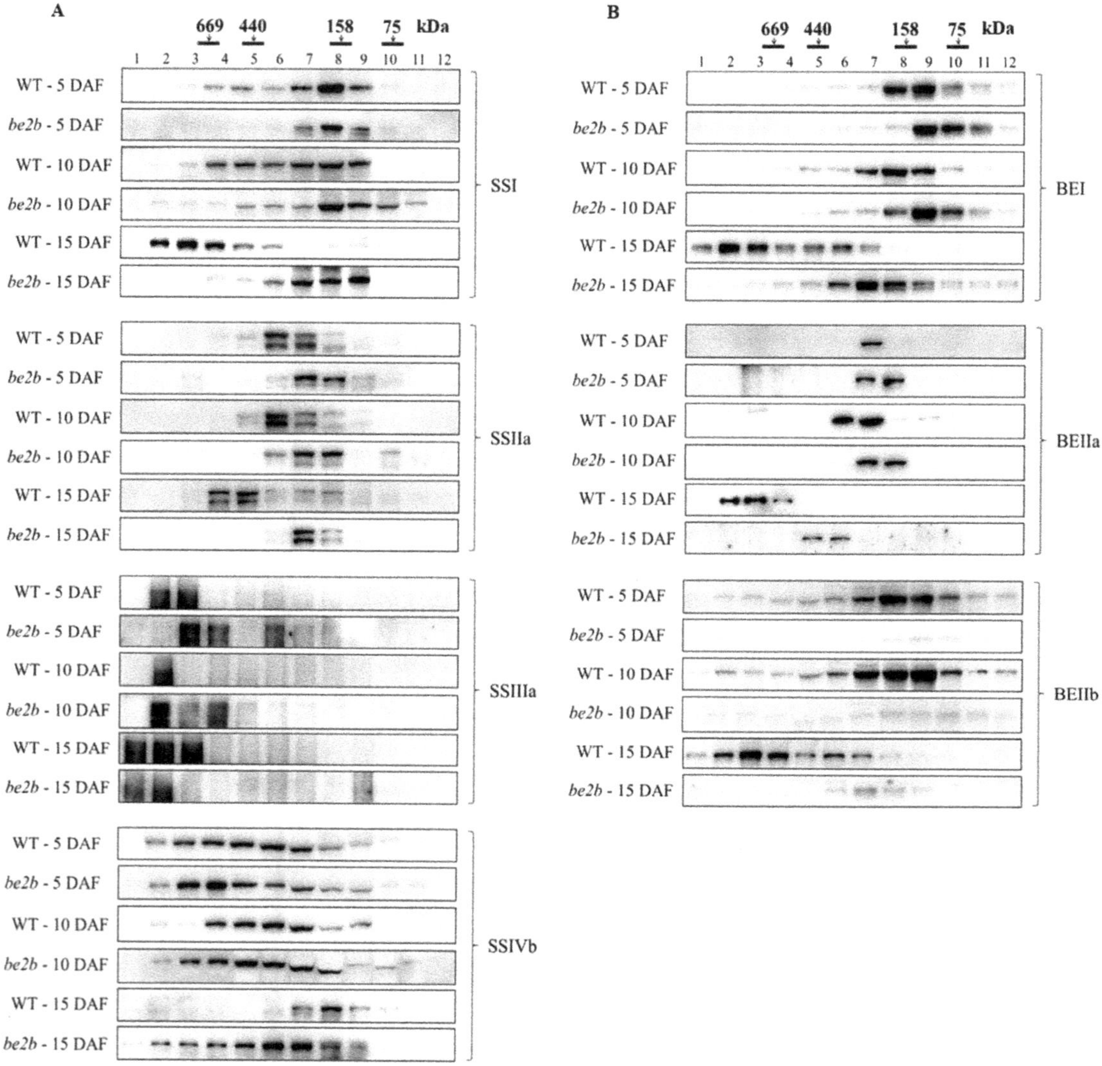

Figure 2. *Cont.*

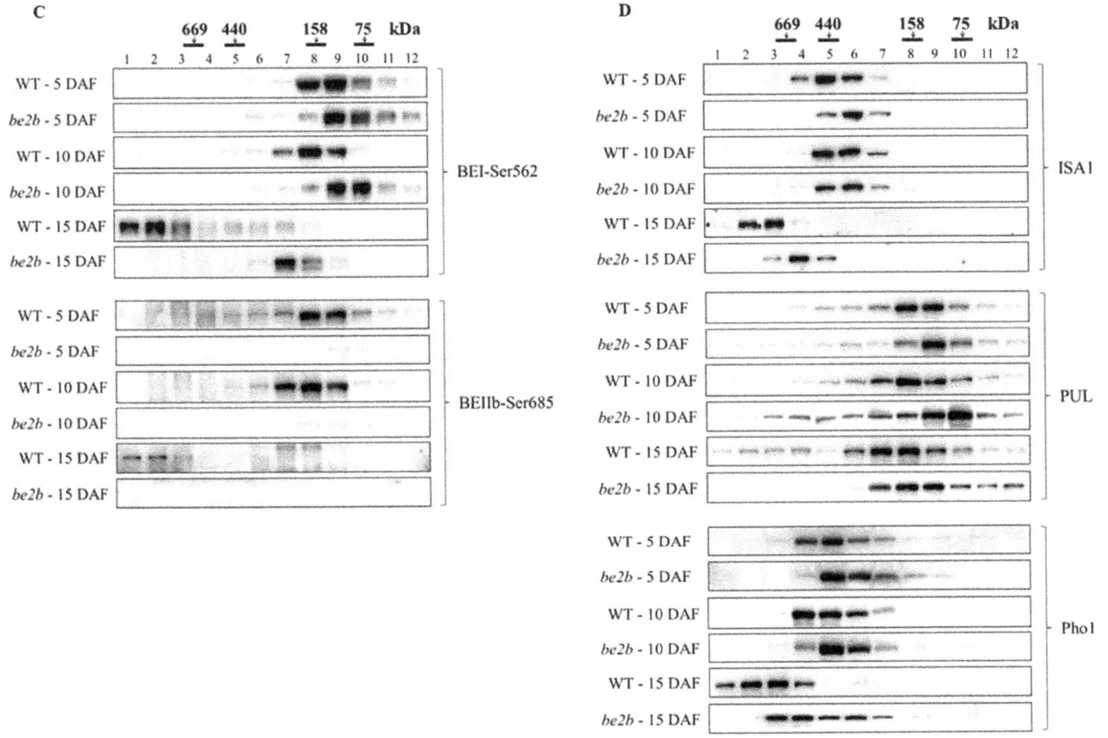

Figure 2. Molecular weight distributions of SSREs during different grain-filling stages (5 DAF, 10 DAF, 15 DAF) in wild type (WT) and *be2b* mutant. Fractions 1–12 obtained after gel filtration chromatography of soluble proteins extracted from the developing seeds were separated by SDS-PAGE prior to Western blotting with indicated isozyme-specific antibodies. Each lane was loaded with 5 μL of 10-fold concentrated fractions. The molecular weight of protein standards is shown at the top (black bars): (**A**) Western blotting of SS isozymes. (**B**) Western blotting of BE isozymes. (**C**) Western blotting of phosphor BE isozymes. (**D**) Western blotting of DBE isozymes and Pho1.

2.3. Co-Immunoprecipitation

In order to investigate the differences in possible interacting partners among SSREs in WT and *be2b* at different developmental stages, soluble proteins were extracted from the developing seeds and co-immunoprecipitation experiments were performed using enzyme-specific antibodies. All of the antibodies used for precipitation (anti-SSI, anti-SSIIa, anti-SSIIIa, anti-SSIVb, anti-BEI, anti-BEIIa, and anti-BEIIb) were able to recognize and precipitate their respective target protein (Figure 3). The results are summarized in Table 1. At the early grain filling stage (5 DAF), no strong pairwise association was detected by reciprocal co-immunoprecipitation in both WT and *be2b*, while weak signals were obtained from reciprocal co-immunoprecipitation experiments for the pairwise interactions BEIIb–SSI, BEIIb–SSIIa, BEIIa–SSIVb, BEIIb–SSIVb, and BEIIb–BEI in WT and only BEIIa–SSIVb in *be2b*. Clear Western blot signals were obtained for only one way in the co-immunoprecipitation experiments in some cases; among such interactions, BEI–SSIIa, SSIIa–SSI, BEIIa–SSI, and BEIIa–SSIIa were commonly observed in WT and *be2b* (first acronym, the antibody used for immunoprecipitation; second acronym, the isozyme subsequently detected by Western blotting).

However, at later stages (10 and 15 DAF), when a higher molecular mass of enzymes was detected in the gel-filtration experiments (Figure 2), strong pairwise associations were obtained by reciprocal co-immunoprecipitation and less weak immunodetections of co-precipitated protein were observed. In WT, BEIIb was clearly co-immunoprecipitated by SSI, SSIIa, SSIIIa, BEI, and BEIIa at 10 DAF and SSI, SSIIa, and BEI at 15 DAF, while no interactions were observed between BEIIb and other enzymes in *be2b* mutant as expected. In addition, interactions among other starch biosynthetic enzymes were reduced in *be2b*, whereas the interaction between BEIIa and BEI was observed by reciprocal co-immunoprecipitation in *be2b* but assessed by a one-side signal in WT at 10 DAF. Pairwise associations detected by reciprocal co-immunoprecipitation were observed for SSIVb–SSIIIa and BEIIa–SSIVb and one-side signals were obtained for SSI–SSIVb, BEI–SSIVb, and BEIIb–SSIVb in WT at 10 DAF. However, at 15 DAF, when SSIVb was eluted in lower molecular weight fractions (Figure 2), only weak pairwise associations for SSIVb–SSIIIa were observed in WT. Interactions of ISA1 with other starch biosynthetic enzymes only occurred between BEIIb and ISA1 at 15 DAF in WT, indicating that ISA1 was not active in multi-enzyme complex formation in rice.

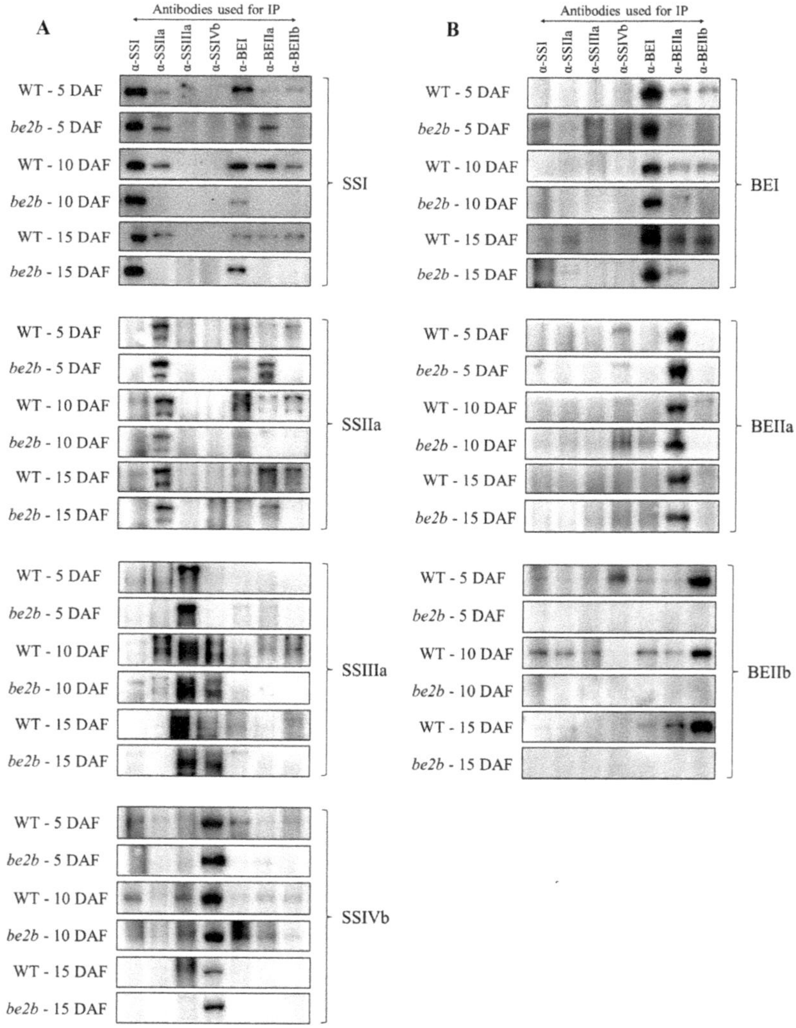

Figure 3. *Cont.*

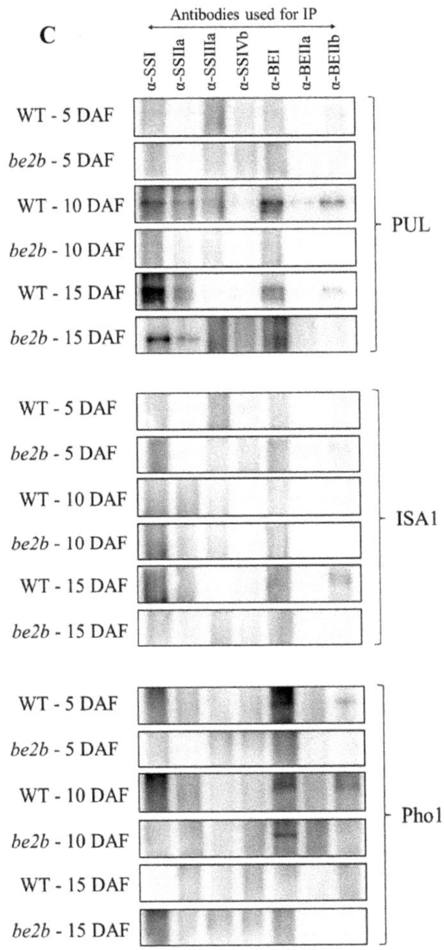

Figure 3. Protein–protein interactions between SSREs during different grain-filling stages (5 DAF, 10 DAF, 15 DAF) in wild type (WT) and *be2b* mutant by co-immunoprecipitation (Co-IP). Soluble proteins extracted from developing seeds were incubated with the indicated isozyme-specific antibodies and protein A magnetic beads. After washing, the captured proteins were released by boiling in a 1× SDS buffer and 5 μL of each supernatant was separated by SDS-PAGE prior to Western blotting. The antibodies used for precipitation as shown in the top panels. The antibodies used for Western blotting are indicated on the right: (**A**) Western blotting of SS isozymes. (**B**) Western blotting of BE isozymes. (**C**) Western blotting of DBE isozymes and Pho1.

Table 1. Comparison of protein-protein interactions among starch synthetic related enzymes during different grain-filling stages (5 DAF, 10 DAF, 15 DAF) in wild type (WT) and *be2b* mutant endosperms determined by co-immunoprecipitation.

Sample		Reciprocal		One Sided [a]	
		Strong Signal [b]	**Weak Signal [b]**	**Strong Signal [b]**	**Weak Signal [b]**
5 DAF	WT		BEIIb–SSI BEIIb–SSIIa **BEIIa–SSIVb** BEIIb–SSIVb BEIIb–BEI	BEI–SSI **BEI–SSIIa** BEI–Pho1	**SSIIa–SSI** **BEIIa–SSI** **BEIIa–SSIIa** SSIVb–SSIIIa SSI–SSIVb BEI–SSIVb BEIIa–BEI BEIIa–BEIIb BEIIb–Pho1
	be2b		**BEIIa–SSIVb**	**SSIIa–SSI** **BEIIa–SSI** **BEIIa–SSIIa**	**BEI–SSIIa** BEI–SSIIIa BEIIa–SSIIIa
10 DAF	WT	BEIIb–SSI BEIIb–SSIIa **SSIVb–SSIIIa** BEIIb–BEI	SSIIa–SSI BEIIb–SSIIIa **BEIIa–SSIVb** BEIIb–BEIIa	**BEI–SSI** BEIIa–SSI BEI–SSIIa **SSIIa–SSIIIa** SSI–SSIVb BEIIa–BEI SSI–PUL BEI–PUL BEIIb–PUL **BEI–Pho1**	**BEIIa–SSIIa** **SSI–SSIIIa** BEIIa–SSIIIa **BEI–SSIVb** BEIIb–SSIVb SSIIa–PUL SSIIIa –PUL BEIIa–PUL BEIIb–Pho1
	be2b	**SSIVb–SSIIIa**	**BEIIa–SSIVb** BEIIa–BEI	**BEI–SSI** **BEI–SSIVb** **BEI–Pho1**	**BEIIa–SSIIa** **SSI–SSIIIa** **SSIIa–SSIIIa**
15 DAF	WT	SSIIa–SSI **BEI–SSI** BEIIb–SSI BEIIb–SSIIa	BEI–SSIIa SSIVb–SSIIIa BEIIb–BEI	BEIIa–SSI **BEIIa–SSIIa** BEIIb–SSIIIa BEIIb–BEIIa **SSI–PUL** BEI–PUL	**BEI–SSIIIa** **BEIIa–BEI** SSIIa–PUL BEIIb–PUL BEIIb–ISA1
	be2b	**BEI–SSI**		SSIVb–SSIIIa **SSI–PUL**	**BEIIa–SSIIa** SSI–SSIIIa **BEI–SSIIIa** SSIIa–BEI **BEIIa–BEI** SSIIa–PUL

[a] The antibody used for immunoprecipitation is shown on the left and the coprecipitated enzyme detected by Western blotting is shown on the right. Same interactions between the wild type and *be2b* mutant are indicated in bold. [b] Protein bands in black or dark grey were defined as strong signals, while those in light gray (but clear) were defined as weak signals.

2.4. Identification of Large Protein Complex Components

To assess dynamic changes in the composition of high-molecular-weight complexes (>400 kDa) in WT and the *be2b* mutant, fractions 1–5 from two samples at three developmental stages separated by GPC were collected, pretreated, digested, and then analyzed by liquid chromatography–tandem mass spectrometry (LC-MS/MS). In all endosperm samples, we obtained 2524 identities representing 2488 unique proteins (Figure 4A; Table S1). Among these proteins detected, only 720 common proteins were detected in all groups, indicating that the components of large protein complexes involved in starch synthesis differ among developmental stages and between WT and mutant. The number of specific proteins identified in WT and *be2b* peaked at 5 DAF (178 and 280, respectively), followed by 10 DAF (86 and 63, respectively) and 15 DAF (168 and 12, respectively). We further analyzed the expression patterns of key enzymes involved in starch synthesis in high-molecular-weight complexes. It is noted that PUL was not detected by LC-MS/MS due to the lack of MSU ID. As shown in Figure 5, most of the enzymes (SSI, SSIIa, BEI, BEIIa, BEIIb, and Pho1) showed higher expression levels in WT compared to *be2b*, and their abundance was significantly enhanced at 15 DAF. The combination of this proteomic information with the results from GPC and Western blot analysis of soluble protein fractions (Figure 2) suggested that more SSREs assemble into large molecular weight protein complexes in WT, especially at 15 DAF.

To detect possible changes in the composition of high-molecular-weight complexes in relation to developmental stages, differentially expressed proteins (DEPs) were identified by RStudio (version 4.1.2). The DEPs were detected between WT and *be2b* in the corresponding period and the differences between 5 and 10 DAF, 10 and 15 DAF, were also compared between WT and *be2b* (Figure 4B: Tables S2 and S3). Both developmental stages and *be2b* mutation showed a large effect on the formed protein complex. From the perspective of endosperm development, a total of 157 DEPs were differentially expressed between 5 and 10 DAF or 10 and 15 DAF in WT and *be2b*. Among these identified DEPs, some proteins were consistently up-regulated or down-regulated in both WT and *be2b*, while a few enzymes displayed converse regulative patterns between WT and *be2b*. One of the most striking DEPs, isopropylmalate synthase (IPMS1, LOC_Os11g04670), was found to be down-regulated from 5 to 10 DAF and up-regulated from 10 to 15 DAF in WT, but inversely up-regulated from 5 to 10 DAF and down-regulated from 10 to 15 DAF in *be2b*. In comparison with WT, the numbers of up-regulated DEPs in *be2b* were 57, 37, and 20 at 5, 10, and 15 DAF, respectively. In contrast, there were 45, 31, and 83 down-regulated DEPs, respectively. Among these DEPs, two proteins of known function, pyruvate orthophosphate dikinase A (PPDKA, LOC_Os03g31750) and PPDKB (also known as FLO4, LOC_Os05g33570), were of particular interest because both of them were dramatically down-regulated in *be2b* at 15 DAF with a fold-change of 7.3 and 14.7, respectively.

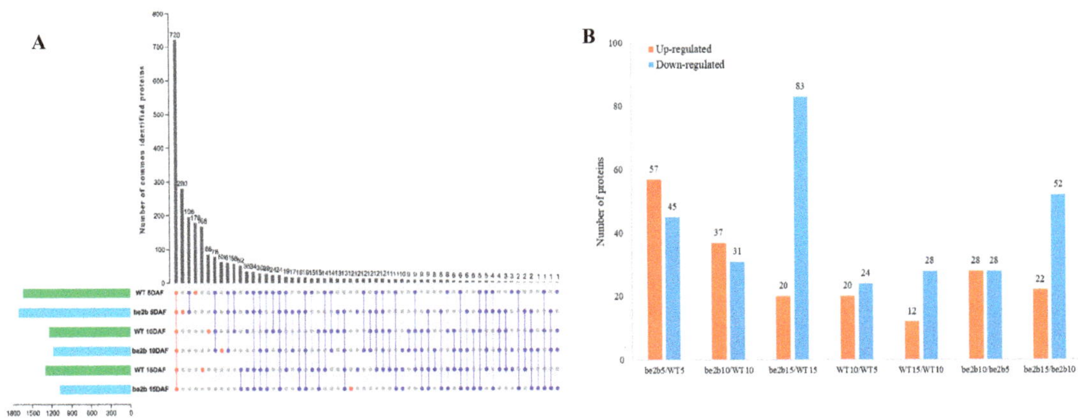

Figure 4. Proteins in high-molecular-weight complexes (>400 kDa) identified by LC-MS/MS during different grain-filling stages (5 DAF, 10 DAF, 15 DAF) in wild type (WT) and *be2b* mutant: (**A**) Upset plot showing the number of proteins that were detected in WT and *be2b* mutant endosperm among the different developmental stages. Red dots indicate proteins identified in all six samples or in only one single sample, blue dots indicate remaining proteins, respectively. (**B**) Number of differentially expressed proteins (DEPs) in relation to *be2b* mutation and developmental stages (3-fold change with p-value < 0.05 and PSM > 2).

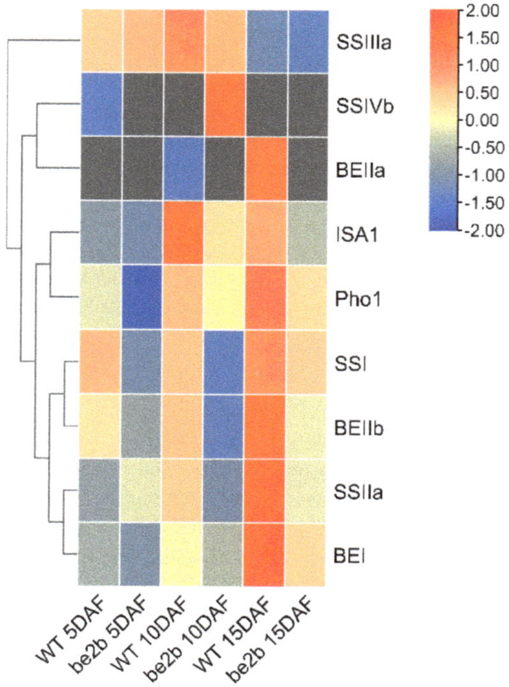

Figure 5. Heat map showing the expression patterns of key enzymes involved in starch synthesis in high-molecular-weight complexes (>400 kDa) identified by LC-MS/MS.

As protein kinases play vital roles in protein complex formation by mediating phosphorylation events, we further delved into these protein kinases identified in high-molecular-weight fractions (>400 kDa) by LC-MS/MS and a total of 29 protein kinases were found (Table S4). Some of them were synchronously present in WT and *be2b*, while other kinases were not. For example, 5′-AMP-activated protein kinase (AMPK) β_1 subunit-related protein (LOC_Os08g29160) existed in both WT and *be2b* at 5 and 15 DAF, and cystathionine β-synthetase (CBS) domain-containing membrane protein (LOC_Os03g63940) existed in both WT and *be2b* at 5 DAF (Figure 6A). Further prediction of protein interaction with STRING (http://String-db.org) revealed the potential association between AMPK β1 subunit-related protein, CBS domain-containing membrane protein, and some starch biosynthetic enzymes in rice (Figure 6B). In addition, the number of protein kinases found in high-molecular-weight fractions in WT and *be2b* peaked at 5 DAF, followed by 10 DAF and 15 DAF (Table S4).

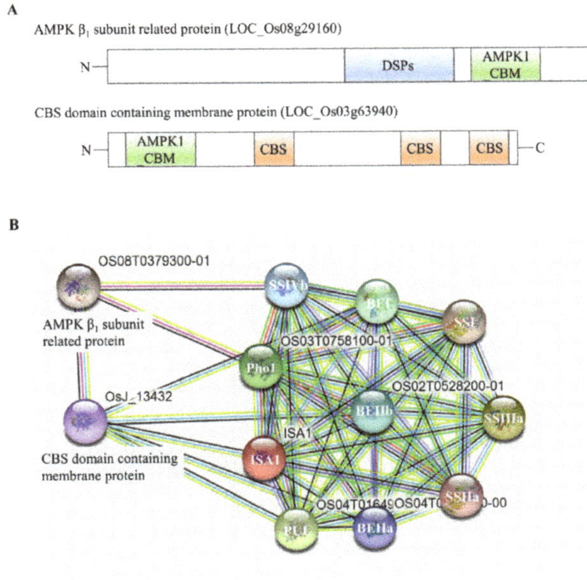

Figure 6. Two protein kinases identified by LC-MS/MS: (**A**) Structure of two protein kinases. (**B**) Potential association networks of two protein kinases and starch synthesis enzymes in rice endosperm from STRING, a database of known and predicted protein interactions. Displayed here is the evidence view, where different line colors represent the types of evidence for the association.

3. Discussion

Starch biosynthetic enzymes in cereal are known to interact with each other and to form protein–protein complexes. The main objectives of the current study were to investigate how starch biosynthetic isozyme complexes were altered during grain-filling stages between BEIIb active and deficient rice. Based on the present and past investigations, possible starch biosynthetic protein complexes in WT (Figure 7A–C) and *be2b* (Figure 7D–F) developing endosperm are schematically illustrated. There are some possibilities that starch biosynthetic enzymes interact with each other through glucan-mediated association. However, it seems unlikely since previous studies showed that pre-incubation of wheat and maize amyloplast extracts with glucan-degrading enzymes to remove glucan polymers did not prevent co-immunoprecipitation of SSs, SBEs, and Pho1 [21,25], and had no effect on the GPC analysis [23], suggesting that the formation of multi-enzyme complexes is due to specific protein–protein interactions.

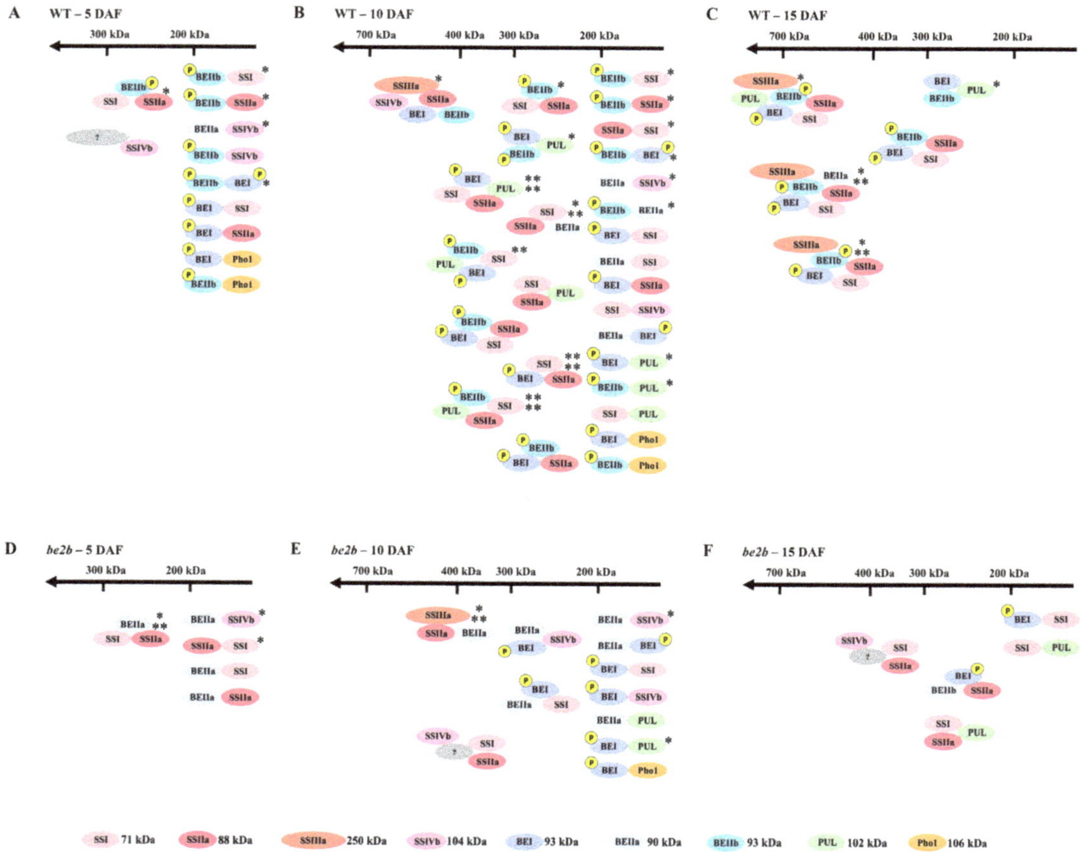

Figure 7. Schematics of speculated starch biosynthetic protein complexes during different grain-filling stages (5 DAF, 10 DAF, 15 DAF) in wild type (WT) and *be2b* mutant: (**A**) WT–5 DAF; (**B**) WT–10 DAF; (**C**) WT–15 DAF; (**D**) *be2b*–5 DAF; (**E**) *be2b*–10 DAF; (**F**) *be2b*–15 DAF. SS isozymes are in red, BE isozymes are in blue, DBE isozymes are in green, Pho1 is in yellow, and phosphate groups are indicated in a small circle. Single, double, triple, and quadruple asterisks indicate the formation of protein complexes confirmed in this study, previously reported by Crofts, et al. [29], Chen and Bao [43], Crofts, et al. [30], and Chen, et al. [44], respectively. Monomeric molecular weights of each isozyme are indicated at the bottom. Note that the schematics do not intend to imply specific details of stoichiometry or direct contacts within the complexes since both direct and indirect physical interactions could be detected by co-immunoprecipitation.

3.1. Components of Multi-Protein Complexes Vary at Different Seed Development Stages

The formation and development of starch undergo a complicated physiological and biochemical process in cereal endosperm [45]. It is clear that the structure and composition of starch granules differ during endosperm development in wheat [46,47], barley [48], maize [49], and rice [41], which can be summarized as an increase in amylose content, a decrease in ordered degree, and changes in the amylopectin chain length distribution. All of these changes in starch structure could be explained by the dynamic expression pattern of genes related to starch biosynthesis [1], giving rise to a different accumulation of enzymes, as described above (Figure 1). These enzymes were demonstrated to display as a complex form through post-translational modifications [50]. Consequently, we attempted to reveal how these protein complexes existed in rice endosperm during seed development, which may facilitate the understanding of starch biosynthetic protein complex formation and the ensuing starch structure and properties.

At the early phase of seed development (5 DAF), when endosperm starch begins to accumulate, most of the enzymes were eluted in low molecular weight fractions in both WT and *be2b* as demonstrated by GPC and Western blot analysis (Figure 2). Accordantly, no strong pairwise association was detected by reciprocal co-immunoprecipitation in rice endosperm at 5 DAF (Figure 3; Table 1), suggesting that proteins weakly associate to form small complexes (Figure 7A,D). However, the immunoblotting signal of SSIVb detected in high-molecular-weight fractions was stronger at 5 DAF compared with later stages (Figure 2A). At the mid-phase of seed development (10 DAF), although no remarkable difference was seen in the position of eluted fractions compared with 5 DAF, the number of proteins in higher molecular-weight fractions was visibly elevated, except for SSIVb (Figure 2). This stage involved active protein–protein interaction in rice endosperm as strong pairwise associations were obtained by reciprocal co-immunoprecipitation and less weak immunodetections of co-precipitated protein were observed (Figures 3 and 7B,E; Table 1). At the late phase of seed development (15 DAF), when endosperm starch rapidly increased, there was a notable shift in the elution patterns of the majority of soluble proteins, suggesting an apparent increase in the aggregation state of these enzymes (Figures 2 and 7C,F); although, no increase was observed in pairwise associations obtained by co-immunoprecipitation (Figure 3; Table 1), which was likely due to the steric inhibition effects emanated from the large molecular mass protein complex. These findings were consistent with previous results from developing wheat endosperm [21] and maize endosperm [24] in which the eluted SS and BE activity in high-molecular-weight fractions could only be detected at the later stages. Although the exact mechanism of SSIVb in rice endosperm starch biosynthesis remains unknown, it showed an opposite elution pattern from other patterns during endosperm development, suggesting its possible role in the interaction with other enzymes and in the formation of large complexes to accomplish its activity at the very early stage of endosperm development (Figure 7A). In fact, SSIV is known to play a significant role in the initiation step of starch synthesis involved in MOSs extension and starch granule control in barley [51], wheat [52], and *Arabidopsis* [53,54]. In rice, the mutation of SSIVb did not show a major impact on the starch structure, but the loss of both SSIVb and SSIIIa resulted in opaque seeds with spherical starch granules, suggesting that SSIVb and SSIIIa are key enzymes affecting starch granule morphology [55]. Intriguingly, in the *be2b* mutant, the elution patterns of SSIVb were similar among different developmental stages (Figure 2A). The increased SSIVb in high-molecular-weight complexes might be regulated by the deficiency of BEIIb (see below). Hence, further investigation of enzyme elution patterns and protein–protein interactions in rice mutant lines lacking SSIVb is necessary to gain a more comprehensive understanding of starch synthesis.

The dynamic changes were further emphasized by proteomic results of high-molecular-weight fractions in which most of the enzymes, including SSI, SSIIa, BEI, BEIIa, BEIIb, and Pho1, showed increased trends from 5 to 15 DAF (Figure 5), indicating these proteins were inclined to form a larger complex at a later developmental stage. IPMS1 was characterized as a DEP in relation to developmental stages, which showed an inverse regulation pattern between WT and *be2b*. IPMS1 was recently identified to regulate seed vigor involved in starch hydrolysis, glycolytic activity, and energy levels [56]. Further work is needed to investigate the effects of BEIIb deficiency on different regulative patterns, which will aid in the understanding of the role of IPMS1 in multi-enzyme complex formation and carbohydrate metabolism.

3.2. *Effects of BEIIb Deficiency on Multi-Enzyme Complex Formation in the Developing Rice Seed*

Protein–protein interactions of starch biosynthetic enzymes are thought to be an important mechanism for efficient starch synthesis [57]. Moreover, studies on mutants that lack starch biosynthetic enzymes have indicated the importance of the formation of starch biosynthetic protein complexes because they could maintain minimal starch biosynthesis by the recruitment of other starch biosynthetic isozymes and the formation of alternative protein complexes; although, the loss of specific starch biosynthetic enzymes may slow starch biosynthesis [33]. Rice mutant seeds with all three major SS activities reduced (*$ss1^L$/$ss2a^L$/ss3a*) complemented the composition of protein complexes within the same enzyme family (SS isozyme) to maximize the storage of photosynthetic products such as starch [32]. The substitution of starch biosynthetic protein complexes in BE deficient mutant of maize [24,25], barley [28], and *japonica* rice [30] could also be a similar phenomenon. Based upon previous experimental results, differences in the component of protein complexes between normal rice (WT) and the BEIIb deficiency mutant (*be2b*) in *indica* background with the active SSIIa isozyme were suspected (Figure 7).

Western blot analysis of SDS-PAGE gels of total proteins from rice endosperm shows that *be2b* used in our study showed a substantial reduction in BEIIb as well as SSI and SSIIa, whilst the protein levels of SSIVb, BEI, ISA1, PUL, and Pho1 were up-regulated compared with WT (Figure 1). Accordingly, the reduction in SSI activity is always accompanied by a deficiency of BEIIb in rice [12,58]. The combined GPC and Western blot analysis of soluble protein involved in amylopectin biosynthesis revealed that the elution pattern of major starch biosynthetic enzymes was altered in *be2b*. In *be2b*, SSI, SSIIa, SSIIIa, BEI, BEIIa, ISA1, PUL, and Pho1 were eluted in lower molecular-weight fractions, and the number of monomeric enzymes (fractions 9–12, <100 kDa) increased compared to WT, especially at 15 DAF (Figure 2). Taking co-immunoprecipitation results into account, associations among these enzymes of *be2b* were weaker than that of WT in the corresponding period (Figure 3; Table 1), indicating that the reduction in BEIIb may either reduce the formation or stability of the protein complex consisting of all these enzymes. Consequently, a considerable decrease in the number of protein complexes at all development stages was hypothesized in *be2b* compared with WT (Figure 7). In addition, no large protein complex containing more than five proteins existed in the *be2b* endosperm, even at 15 DAF (Figure 7F), implying a significant role of BEIIb in protein complex formation. By contrast, SSIVb in *be2b* rice was eluted in earlier fractions than WT and was present in a broader molecular weight range. In addition, the interactions of BEI–SSIVb and BEIIa–SSIVb were stronger in *be2b*, implying alternative protein complexes may form to compensate for BEIIb deficiency. These observations were different from previous studies in *ae^-* mutants of maize [24] and *japonica* rice [30], which reported the SSI–SSIIa–BEIIb trimeric protein was substituted by the compensatory effects of BEI/BEIIa/Pho1 and BEIIa, respectively, to form the altered complexes. Maize BEI and *japonica* rice BEIIa showed similar changed elution patterns in *be2b* with an increased ratio in the 200–300 kDa fraction (the trimeric protein complex elutes) and broader eluded molecular weight range [24,30]. However, neither BEI nor BEIIa in *be2b* investigated in our study displayed this alteration in molecular weight distribution. Taken together, although BEIIb deficiency leads to a similar altered amylopectin fine structure with less amylopectin short chains and more amylopectin long chains, giving rise to increased gelatinization temperature and amylose content as a result of reduced amylopectin biosynthesis in maize [25], *japonica* rice [58], and *indica* rice [41], the underlying mechanism involved in the synthesis of such *ae* starch may differ as judged by distinct compensatory effects. Our present results suggested that SSIVb likely complemented the role of BEIIb in *be2b* as it was observed in higher molecular-weight protein complexes and had enhanced Western blot signals by co-immunoprecipitation (Figures 2A and 3A), which were analogous to that of BEI/BEIIa/Pho1 in the maize *ae^-* mutant [24] and BEIIa in the *japonica* rice *ae^-* mutant [30]. However, neither SSI nor SSIIa associated with SSIVb in *be2b* to form the trimeric protein complex substituted for SSI–SSIIa–BEIIb. The likely explanation is that other potential proteins might bridge the binding of SSIVb to SSI–SSIIa to form a larger alternative protein complex and enhance the association among other starch biosynthetic enzymes to maintain starch biosynthesis in *be2b* (Figure 7E,F). Several enzymes and novel non-enzymatic proteins, including MFP1 [59], PII1 [60], PHS1, and PTST2 [53], were identified to interact with SSIV in *Arabidopsis*. Recently, Zhang, et al. [61] found that carbohydrate-binding module 48 (CBM48) domain-containing protein, FLO6, had a physical association with SSIVb in rice. In addition, correlation analysis between model fitting parameters of amylose and amylopectin chain-length distributions and ratios of the protein content of enzyme pairs revealed that rice SSIVb functionally interacted with SSI, SSIIa, BEI, BEIIb, ISA1, and PUL [62]. Further analyses of the possible interacting partners of SSIVb, the major SSIV isozymes in rice endosperm, will give a better understanding of protein complex formation and starch biosynthesis.

Proteomic evidence of high-molecular-weight fractions from GPC showed that most of the enzyme levels (SSI, SSIIa, BEI, BEIIa, BEIIb, and Pho1) significantly decreased in *be2b* compared with WT (Figure 5), which supported the conclusion that those enzymes are present in a complex with BEIIb. Both PPDKA and PPDKB were significantly down-regulated in the *be2b* mutant at 15 DAF. PPDK catalyzes the formation of the CO_2 acceptor phosphoenolpyruvate (PEP) from pyruvate, which is most well known as a photosynthetic enzyme in C_4 plants [63], and the PPDKB deficient mutant caused by T-DNA insertion showed a white-core endosperm, suggesting that the essential function of PPDKB in modulating the carbon flow during grain filling [64]. It was reported that PPDK1 and/or PPDK2 existed in high molecular mass forms that require multiple starch biosynthetic enzymes as both of them were present in the partially purified C670 fraction and identified in the eluate from the SSIIIHD affinity column [23]. In addition, PPDK was identified by nano-LC-MS/MS in rice starch granule-bound proteins whose composition was thought to reflect the composition of the starch biosynthetic protein complex [30]. These findings, taken together with the proteomic investigations of high-molecular-weight fractions from GPC, suggest that PPDKA and PPDKB might also be assembled into large molecular-weight protein complexes in rice for starch biosynthesis. It can be speculated that PPDKA and PPDKB may participate in the complex with the combination of BEIIb because both of them were remarkably down-regulated in *be2b* as described above. Further analyses are required to detect the direct evidence of such protein–protein interactions using antibodies against PPDK, to validate the prediction of the composition of protein complexes.

3.3. The Essential Role of Protein Phosphorylation and Protein Kinases in Multi-Protein Complexes Formation

It is now well accepted that the ability of some key enzymes to form physical interactions with other proteins and their catalytic activity is modulated by protein phosphorylation [50,65]. The phosphorylation status of BEs was also confirmed in our study (Figure 2C), and their narrower eluted molecular weight ranges compared with BEI and BEIIb reflected a regulatory mechanism of protein phosphorylation in rice endosperm.

Protein phosphorylation is a reversible process regulated by a series of protein kinases [57]; however, the kinases present in starch biosynthetic enzyme complexes remain to be defined. One of the protein kinases eluted from high-molecular-weight fractions and characterized by LC-MS/MS in the present study was AMPK β_1 subunit-related protein, which contains a dual specificity protein phosphatases (DSPs) domain and an AMPK1_carbohydrate-binding module (AMPK1_CBM) domain (Figure 6A). DSPs regulated the activity of their substrates by dephosphorylating threonine/serine and/or tyrosine residues [66]. The AMPK1_CBM domain showed high similarity to CBM20, CBM48, and CBM53 [67], whose surface revealed a carbohydrate-binding pocket to help the kinases, as well as AMPK1_CBM-associated enzymes, bind to starch. AMPK is a ubiquitously expressed, highly conserved heterotrimeric kinase complex with an α (catalytic) subunit and regulatory β and γ subunits in eukaryotic animal cells, which works as a cellular energy sensor in glucose and lipid metabolism [68], while the exact role of AMPK in plants and fungi has not been declared yet. Another kinase was CBS domain-containing membrane protein, which also possessed an AMPK1_CBM domain (Figure 6A). We boldly speculated that this kinase might work as a γ subunit of AMPK in rice because it was composed of three tandem repeats of the CBS domain (four CBS in γ subunit), which bound AMP, ADP, or ATP in a competitive manner according to the changes in energy to modulate the activity of AMPK [69]. Consistent with our hypothesis, the prediction of functional protein association networks showed that the protein interaction of two identified kinases can be formed in the rice endosperm (Figure 6B). Interestingly, both of them also associated with several starch biosynthetic enzymes (Figure 6B), indicating they might be essential components of multi-enzyme complexes in starch biosynthesis. It is difficult to verify such associations by conventional methods because the interaction between kinase and starch biosynthetic enzymes was weak and transient, and the investigation might suffer from the low water solubility of protein. Therefore, the employment of emerging approaches such as TurboID-based proximity labeling technology [70], will shed new light on the key roles of regulatory kinases in multi-protein complexes formation and starch synthesis.

In conclusion, the evidence presented in this article clearly shows that the components of multi-protein complexes changed among different developmental stages, as well as between BEIIb active and deficient rice. With the development of endosperm, most of the enzymes tend to form larger protein complexes except for SSIVb. The converse distribution pattern of SSIVb may be attributed to its vital role in the initiation step of starch synthesis. However, in the *be2b* mutant, BEIIb is not present in any complex, which results in the reduced molecular weight of protein complexes, and a considerable decrease in the number of protein–protein interactions. The large and coordinated protein complexes formed in normal rice endosperm coincide with the fact that starch biosynthesis becomes accelerative during seed development. At the late stage, the diffusion rates of enzymes and substrates are expected to decelerate compared with the relatively aqueous environments during the early stages of seed development, so more efficient machinery is necessary. However, the massive changes in the formation of multi-enzyme complexes in *be2b* endosperm hindered efficient starch synthesis and resulted in the production of modified amylopectin with reduced branching frequency, and longer DP of branches during seed development [41], since other enzymes could only supplement but could not substitute for the crucial role of BEIIb. Again, our results emphasized the importance of the temporal and spatial coordination of multiple starch biosynthetic enzymes for efficient starch synthesis [29,33]. The loss of any components of protein complexes at any stage of seed development will potentially endow starches with altered structure. Hence, the identification of novel proteins involved in multi-enzyme complexes will not only be required to gain further insight into the mechanisms responsible for starch synthesis, but also to provide new targets for improving the quality and yield of rice grains.

4. Materials and Methods

4.1. Plant Materials and Growth Conditions

Oryza sativa subsp. *indica* cv. IR36 (wild type, WT) and a BEIIb-deficient mutant (*be2b*) were used in this study [41]. Both of them were planted at the experimental farm of Zhejiang University, Hangzhou, China, during the summer months under natural conditions. Individual panicles were labeled during flowering and developing seeds were handpicked from fresh plants at 5, 10, and 15 DAF, then immediately frozen in liquid nitrogen and stored at −80 °C. Husks and seed coats were removed before further analysis.

4.2. Protein Extraction

Total protein and soluble proteins were extracted on ice as previously described [29]. After extraction, samples were centrifuged at 14,000× g at 4 °C for 40 min. The supernatant was collected, and the protein concentration was estimated using a NANODROP 2000 spectrophotometer (Thermo, San Jose, CA, USA) before further analysis.

4.3. Gel Permeation Chromatography

Soluble protein was filtered through a 0.22 µm syringe filter to remove large particles and injected into a 500 µL sample loop, prior to fractionation by gel permeation chromatography (GPC) using Superdex 200 resin packed in a 10/300 column connected to an ÄKTA™ *prime plus* chromatography system (GE Healthcare, Chicago, IL, USA). The column was routinely calibrated using commercial gel filtration calibration kits from 75 to 669 kDa (GE

Healthcare, USA) and equilibrated with 10 mM HEPES-KOH, pH 7.5, 100 mM NaCl, at a flow rate of 0.4 mL min^{-1}. Fractions of 0.8 mL were collected when elution volume was 6.6 mL and concentrated 10-fold using an Amicon Ultra 30K centrifugal filter unit (Merck Millipore, Darmstadt, Germany). Concentrated samples were further supplemented with SDS–PAGE sample loading buffer (Beyotime Biotechnology, Haimen, China) following the manufacturer's instructions before SDS-PAGE and Western blotting.

4.4. Co-Immunoprecipitation

Co-immunoprecipitation experiments were conducted as described in Crofts, et al. [29] with some modifications. Soluble proteins, extracted as described above, were incubated with isozyme-specific antibodies and protein A magnetic beads (New England Biolabs, Ipswich, MA, USA). After extensive washing with phosphate-buffered saline (PBS) (137 mM NaCl, 10 mM Na_2HPO_4, 2.7 mM KCl, and 1.8 mM KH_2PO_4 at pH 7.4), bound proteins were released by boiling in 1× sodium dodecyl sulfate (SDS) buffer (Beyotime Biotechnology, China) and 5 μL of each supernatant was analyzed by Western blotting.

4.5. Western Blotting

Proteins were resolved by 8% SDS-PAGE (SDS-polyacrylamide gel electrophoresis), and transferred onto polyvinylidene fluoride (PVDF) membranes using a transblotter. Western blotting procedure was carried out according to the method of Crofts, et al. [71]. Anti-rice GBSSI, SSI [72], SSIIa, SSIIIa [71], SSIVb [29], BEI, BEIIb [73], PUL [74], and Pho1 [16] antibodies were kindly gifted by Prof. Naoko Fujita (Akita Prefectural University, Akita, Japan). Anti-rice BEIIa (the polypeptide CAGAPGKVLVPG) and ISA1 (the polypeptide CEPLVDTGKPAPYD) [29] antibodies were produced by company (HuaAn Biotechnology Co., Ltd., Hangzhou, China). Site-specific phosphopeptide antibodies (BEI-Ser562, BEIIb-Ser685) were produced as previously described [50]. Anti-beta-actin antibody was purchased from Sigma-Aldrich (St. Louis, MO, USA). Western blot results were quantitated using the Image J software [75].

4.6. Protein Preparation, Digestion, LC-MS/MS, and Data Analysis

High-molecular-weight fractions (fractions 1–5, >400 kDa) of 300 μL separated by GPC were mixed together and concentrated 30-fold using an Amicon Ultra 30K centrifugal filter unit (Merck Millipore, Germany). FASP digestion and LC-MS/MS analysis were performed according to the method described by Pang, et al. [50]. Raw data were analyzed with Proteome Discoverer (version 2.4) and were compared with the rice database. Searches were performed using a fragment tolerance of 0.10 Da, and a parent tolerance of 20 ppm, with carbamidomethyl of cysteine as a fixed and oxidation of methionine as variable modifications. Trypsin/P was specified as the enzyme, with maximum missed cleavages allowed of up to 2. Protein identifications were accepted if they achieved a minimum of 1 peptide per protein and a false discovery rate (FDR) of <1%.

4.7. Statistical and Bioinformatic Analyses

For differentially expressed proteins (DEPs) analysis, quantified protein abundances were integrated and then normalized by the DEP (Differential Enrichment analysis of Proteomics data) package (version 1.8.0) in RStudio (version 4.1.2) as the input for statistical analysis. Candidates that met the following criteria were selected: fold-change > 3, $p < 0.05$, and peptide-spectrum match (PSM) >2. The least significant difference (LSD) multiple range test was conducted for comparison of the mean of samples at $p < 0.05$.

Supplementary Materials: The following supporting information can be downloaded at: https://www.mdpi.com/article/10.3390/ijms231810714/s1.

Author Contributions: J.B. and Y.Y. conceived the original research plans; J.B. and F.X. supervised the experiments; Y.Y., Z.Z. and P.T. performed the experiments; Y.Y. wrote the manuscript. All authors have read and agreed to the published version of the manuscript.

Funding: The work was financially supported by the National Natural Science Foundation of China (32201817) and the Zhejiang Provincial Natural Science Foundation (Grant No. LZ21C130003).

Institutional Review Board Statement: Not applicable.

Informed Consent Statement: Not applicable.

Data Availability Statement: The data supporting the findings of this study are available within the article and its supplementary materials.

Acknowledgments: We sincerely thank Naoko Fujita, Akita Prefectural University, Akita, Japan, for kindly providing some antibodies used in this study.

Conflicts of Interest: The authors declare no conflict of interest.

References

1. Ohdan, T.; Francisco, P.B., Jr.; Sawada, T.; Hirose, T.; Terao, T.; Satoh, H.; Nakamura, Y. Expression profiling of genes involved in starch synthesis in sink and source organs of rice. *J. Exp. Bot.* **2005**, *56*, 3229–3244. [CrossRef]
2. Ishimaru, T.; Matsuda, T.; Ohsugi, R.; Yamagishi, T. Morphological development of rice caryopses located at the different positions in a panicle from early to middle stage of grain filling. *Funct. Plant Biol.* **2003**, *30*, 1139–1149. [CrossRef]
3. Xu, S.B.; Li, T.; Deng, Z.Y.; Chong, K.; Xue, Y.; Wang, T. Dynamic proteomic analysis reveals a switch between central carbon metabolism and alcoholic fermentation in rice filling grains. *Plant Physiol.* **2008**, *148*, 908–925. [CrossRef]
4. Rösti, S.; Fahy, B.; Denyer, K. A mutant of rice lacking the leaf large subunit of ADP-glucose pyrophosphorylase has drastically reduced leaf starch content but grows normally. *Funct. Plant Biol.* **2007**, *34*, 480–489. [CrossRef] [PubMed]
5. Lee, S.K.; Eom, J.S.; Hwang, S.K.; Shin, D.; An, G.; Okita, T.W.; Jeon, J.S. Plastidic phosphoglucomutase and ADP-glucose pyrophosphorylase mutants impair starch synthesis in rice pollen grains and cause male sterility. *J. Exp. Bot.* **2016**, *67*, 5557–5569. [CrossRef] [PubMed]
6. Cook, F.R.; Fahy, B.; Trafford, K. A rice mutant lacking a large subunit of ADP-glucose pyrophosphorylase has drastically reduced starch content in the culm but normal plant morphology and yield. *Funct. Plant Biol.* **2012**, *39*, 1068–1078. [CrossRef] [PubMed]
7. Okamura, M.; Hirose, T.; Hashida, Y.; Yamagishi, T.; Ohsugi, R.; Aoki, N. Starch reduction in rice stems due to a lack of OsAGPL1 or OsAPL3 decreases grain yield under low irradiance during ripening and modifies plant architecture. *Funct. Plant Biol.* **2013**, *40*, 1137–1146. [CrossRef] [PubMed]
8. Irshad, A.; Guo, H.; Rehman, S.U.; Wang, X.; Wang, C.; Raza, A.; Zhou, C.; Li, Y.; Liu, L. Soluble starch synthase enzymes in cereals: An updated review. *Agronomy* **2021**, *11*, 1983. [CrossRef]
9. Pandey, M.K.; Rani, N.S.; Madhav, M.S.; Sundaram, R.M.; Varaprasad, G.S.; Sivaranjani, A.K.; Bohra, A.; Kumar, G.R.; Kumar, A. Different isoforms of starch-synthesizing enzymes controlling amylose and amylopectin content in rice (*Oryza sativa* L.). *Biotechnol. Adv.* **2012**, *30*, 1697–1706. [CrossRef]
10. Nakamura, Y.; Utsumi, Y.; Sawada, T.; Aihara, S.; Utsumi, C.; Yoshida, M.; Kitamura, S. Characterization of the reactions of starch branching enzymes from rice endosperm. *Plant Cell Physiol.* **2010**, *51*, 776–794. [CrossRef]
11. Satoh, H.; Nishi, A.; Yamashita, K.; Takemoto, Y.; Tanaka, Y.; Hosaka, Y.; Sakurai, A.; Fujita, N.; Nakamura, Y. Starch-branching enzyme I-deficient mutation specifically affects the structure and properties of starch in rice endosperm. *Plant Physiol.* **2003**, *133*, 1111–1121. [CrossRef] [PubMed]
12. Nishi, A.; Nakamura, Y.; Tanaka, N.; Satoh, H. Biochemical and genetic analysis of the effects of *amylose-extender* mutation in rice endosperm. *Plant Physiol.* **2001**, *127*, 459–472. [CrossRef]
13. Utsumi, Y.; Utsumi, C.; Sawada, T.; Fujita, N.; Nakamura, Y. Functional diversity of isoamylase oligomers: The ISA1 homo-oligomer is essential for amylopectin biosynthesis in rice endosperm. *Plant Physiol.* **2011**, *156*, 61–77. [CrossRef] [PubMed]
14. Yun, M.-S.; Umemoto, T.; Kawagoe, Y. Rice debranching enzyme isoamylase3 facilitates starch metabolism and affects plastid morphogenesis. *Plant Cell Physiol.* **2011**, *52*, 1068–1082. [CrossRef] [PubMed]
15. Nakamura, Y.; Kubo, A.; Shimamune, T.; Matsuda, T.; Harada, K.; Satoh, H. Correlation between activities of starch debranching enzyme and α-polyglucan structure in endosperms of sugary-1 mutants of rice. *Plant J.* **1997**, *12*, 143–153. [CrossRef]
16. Satoh, H.; Shibahara, K.; Tokunaga, T.; Nishi, A.; Tasaki, M.; Hwang, S.K.; Okita, T.W.; Kaneko, N.; Fujita, N.; Yoshida, M.; et al. Mutation of the plastidial alpha-glucan phosphorylase gene in rice affects the synthesis and structure of starch in the endosperm. *Plant Cell* **2008**, *20*, 1833–1849. [CrossRef]
17. Hwang, S.-K.; Nishi, A.; Satoh, H.; Okita, T. Rice endosperm-specific plastidial α-glucan phosphorylase is important for synthesis of short-chain malto-oligosaccharides. *Arch. Biochem. Biophys.* **2010**, *495*, 82–92. [CrossRef]
18. Hwang, S.-K.; Koper, K.; Satoh, H.; Okita, T.W. Rice endosperm starch phosphorylase (Pho1) assembles with disproportionating enzyme (Dpe1) to form a protein complex that enhances synthesis of malto-oligosaccharides. *J. Biol. Chem.* **2016**, *291*, 19994–20007. [CrossRef]
19. Zhang, G.; Cheng, Z.; Zhang, X.; Guo, X.; Su, N.; Jiang, L.; Mao, L.; Wan, J. Double repression of soluble starch synthase genes SSIIa and SSIIIa in rice (*Oryza sativa* L.) uncovers interactive effects on the physicochemical properties of starch. *Genome* **2011**, *54*, 448–459. [CrossRef]

20. Tetlow, I.J.; Wait, R.; Lu, Z.; Akkasaeng, R.; Bowsher, C.G.; Esposito, S.; Kosar-Hashemi, B.; Morell, M.K.; Emes, M.J. Protein phosphorylation in amyloplasts regulates starch branching enzyme activity and protein-protein interactions. *Plant Cell* **2004**, *16*, 694–708. [CrossRef]
21. Tetlow, I.J.; Beisel, K.G.; Cameron, S.; Makhmoudova, A.; Liu, F.; Bresolin, N.S.; Wait, R.; Morell, M.K.; Emes, M.J. Analysis of protein complexes in wheat amyloplasts reveals functional interactions among starch biosynthetic enzymes. *Plant Physiol.* **2008**, *146*, 1878–1891. [CrossRef] [PubMed]
22. Hennen-Bierwagen, T.A.; Liu, F.; Marsh, R.S.; Kim, S.; Gan, Q.; Tetlow, I.J.; Emes, M.J.; James, M.G.; Myers, A.M. Starch biosynthetic enzymes from developing maize endosperm associate in multisubunit complexes. *Plant Physiol.* **2008**, *146*, 1892–1908. [CrossRef] [PubMed]
23. Hennen-Bierwagen, T.A.; Lin, Q.; Grimaud, F.; Planchot, V.; Keeling, P.L.; James, M.G.; Myers, A.M. Proteins from multiple metabolic pathways associate with starch biosynthetic enzymes in high molecular weight complexes: A model for regulation of carbon allocation in maize amyloplasts. *Plant Physiol.* **2009**, *149*, 1541–1559. [CrossRef] [PubMed]
24. Liu, F.; Makhmoudova, A.; Lee, E.A.; Wait, R.; Emes, M.J.; Tetlow, I.J. The amylose extender mutant of maize conditions novel protein-protein interactions between starch biosynthetic enzymes in amyloplasts. *J. Exp. Bot.* **2009**, *60*, 4423–4440. [CrossRef] [PubMed]
25. Liu, F.; Ahmed, Z.; Lee, E.A.; Donner, E.; Liu, Q.; Ahmed, R.; Morell, M.K.; Emes, M.J.; Tetlow, I.J. Allelic variants of the amylose extender mutation of maize demonstrate phenotypic variation in starch structure resulting from modified protein-protein interactions. *J. Exp. Bot.* **2012**, *63*, 1167–1183. [CrossRef]
26. Makhmoudova, A.; Williams, D.; Brewer, D.; Massey, S.; Patterson, J.; Silva, A.; Vassall, K.A.; Liu, F.; Subedi, S.; Harauz, G.; et al. Identification of multiple phosphorylation sites on maize endosperm starch branching enzyme IIb, a key enzyme in amylopectin biosynthesis. *J. Biol. Chem.* **2014**, *289*, 9233–9246. [CrossRef]
27. Liu, F.; Romanova, N.; Lee, E.A.; Ahmed, R.; Evans, M.; Gilbert, E.P.; Morell, M.K.; Emes, M.J.; Tetlow, I.J. Glucan affinity of starch synthase IIa determines binding of starch synthase I and starch-branching enzyme IIb to starch granules. *Biochem. J.* **2012**, *448*, 373–387. [CrossRef]
28. Ahmed, Z.; Tetlow, I.J.; Ahmed, R.; Morell, M.K.; Emes, M.J. Protein–protein interactions among enzymes of starch biosynthesis in high-amylose barley genotypes reveal differential roles of heteromeric enzyme complexes in the synthesis of A and B granules. *Plant Sci.* **2015**, *233*, 95–106. [CrossRef]
29. Crofts, N.; Abe, N.; Oitome, N.F.; Matsushima, R.; Hayashi, M.; Tetlow, I.J.; Emes, M.J.; Nakamura, Y.; Fujita, N. Amylopectin biosynthetic enzymes from developing rice seed form enzymatically active protein complexes. *J. Exp. Bot.* **2015**, *66*, 4469–4482. [CrossRef]
30. Crofts, N.; Iizuka, Y.; Abe, N.; Miura, S.; Kikuchi, K.; Matsushima, R.; Fujita, N. Rice mutants lacking starch synthase I or branching enzyme IIb activity altered starch biosynthetic protein complexes. *Front. Plant Sci.* **2018**, *9*, 1817. [CrossRef]
31. Miura, S.; Crofts, N.; Saito, Y.; Hosaka, Y.; Oitome, N.F.; Watanabe, T.; Kumamaru, T.; Fujita, N. Starch synthase IIa-deficient mutant rice line produces endosperm starch with lower gelatinization temperature than japonica rice cultivars. *Front. Plant Sci.* **2018**, *9*, 645. [CrossRef] [PubMed]
32. Hayashi, M.; Crofts, N.; Oitome, N.F.; Fujita, N. Analyses of starch biosynthetic protein complexes and starch properties from developing mutant rice seeds with minimal starch synthase activities. *BMC Plant Biol.* **2018**, *18*, 59. [CrossRef] [PubMed]
33. Crofts, N.; Nakamura, Y.; Fujita, N. Critical and speculative review of the roles of multi-protein complexes in starch biosynthesis in cereals. *Plant Sci.* **2017**, *262*, 1–8. [CrossRef] [PubMed]
34. Nakamura, Y. Towards a Better Understanding of the metabolic system for amylopectin biosynthesis in plants: Rice endosperm as a model Tissue. *Plant Cell Physiol.* **2002**, *43*, 718–725. [CrossRef] [PubMed]
35. Yamanouchi, H.; Nakamura, Y. Organ specificity of isoforms of starch branching enzyme (Q-enzyme) in rice. *Plant Cell Physiol.* **1992**, *33*, 985–991.
36. Tanaka, N.; Fujita, N.; Nishi, A.; Satoh, H.; Hosaka, Y.; Ugaki, M.; Kawasaki, S.; Nakamura, Y. The structure of starch can be manipulated by changing the expression levels of starch branching enzyme IIb in rice endosperm. *Plant Biotechnol. J.* **2004**, *2*, 507–516. [CrossRef]
37. Kubo, A.; Akdogan, G.; Nakaya, M.; Shojo, A.; Suzuki, S.; Satoh, H.; Kitamura, S. Structure, physical, and digestive properties of starch from wx ae double-mutant rice. *J. Agric. Food Chem.* **2010**, *58*, 4463–4469. [CrossRef]
38. Tsuiki, K.; Fujisawa, H.; Itoh, A.; Sato, M.; Fujita, N. Alterations of starch structure lead to increased resistant starch of steamed rice: Identification of high resistant starch rice lines. *J. Cereal Sci.* **2016**, *68*, 88–92. [CrossRef]
39. Sawada, T.; Itoh, M.; Nakamura, Y. Contributions of three starch branching enzyme isozymes to the fine structure of amylopectin in rice endosperm. *Front. Plant Sci.* **2018**, *9*, 1536. [CrossRef]
40. Nakamura, Y.; Francisco, P.B.; Hosaka, Y.; Sato, A.; Sawada, T.; Kubo, A.; Fujita, N. Essential amino acids of starch synthase IIa differentiate amylopectin structure and starch quality between *japonica* and *indica* rice varieties. *Plant Mol. Biol.* **2005**, *58*, 213–227. [CrossRef]
41. Ying, Y.; Zhang, Z.; Tappiban, P.; Xu, F.; Deng, G.; Dai, G.; Bao, J. Starch fine structure and functional properties during seed development in BEIIb active and deficient rice. *Carbohydr. Polym.* **2022**, *292*, 119640. [CrossRef] [PubMed]
42. Pang, Y.; Zhou, X.; Chen, Y.; Bao, J. Comparative phosphoproteomic analysis of the developing seeds in two indica rice (*Oryza sativa* L.) cultivars with different starch quality. *J. Agric. Food Chem.* **2018**, *66*, 3030–3037. [CrossRef] [PubMed]

43. Chen, Y.; Bao, J. Underlying Mechanisms of zymographic diversity in starch synthase I and pullulanase in rice-developing endosperm. *J. Agric. Food Chem.* **2016**, *64*, 2030–2037. [CrossRef] [PubMed]
44. Chen, Y.; Pang, Y.; Bao, J. Expression profiles and protein complexes of starch biosynthetic enzymes from white-core and waxy mutants induced from high amylose *indica* rice. *Rice Sci.* **2020**, *27*, 152–161.
45. Qu, J.; Xu, S.; Tian, X.; Li, T.; Wang, L.; Zhong, Y.; Xue, J.; Guo, D. Comparative transcriptomics reveals the difference in early endosperm development between maize with different amylose contents. *PeerJ* **2019**, *7*, e7528. [CrossRef]
46. Waduge, R.N.; Xu, S.; Seetharaman, K. Iodine absorption properties and its effect on the crystallinity of developing wheat starch granules. *Carbohydr. Polym.* **2010**, *82*, 786–794. [CrossRef]
47. Wei, C.; Zhang, J.; Chen, Y.; Zhou, W.; Xu, B.; Wang, Y.; Chen, J. Physicochemical properties and development of wheat large and small starch granules during endosperm development. *Acta Physiol. Plant.* **2010**, *32*, 905–916. [CrossRef]
48. Källman, A.; Bertoft, E.; Koch, K.; Sun, C.; Åman, P.; Andersson, R. Starch structure in developing barley endosperm. *Int. J. Biol. Macromol.* **2015**, *81*, 730–735. [CrossRef]
49. Li, L.; Blanco, M.; Jane, J.-L. Physicochemical properties of endosperm and pericarp starches during maize development. *Carbohydr. Polym.* **2007**, *67*, 630–639. [CrossRef]
50. Pang, Y.; Hu, Y.; Bao, J. Comparative phosphoproteomic analysis reveals the response of starch metabolism to high-temperature stress in rice endosperm. *Int. J. Mol. Sci.* **2021**, *22*, 10546. [CrossRef]
51. Cuesta-Seijo, J.A.; Nielsen, M.M.; Ruzanski, C.; Krucewicz, K.; Beeren, S.R.; Rydhal, M.G.; Yoshimura, Y.; Striebeck, A.; Motawia, M.S.; Willats, W.G.; et al. In vitro biochemical characterization of all barley endosperm starch synthases. *Front. Plant Sci.* **2015**, *6*, 1265. [CrossRef] [PubMed]
52. Guo, H.; Liu, Y.; Li, X.; Yan, Z.; Xie, Y.; Xiong, H.; Zhao, L.; Gu, J.; Zhao, S.; Liu, L. Novel mutant alleles of the starch synthesis gene TaSSIVb-D result in the reduction of starch granule number per chloroplast in wheat. *BMC Genom.* **2017**, *18*, 358. [CrossRef] [PubMed]
53. Malinova, I.; Qasim, H.M.; Brust, H.; Fettke, J. Parameters of starch granule genesis in chloroplasts of *Arabidopsis thaliana*. *Front. Plant Sci.* **2018**, *9*, 761. [CrossRef] [PubMed]
54. Seung, D.; Smith, A.M. Starch granule initiation and morphogenesis-progress in Arabidopsis and cereals. *J. Exp. Bot.* **2019**, *70*, 771–784. [CrossRef] [PubMed]
55. Toyosawa, Y.; Kawagoe, Y.; Matsushima, R.; Crofts, N.; Ogawa, M.; Fukuda, M.; Kumamaru, T.; Okazaki, Y.; Kusano, M.; Saito, K.; et al. Deficiency of starch synthase IIIa and IVb alters starch granule morphology from polyhedral to spherical in rice endosperm. *Plant Physiol.* **2016**, *170*, 1255–1270. [CrossRef]
56. He, Y.; Cheng, J.; He, Y.; Yang, B.; Cheng, Y.; Yang, C.; Zhang, H.; Wang, Z. Influence of isopropylmalate synthase OsIPMS1 on seed vigour associated with amino acid and energy metabolism in rice. *Plant Biotechnol. J.* **2019**, *17*, 322–337. [CrossRef]
57. Tappiban, P.; Ying, Y.; Xu, F.; Bao, J. Proteomics and Post-Translational Modifications of Starch Biosynthesis-Related Proteins in Developing Seeds of Rice. *Int. J. Mol. Sci.* **2021**, *22*, 5901. [CrossRef]
58. Itoh, Y.; Crofts, N.; Abe, M.; Hosaka, Y.; Fujita, N. Characterization of the endosperm starch and the pleiotropic effects of biosynthetic enzymes on their properties in novel mutant rice lines with high resistant starch and amylose content. *Plant Sci.* **2017**, *258*, 52–60. [CrossRef]
59. Seung, D.; Schreier, T.B.; Bürgy, L.; Eicke, S.; Zeeman, S.C. Two plastidial coiled-coil proteins are essential for normal starch granule initiation in *Arabidopsis*. *Plant Cell* **2018**, *30*, 1523–1542. [CrossRef]
60. Vandromme, C.; Spriet, C.; Dauvillée, D.; Courseaux, A.; Putaux, J.L.; Wychowski, A.; Krzewinski, F.; Facon, M.; D'Hulst, C.; Wattebled, F. PII1: A protein involved in starch initiation that determines granule number and size in Arabidopsis chloroplast. *New Phytol.* **2019**, *221*, 356–370. [CrossRef]
61. Zhang, L.; Li, N.; Zhang, J.; Zhao, L.; Qiu, J.; Wei, C. The CBM48 domain-containing protein FLO6 regulates starch synthesis by interacting with SSIVb and GBSS in rice. *Plant Mol. Biol.* **2022**, *108*, 343–361. [CrossRef] [PubMed]
62. Zhang, Z.; Tappiban, P.; Ying, Y.; Hu, Y.; Bao, J. Functional interactions between enzymes involved in amylose and amylopectin biosynthesis in rice based on mathematical models. *Biomacromolecules* **2022**, *23*, 1443–1452. [CrossRef] [PubMed]
63. Chastain, C.J.; Failing, C.J.; Manandhar, L.; Zimmerman, M.A.; Lakner, M.; Nguyen, T. Functional evolution of C(4) pyruvate, orthophosphate dikinase. *J. Exp. Bot.* **2011**, *62*, 3083–3091. [CrossRef] [PubMed]
64. Kang, H.G.; Park, S.; Matsuoka, M.; An, G. White-core endosperm floury endosperm-4 in rice is generated by knockout mutations in the C-type pyruvate orthophosphate dikinase gene (OsPPDKB). *Plant J.* **2005**, *42*, 901–911. [CrossRef]
65. Subasinghe, R.M.; Liu, F.; Polack, U.C.; Lee, E.A.; Emes, M.J.; Tetlow, I.J. Multimeric states of starch phosphorylase determine protein-protein interactions with starch biosynthetic enzymes in amyloplasts. *Plant Physiol. Biochem.* **2014**, *83*, 168–179. [CrossRef]
66. Prabhakar, S.; Asuthkar, S.; Tsung, A.J.; Velpula, K.K. Dual-specificity Protein Phosphatases. In *Encyclopedia of Signaling Molecules*; Choi, S., Ed.; Springer: New York, NY, USA, 2017; pp. 1–9.
67. Zhong, Y.; Blennow, A.; Kofoed-Enevoldsen, O.; Jiang, D.; Hebelstrup, K.H. Protein Targeting to Starch 1 is essential for starchy endosperm development in barley. *J. Exp. Bot.* **2019**, *70*, 485–496. [CrossRef]
68. Hardie, D.G.; Ross, F.A.; Hawley, S.A. AMPK: A nutrient and energy sensor that maintains energy homeostasis. *Nat. Rev. Mol. Cell Biol.* **2012**, *13*, 251–262. [CrossRef]
69. Ross, F.A.; Thomas, E.J.; Hardie, D.G. Differential regulation by AMP and ADP of AMPK complexes containing different γ subunit isoforms. *Biochem. J.* **2016**, *473*, 189–199. [CrossRef]

70. Xu, F.; Jia, M.; Li, X.; Tang, Y.; Jiang, K.; Bao, J.; Gu, Y. Exportin-4 coordinates nuclear shuttling of TOPLESS family transcription corepressors to regulate plant immunity. *Plant Cell* **2021**, *33*, 697–713. [CrossRef]
71. Crofts, N.; Abe, K.; Aihara, S.; Itoh, R.; Nakamura, Y.; Itoh, K.; Fujita, N. Lack of starch synthase IIIa and high expression of granule-bound starch synthase I synergistically increase the apparent amylose content in rice endosperm. *Plant Sci.* **2012**, *193–194*, 62–69. [CrossRef]
72. Fujita, N.; Yoshida, M.; Asakura, N.; Ohdan, T.; Miyao, A.; Hirochika, H.; Nakamura, Y. Function and Characterization of Starch Synthase I Using Mutants in Rice. *Plant Physiol.* **2006**, *140*, 1070–1084. [CrossRef] [PubMed]
73. Nakamura, Y.; Takeichi, T.; Kawaguchi, K.; Yamanouchi, H. Purification of two forms of starch branching enzyme (Q-enzyme) from developing rice endosperm. *Physiol. Plant.* **1992**, *84*, 329–335. [CrossRef]
74. Nakamura, Y.; Umemoto, T.; Ogata, N.; Kuboki, Y.; Yano, M.; Sasaki, T. Starch debranching enzyme (R-enzyme or pullulanase) from developing rice endosperm: Purification, cDNA and chromosomal localization of the gene. *Planta* **1996**, *199*, 209–218. [CrossRef] [PubMed]
75. Rueden, C.T.; Schindelin, J.; Hiner, M.C.; DeZonia, B.E.; Walter, A.E.; Arena, E.T.; Eliceiri, K.W. ImageJ2: ImageJ for the next generation of scientific image data. *BMC Bioinform.* **2017**, *18*, 529. [CrossRef] [PubMed]

International Journal of *Molecular Sciences*

MDPI

Article

Chromosome-Level Genome Assembly of a Fragrant *Japonica* Rice Cultivar 'Changxianggeng 1813' Provides Insights into Genomic Variations between Fragrant and Non-Fragrant *Japonica* Rice

Ruisen Lu [1], Jia Liu [1], Xuegang Wang [2], Zhao Song [3], Xiangdong Ji [2], Naiwei Li [1,*], Gang Ma [2] and Xiaoqin Sun [1,*]

[1] Institute of Botany, Jiangsu Province and Chinese Academy of Sciences, Nanjing 210014, China
[2] Changshu Agricultural Science Research Institute, Changshu 215500, China
[3] Guangdong Academy of Forestry, Guangzhou 510520, China
* Correspondence: linaiwei@jib.ac.cn (N.L.); xiaoqinsun@cnbg.net (X.S.)

Abstract: East Asia has an abundant resource of fragrant *japonica* rice that is gaining increasing interest among both consumers and producers. However, genomic resources and in particular complete genome sequences currently available for the breeding of fragrant *japonica* rice are still scarce. Here, integrating Nanopore long-read sequencing, Illumina short-read sequencing, and Hi-C methods, we presented a high-quality chromosome-level genome assembly (~378.78 Mb) for a new fragrant *japonica* cultivar 'Changxianggeng 1813', with 31,671 predicated protein-coding genes. Based on the annotated genome sequence, we demonstrated that it was the *badh2-E2* type of deletion (a 7-bp deletion in the second exon) that caused fragrance in 'Changxianggeng 1813'. Comparative genomic analyses revealed that multiple gene families involved in the abiotic stress response were expanded in the 'Changxianggeng 1813' genome, which further supported the previous finding that no generalized loss of abiotic stress tolerance associated with the fragrance phenotype. Although the 'Changxianggeng 1813' genome showed high genomic synteny with the genome of the non-fragrant *japonica* rice cultivar Nipponbare, a total of 289,970 single nucleotide polymorphisms (SNPs), 96,093 small insertion-deletion polymorphisms (InDels), and 8690 large structure variants (SVs, >1000 bp) were identified between them. Together, these genomic resources will be valuable for elucidating the mechanisms underlying economically important traits and have wide-ranging implications for genomics-assisted breeding in fragrant *japonica* rice.

Keywords: *BADH2*; 'Changxianggeng 1813'; fragrant rice; genome assembly; genomic variations; *japonica* cultivar

check for updates

Citation: Lu, R.; Liu, J.; Wang, X.; Song, Z.; Ji, X.; Li, N.; Ma, G.; Sun, X. Chromosome-Level Genome Assembly of a Fragrant *Japonica* Rice Cultivar 'Changxianggeng 1813' Provides Insights into Genomic Variations between Fragrant and Non-Fragrant *Japonica* Rice. *Int. J. Mol. Sci.* **2022**, *23*, 9705. https://doi.org/10.3390/ijms23179705

Academic Editors: Jinsong Bao and Jianhong Xu

Received: 1 August 2022
Accepted: 24 August 2022
Published: 26 August 2022

Publisher's Note: MDPI stays neutral with regard to jurisdictional claims in published maps and institutional affiliations.

Copyright: © 2022 by the authors. Licensee MDPI, Basel, Switzerland. This article is an open access article distributed under the terms and conditions of the Creative Commons Attribution (CC BY) license (https://creativecommons.org/licenses/by/4.0/).

1. Introduction

Fragrant rice (*Oryza sativa* L.), well-known for its pleasant and subtle aroma, is widely preferred among rice consumers and fetches a higher price than non-fragrant rice in both domestic and international markets [1,2]. At present, Basmati rice from India and Pakistan and Jasmine rice from Thailand are the two most popular fragrant rice cultivars in the world [3,4]. It is, however, noteworthy that both of these two fragrant rice cultivars belong to the *indica* subspecies, with fluffy and dry cooked rice, while consumers from East Asia, including China, Japan, and Korea tend to prefer *japonica* rice that becomes sticky and soft when cooked [4]. Although East Asia has diverse and rich germplasm resources of fragrant *japonica* rice, none of them have been fully commercially utilized [5]. Thus, breeding and cultivation of fragrant *japonica* rice has become one of the most important jobs in modern rice breeding projects, especially in East Asia [6].

Hundreds of volatile compounds have been detected in fragrant rice, but the key compound responsible for the characteristic fragrance is 2-acetyl-1-pyrroline (2AP) [2,7].

Investigations into the genetic basis of rice fragrance have demonstrated that the fragrance phenotype is largely controlled by a recessive *betaine aldehyde dehydrogenase 2* (*BADH2*) gene, which comprises 15 exons and 14 introns with approximately 7 kilobase pairs in length [7,8]. The dominant *BADH2* gene encoding the active BADH2 catalyzes the oxidation of γ-aminobutyraldehyde (AB-ald, a 2AP precursor), while the recessive *BADH2* gene encoding the inactive BADH2 results in the accumulation of both AB-ald and its cyclic form Δ^1pyrroline, and finally acetylates 2AP through enzymatic or non-enzymatic reactions [7,9–11]. To date, multiple types of loss-of-function mutations in the *BADH2* gene responsible for rice fragrance have been reported, e.g., an 8-bp deletion and three single nucleotide polymorphisms (SNPs) in the seventh exon (designated as *badh2-E7* or *badh2.1*), a 7 bp deletion in the second exon (*badh2-E2* or *badh2.2*), and an 803 bp deletion between the fourth and fifth exons (*badh2-E4/5*) [6,7,12–14]. Based on the above information, functional molecular markers have also been developed for various SNPs and small insertion-deletion polymorphisms (InDels) on different exons of *BADH2*, improving the efficiency of selection and breeding of fragrant rice, e.g., [6,13]. However, these molecular markers were usually developed based on old conventional fragrant rice varieties, most of which have relatively low yields and demonstrate inferior agronomic performance, such as weak disease resistance and low tolerance to climatic stresses [3]. Thus, excluding inferior agronomic traits has become a major challenge during introgression of fragrance alleles from old conventional fragrant rice cultivars into modern rice cultivars [3].

The new fragrant *japonica* rice cultivar 'Changxianggeng 1813' (2AP content: ~310 ug/kg, unpublished data), derived from a cross between '93-63/wuyungeng 20' and 'wuyungeng 31', was developed by the Changshu Institute of Agricultural Sciences (Changshu, Jiangsu, China) and licensed for release in Jiangsu Province, China in 2020 [15]. In contrast to old conventional fragrant *japonica* rice cultivars, 'Changxianggeng 1813' shows high resistances to lodging and blast, with both high yield and good quality [15], which is not only suitable for being widely planted in the South Yangtze River regions, but also could be used as a parental line to develop new fragrant *japonica* rice cultivars. Therefore, the construction of a high-quality genome of 'Changxianggeng 1813' is essential for further improvement of this cultivar or its progenies, as well as accelerating the process of fragrant *japonica* rice breeding, by providing genomic resources that could be directly applied to fragrant *japonica* rice cultivars.

With the rapid progress in next-generation sequencing technologies, unprecedented amounts of genomic data for wild and cultivated rice are currently available, providing important resources for investigation of the genetic basis behind rice domestication and improvement [16–29]. Within cultivated rice, however, genome assemblies for most cultivars were based on short-read sequencing data, which often showed higher levels of incompleteness than those generated from long-read sequences, e.g., [16,30–32]. Moreover, the information from highly polymorphic regions, especially for large structural variations (SVs), would often be inevitably lost by direct mapping of short sequencing reads onto a single reference genome (typically, *O. sativa japonica* Nipponbare) [23,33]. Thus, high-quality, chromosome-level genome assemblies for different rice cultivars are still needed to comprehensively capture the genomic variations in rice.

In this study, we generated a high-quality, chromosome-level genome sequence of the fragrant *japonica* rice cultivar 'Changxianggeng 1813', based on Oxford Nanopore, Illumina, and Hi-C sequencing technologies. Then, we aligned the *BADH2* gene in 'Changxianggeng 1813' to previously described *BADH2* haplotypes to verify the presence/absence of the mutations associated with fragrance and determine their phylogenetic relationships. We also carried out comparative genomic analyses to provide insights into the evolution and adaptation of this cultivar. Finally, we performed a pairwise genome comparison between the fragrant *japonica* cultivar 'Changxianggeng 1813' and the non-fragrant *japonica* cultivar Nipponbare to identify genomic variations (SNPs, InDels, SVs). Of note, this is the first high-quality de novo assembly genome sequence for fragrant *japonica* rice published to date, and is expected to have a lasting direct impact on molecular breeding and improvement of fragrant *japonica* rice.

2. Results and Discussion

2.1. Genome Sequencing and De Novo Assembly

With the rapid development of genome sequencing methods, long-read sequencing technologies such as Oxford Nanopore Technology and Pacific Biosciences combined with Illumina short-read sequencing and chromosome conformation capture (Hi-C) technologies have become a common standard protocol to generate high-quality assemblies of plant genomes [34–36]. In this study, the genome of 'Changxianggeng 1813' was sequenced and de novo assembled by a hybrid strategy combining Oxford Nanopore, Illumina, and Hi-C technologies. A total of ~51.59 Gb Nanopore long reads, ~28.21 Gb Illumina short reads, and ~41.45 Gb Hi-C reads were generated, respectively, after filtering (Table S1). Using *k*-mer analysis with Illumina clean reads, the genome size of 'Changxianggeng 1813' was estimated to be approximately 394.39 Mb, with a heterozygosity rate of 0.08% (Table S2).

The 'Changxianggeng 1813' genome was preliminarily assembled based on Nanopore long reads, followed by two rounds of assembly corrections using both of Nanopore and Illumina sequencing data, which produced an assembled genome (scaffold level) with a total length of ~378.78 Mb, a GC content of 43.55%, and a surprisingly long scaffold N50 of 29.83 Mb (Table 1). Despite the super-long scaffolds generated, Hi-C data were employed to further improve assembly contiguity and obtain a high-quality reference genome of 'Changxianggeng 1813'. Approximately 62.20 million valid interaction pairs (~18.65 Gb Hi-C data), accounting for 82.55% of the unique mapped read pairs, were used for the Hi-C assembly. Consequently, all ~378.78 Mb (100%) data in 20 scaffolds were anchored and orientated onto 12 chromosomes by agglomerative hierarchical clustering, with their lengths ranging from 22.66 to 43.60 Mb (Figure 1a,b; Table S3). The 12 chromosomes could be distinguished obviously, and the near-diagonal interaction signals were considerably stronger than that of other positions within each chromosome, which illustrated that Hi-C scaffolding was reliable and robust (Figure 1a).

Table 1. Statistics of the genome assembly of 'Changxianggeng 1813'.

Genome Assembly	'Changxianggeng 1813'
Total number of scaffolds	20
GC content (%)	43.55
Scaffold N50 (bp)	29,831,576
Scaffold N90 (bp)	16,183,542
Longest Scaffold	43,598,337
Total length (bp)	378,775,865
Mapping efficiency (%)	99.11%
Complete BUSCO (%)	96.16%

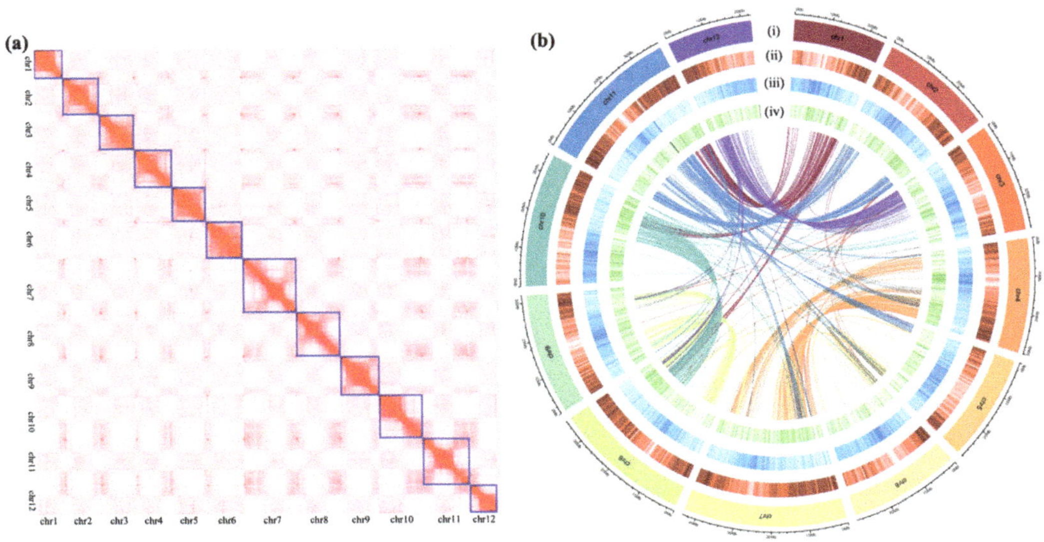

Figure 1. Basic characteristics of the 'Changxianggeng 1813' genome. (**a**) Genome-wide Hi-C heat map of the 'Changxianggeng 1813' genome showing chromatin interactions among the 12 chromosomes. Darker red color indicates higher contact probability. The blue boxes show the location of the chromosomes. (**b**) Circos plot of the multidimensional topography of the 12 chromosomes in the 'Changxianggeng 1813' genome. Concentric circles, from outermost to innermost, show (i) the chromosome, (ii) gene density, (iii) percentage of repeats, and (iv) GC content. The three metrics were calculated in 500 kb sliding windows. In the innermost circle, each line shows the syntenic relationship between different chromosomes, indicating the existence of large episodic duplications derived from the ancient whole-genome duplication in rice.

The accuracy and completeness of the genome assembly were first assessed by mapping the Illumina reads back to the reference genome, which revealed a mapping efficiency of 99.11% (Table 1). Furthermore, 1552 (96.16%) of 1614 conserved BUSCO (Benchmarking Universal Single-Copy Orthologs) genes, including 1514 (93.8%) complete and single-copy BUSCOs and 38 (2.4%) complete and duplicated BUSCOs (Tables 1 and S4), were completely recalled in our assembly. Taken together, these results implied that the genome assembly of 'Changxianggeng 1813' was performed well and in high completeness. In a word, the assembled genome of 'Changxianggeng 1813' was at the chromosomal level, with a longer scaffold N50 length than in most de novo assemblies of *Oryza* genomes e.g., [16,31,37,38], which provides good quality, high-resolution resources for associating traits of interest with genetic variations and identifying the genes controlling those important economical traits in fragrant *japonica* rice.

2.2. Genome Annotation

Repetitive sequences constitute large proportions of plant genomes and often play key roles in plant genome evolution due to their roles in both genome size variation and functional adaption [39,40]. Using a combination of homology-based and de novo approaches, about 50.52% of the 'Changxianggeng 1813' genome was identified as transposable elements (TEs; Table S5). Of these TEs, DNA transposons were the most abundant, occupying 24.47% of the genome, followed by long terminal repeats (LTRs; 24.15%), while long interspersed nuclear elements (LINEs) and short interspersed nuclear elements (SINEs) accounted for 1.78% and 0.12%, respectively (Table S5). Additionally, approximately 0.97% of the 'Changxianggeng 1813' genome was identified as tandem repeats (Table S5). Indeed, among various types of repetitive sequences, LTRs are one of the most important contributors to the genome size variation across the *Oryza* genus [41,42]. It was thus speculated that the genome size of 'Changxianggeng 1813' (~379 Mb), nearly half of that in *O. granulate* (~777 Mb), is largely due to the differences of the proportion of LTRs between them (24.15% for 'Changxianggeng 1813' and 59.33% for *O. granulate*) [24].

A total of 32,165 protein-coding genes were predicted by integrating protein-based homology, de novo and transcriptome-based prediction approaches, with average gene and coding sequence lengths of 4244 and 1224 bp, respectively, and an average of 4.62 exons per gene (Table 2). Among these protein-coding genes, 98.46% (31,671) could be annotated by at least one of the six functional databases employed, including Uniprot, Pfam, GO (Gene Ontology), KEGG (Kyoto Encyclopedia of Genes and Genomes), and NR (Non-redundant) (Table S6). In addition, 3524 RNAs were identified as potential noncoding RNAs, including 1756 microRNAs (miRNAs), 715 transfer RNAs (tRNAs), 322 ribosomal RNAs (rRNAs), and 731 small nuclear RNAs (snRNAs) (Table S7).

Table 2. Prediction of protein-coding gene models in the 'Changxianggeng 1813' genome.

Method	Software/Gene Set	Gene Number	Average Gene Length (bp)	Average CDS Length (bp)	Average Exon per Gene	Average Exon Length (bp)	Average Intron Length (bp)
De novo	Augustus	55,473	5539.04	2107.58	3.38	623.88	1442.88
	GlimmerHMM	35,280	2841.15	1183.67	4.81	246.27	435.45
Homolog	*Oryza brachyantha*	46,486	11,708.76	911.19	3.79	240.46	3871.05
	Oryza sativa japonica	55,308	10,634.66	964.91	3.6	268.35	3725.27
	Aegilops tauschii	55,839	13,039.16	939.32	3.55	264.97	4754.30
	Panicum hallii	48,315	11,231.61	911.68	3.66	249.38	3885.87
RNA-Seq	TransDecoder	12,400	4698.81	1203.90	6.73	329.13	433.89
Final set	EvidenceModeler	32,165	4243.68	1224.17	4.87	434.61	546.96

2.3. Characterization and Evolutionary Analysis of BADH2 Gene

Since 'Changxianggeng 1813' has been identified to be a fragrant rice (2AP content: ~310 ug/kg, unpublished data), we checked whether it indeed carries a recessive *BADH2* gene and investigated its allelic variation. By comparing to the non-fragrant rice cultivar Nipponbare, a 7 bp deletion (5′-CGGGCGC-3′) in the second exon was observed at the *BADH2* allele (*badh2-E2*), which generated a premature stop codon that disabled the BADH2 enzyme (Figure S1a), thereby promoting the accumulation of 2AP in 'Changxianggeng 1813'. The *badh2-E2* allele carried by 'Changxianggeng 1813' was consistent with that in a number of Chinese fragrant *japonica* rice cultivars, e.g., 'Wuxiang 9915', 'Xiangjing 111', 'Zhenxiangjing 5', suggesting that this allele is common by descent in Chinese fragrant rice cultivars [8,13]. Phylogenetic analysis of *BADH2* haplotype data (Figure S1b) further showed that the *badh2-E2* allele in 'Changxianggeng 1813' clustered together with a previously identified haplotype sequence endemic to two cultivated *japonica* rice (see [2] for full details). Taken together, these findings provided additional support for the previous studies indicating that *badh2-E2* may arise and become fixed in the *japonica* gene pool [7,8,13]. However, it is worth noting that the sample size analyzed to date is still inadequate to comprehensively detect the origin and evolution of *badh2-E2* in fragrant rice.

2.4. Genome Synteny

Comparisons of genome synteny within and between species have provided a framework to reveal evolutionary processes that lead to diversity of genome structure and function in many lineages [43]. Nowadays, genome synteny analysis has become an integral part of comparative genomics for almost every new published genome. Using the MCScan toolkit, a total of 19,912 and 24,975 gene pairs were identified in the intergenomic comparisons of the fragrant cultivar 'Changxianggeng 1813' vs. the non-fragrant cultivar Nipponbare (Figure 2a), and 'Changxianggeng 1813' vs. the common wild rice *O. rufipogon* (Figure 2b), respectively. In general, extremely high degrees of collinearity were observed in these two comparisons; each chromosome of 'Changxianggeng 1813' corresponded to one chromosome of Nipponbare and *O. rufipogon*, respectively, although some interchromosomal rearrangement events were detected (Figure 2c). It was also found that there were fewer scattered points in the comparison of 'Changxianggeng 1813' vs. Nipponbare, than in 'Changxianggeng 1813' vs. *O. rufipogon* (Figure 2a,b), suggesting a close relationship between 'Changxianggeng 1813' and Nipponbare.

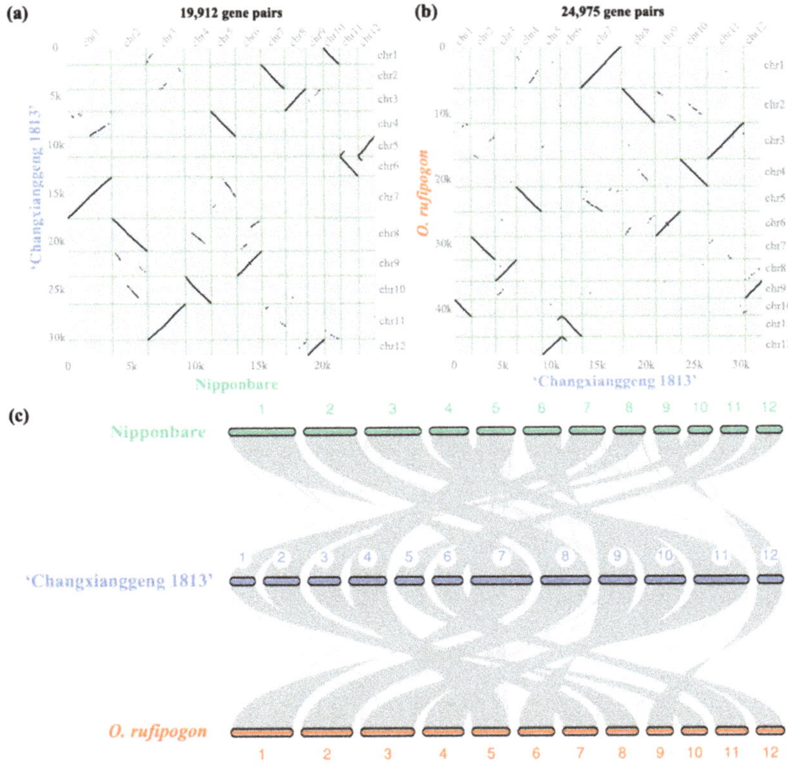

Figure 2. Chromosome synteny between the fragrant *japonica* cultivar 'Changxianggeng 1813' and its close relatives, i.e., the non-fragrant *japonica* cultivar Nipponbare and the common wild rice *O. rufipogon*. (**a**,**b**) Syntenic dot plots for intergenomic comparisons of (**a**) 'Changxianggeng 1813' vs. Nipponbare, and (**b**) 'Changxianggeng 1813' vs. *O. rufipogon*. (**c**) Macrosyntenic relationship pattern between 'Changxianggeng 1813' and its two close relatives (Nipponbare and *O. rufipogon*).

2.5. Gene Family Evolution and Phylogenetic Relationships

Of the 32,165 protein-coding genes identified in the 'Changxianggeng 1813' genome, 11,413 were classified as single-copy orthologs, 9531 as multiple-copy orthologs, 2899 as unique paralogs, and 17,837 as other paralogs (Figure 3a). All the 32,165 protein-coding genes were clustered into 26,658 gene families, of which 1576 (5.91%) were unique in the 'Changxianggeng 1813' genome (Table S8). A total of 7658 single-copy orthologous genes shared among the six *Oryza* genomes were identified and used for phylogenetic analysis. Phylogenetic analysis strongly supported that the fragrant cultivar 'Changxianggeng 1813' and the non-fragrant cultivar Nipponbare, both of which belong to the *japonica* subspecies, were sister to each other, and jointly sister to the common wild rice *O. rufipogon* (Figure 3b). The divergence time of 'Changxianggeng 1813' and Nipponbare was estimated to be 0.5 (0.4–0.6) million years ago (Ma; Figure 3b), unambiguously older than the date of domestication of the rice (10,000 years ago). One possible explanation for this is that the divergence between these two cultivars from two different subpopulations (temperate *japonica* and *aromatic*) is in part due to differentiation of their ancestral populations in different locations and/or at different times. Furthermore, although our estimated divergence time is slightly older than the date for *japonica* and *indica* (about 0.44 Ma), this estimate conformed generally with the previous findings that (i) genomic variation in the rice is deeply partitioned and that divergent haplotypes can be readily associated with major varietal groups and subpopulations, and (ii) rice domestication proceeded from multiple predifferentiated ancestral pools much earlier than the beginning of agriculture in Asia [37,44].

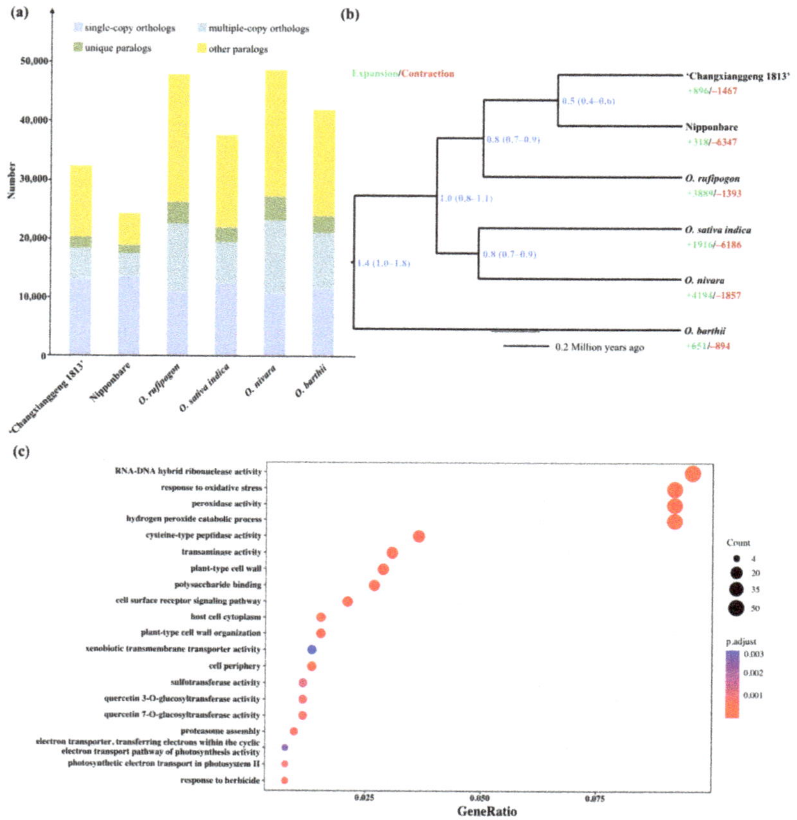

Figure 3. (**a**) Comparison of copy numbers in gene clusters residing in the genomes of 'Changxianggeng 1813' and five other members of *Oryza*. (**b**) Phylogenetic tree inferred from single-copy orthogroups. Numbers near each node refer to divergence times (in million years ago, Ma). Bootstrap values are all 100. Numbers marked in green and red represent gene family expansions and contractions, respectively. (**c**) Visualization of results from GO enrichment analysis of significantly expanded gene families in 'Changxianggeng 1813'. The top 20 GO terms were selected for display after using the Benjamini–Hochberg multiple test correction for p-value adjustment (adjusted p-value < 0.01).

Gene family expansion and contraction are generally considered important evolutionary mechanisms that contribute to evolutionary adaption to the environment [45,46]. To reveal gene family expansion and contraction related to environmental stress in 'Changxianggeng 1813', we undertook a computational analysis of gene family sizes among different members of *Oryza*. Our results indicated that 896 gene families in 'Changxianggeng 1813' genome underwent expansion, while 1467 genes families underwent contraction (Figure 3b). Functional enrichment analysis of expanded gene families revealed 25 GO terms that were significantly enriched (*p*.adjust < 0.01). The expanded gene families were mainly enriched in genes associated with RNA-DNA hybrid ribonuclease activity (GO:0004523, *p*.adjust = 2.6×10^{-30}), hydrogen peroxide catabolic process (GO:0042744, *p*.adjust = 1.71×10^{-37}), peroxidase activity (GO:0004601, *p*.adjust = 5.05×10^{-35}), and response to oxidative stress (GO:0006979, *p*.adjust = 8.24×10^{-32}) (Figure 3c). It needs to be emphasized here that oxidative stress is regarded as a major damaging factor in plants exposed to a variety of abiotic stresses [47]. Thus, these expanded oxidative stress response genes may have a role in conferring enhanced stress tolerance to 'Changxianggeng 1813' during periods of rapid climate change. This result also supported the previous findings that *BADH2* does not play a role in abiotic stress tolerance in rice, and no generalized loss of abiotic stress tolerance associated with the fragrance phenotype [48].

2.6. Genomic Variations between 'Changxianggeng 1813' and Nipponbare

Since large-scale genome sequencing has been undertaken in rice, a substantial number of genetic variations, such as single nucleotide polymorphisms (SNPs) and small insertion-deletion polymorphisms (InDels), have become available across the rice genome, e.g., [26,49]. However, few recent studies have been concentrated on fragrant *japonica* rice, resulting in a severe lack of knowledge for valuable fragrant *japonica* rice, especially in East Asia. Although the genome assembly of the fragrant *japonica* cultivar 'Changxianggeng 1813' very closely matched the genome of non-fragrant *japonica* cultivar Nipponbare (Figure 2a,c), a total of 289,970 SNPs and 96,093 InDels were identified in the 'Changxianggeng 1813' genome when compared to the Nipponbare genome, with an average density of 0.76 SNPs and 0.25 InDels per kb, respectively (Figure 4; Tables S9 and S10). The number of SNPs and InDels per 1 Mb varied considerably across each chromosome. In particular, chromosome 9 had the highest density of both SNPs (208.3 Mb^{-1}) and InDels (49.0 Mb^{-1}), while chromosome 4 had the lowest SNP (19.6 Mb^{-1}) and Indel (1.2 Mb^{-1}) densities (Figure 4a,b; Tables S9 and S10). The distribution of SNPs and InDels was also uneven within a chromosome. For example, on chromosome 1, SNPs and InDels were dense from 11.9 to 12.7 Mb, but sparse from the regions of 9.8–10.5 and 17.6–20.0 Mb (Figure 4a,b). The distributions of SNPs and InDels were positively correlated, and both were more abundant in intergenic spacer (IGS) regions. More specifically, about 67.12% (13,142/19,580, chromosome 10) to 80.07% (4728/5905, chromosome 4) of SNPs and 67.62% (6233/9217, chromosome 10) to 79.20% (2856/3606, chromosome 4) of InDels were located in the IGS regions (Figure 4c,d; Tables S9 and S10). The distributions of the SNPs and InDels in the genomic regions were also examined, which indicated that most of them were in the introns (SNPs: 57.36% on chromosome 8 to 70.20% on chromosome 11; InDels: 61.53% on chromosome 10 to 76.89% on chromosome 12), while 5′ UTRs, 3′ UTRs, and CDS contained only a small fraction (Figure 4c,d; Tables S9 and S10). The information described here can be exploited in future studies to provide novel perspectives on genetics and breeding of fragrant *japonica* rice.

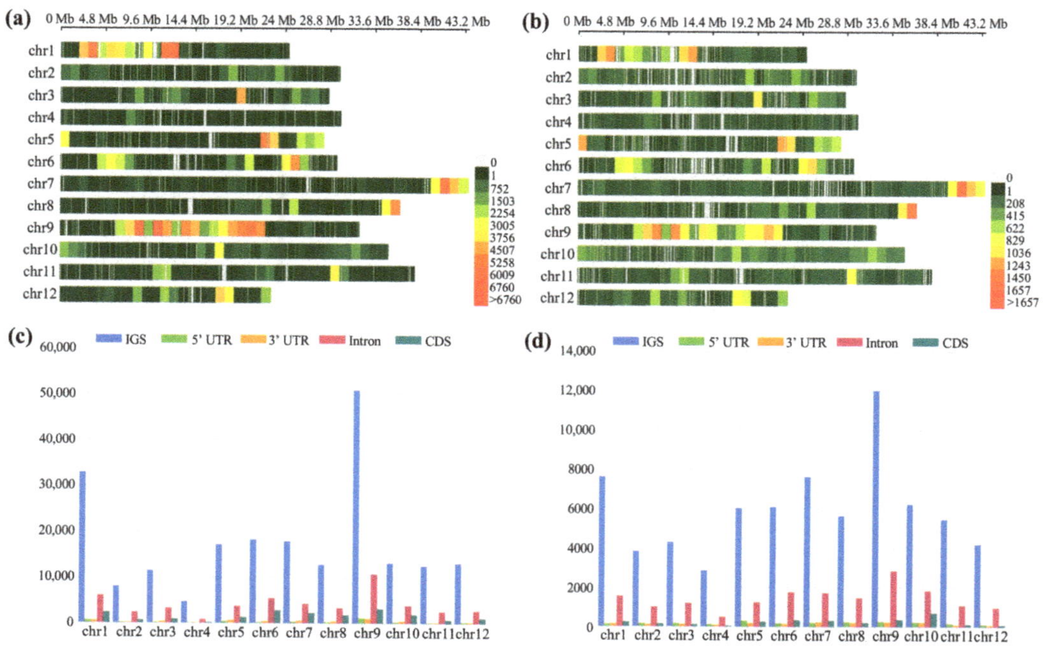

Figure 4. (**a**,**b**) Distribution patterns of SNPs (**a**) and InDels (**b**) across the 'Changxianggeng 1813' genome by comparing to the Nipponbare genome. (**c**,**d**) The distribution of (**c**) SNPs and (**d**) InDels in different genomic regions, including intergenic spacer regions (IGS), 5′ untranslated regions (UTR), 3′ UTR, intron and protein coding regions (CDS).

It is also noteworthy that SNPs and small InDels do not capture all the meaningful genomic variations that underlie crop improvement, and that structure variants (SVs) also play an important role in plant evolution and agriculture [50,51]. SVs typically defined as genomic variations that involve segments of DNA larger than 1 kb in length, hence detecting SVs with short-read sequencing is a challenging problem, leaving the vast majority of SVs poorly resolved in rice [26,33]. Nowadays, the recent development of high-throughput Oxford Nanopore long-read sequencing has enabled us to take a broad survey on previously hidden SVs in rice genomes [16]. In this study, establishing a high-quality de novo genome assembly for 'Changxianggeng 1813' allowed us to resolve large SVs between fragrant and non-fragrant *japonica* rice. A total of 8690 large SVs were identified between the genomes of 'Changxianggeng 1813' and Nipponbare through direct genome comparison (Figure 5a, Table S11). Of these SVs, the dominant type was DUP (gap between two mutually consistent alignments), accounting for 81.51% (7083/8650) of all identified SVs, followed by BRK (other inserted sequence) (11.09%, 964/8650) and GAP (gap between two mutually consistent alignments) (4.10%, 356/8650), while the JMP (rearrangement) (1.31%, 114/8650), SEQ (rearrangement with another sequence) (1.13%, 98/8650), and INV (rearrangement with inversion) (0.86%, 75/8650) were least abundant (Figure 5a; Table S11).

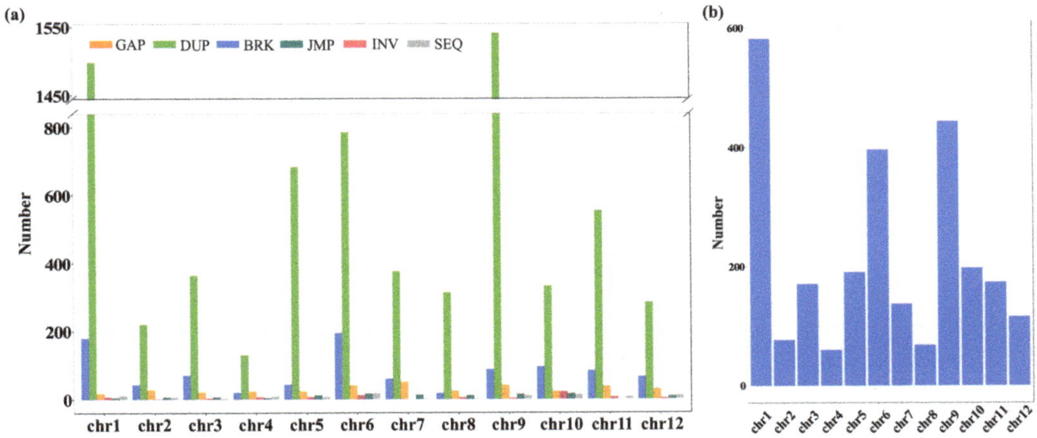

Figure 5. (**a**) SV types and numbers across 12 chromosomes of the 'Changxianggeng 1813' genome. (**b**) Total counts of SVs overlapping genes for each chromosome in the 'Changxianggeng 1813' genome. GAP, gap between two mutually consistent alignments; DUP, inserted duplication; BRK, other inserted sequence; JMP, rearrangement; INV, rearrangement with inversion; SEQ, rearrangement with another sequence.

The total number of SVs detected also varied across different chromosomes. To be specific, the highest number of SVs (Total: 1718; GAP: 17, DUP: 1598, BRK: 180, JMP: 5, INV: 7, SEQ: 11) was observed on chromosome 1, while chromosome 4 had the lowest number of SVs (Total: 192; GAP: 23, DUP: 130, BRK: 20, JMP: 4, INV: 6, SEQ: 9) (Figure 5a, Table S11). Because SVs overlapping genes can impact gene functions and expression, and those in noncoding genes have a disproportionate impact on gene expression of nearby genes [51,52], we examined the distributions of SVs in different genomic regions. Our results indicated that a majority (~70%) of SVs located in noncoding regions, notably higher than the proportion (~30%) in gene regions (Figure 5b, Table S11). As expected, SVs overlapping genes were also distributed unevenly on each chromosome, ranging from 60 SVs on chromosome 4 to 582 SVs on chromosome 1 (Figure 5b, Table S11). This result suggested that some regions might be conserved and share a common ancestral gene pool between the two *japonica* cultivars ('Changxianggeng 1813' and Nipponbare).

3. Materials and Methods

3.1. Plant Materials and DNA Extraction

Genomic DNA was extracted from fresh leaves of 15-day-old seedlings of the fragrant *japonica* cultivar 'Changxianggeng 1813' using the DNAsecure Plant Kit (Tiangen Biotech, Beijing, China) according to the manufacturer's protocol. The quality and integrity of the DNA products were assessed using agarose gel electrophoresis, NanoDrop spectrophotometry (NanoDrop Technologies, Wilmington, DE, USA), and Qubit fluorometry (Thermo Fisher Scientific, Waltham, MA, USA). The genomic DNA that met the quality and quantity standards was used to construct Illumina and Nanopore libraries.

3.2. Genome and Transcriptome Sequencing

For Illumina sequencing, a short-insert (350 bp) genomic library was performed using the NEBNext Ultra DNA Library Prep Kit (New England Biolabs, Beverly, MA, USA), and sequenced on the Illumina NovaSeq 6000 platform using a paired-end sequencing strategy. To reduce the effect of sequencing errors, we discarded those reads that met either of the following criteria: (i) reads with adapters; (ii) reads having more than 50% bases with Phred quality < 5; (iii) reads with N bases more than 5%; and (iv) PCR duplicated reads. All the obtained clean reads were prepared to carry out genome size estimation, genome assembly correction and evaluation.

For Nanopore sequencing, approximately 10 μg of genomic DNA was size-selected (10–50 kb) with the BluePippin System (Sage Science, Beverly, MA, USA), and then the DNA was subjected to a 30 μL end-repair/dA-tailing reaction using the NEBNext Ultra End Repair/dA-Tailing module (New England Biolabs, Beverly, MA, USA). The sequencing adaptors were further ligated using the Ligation Sequencing Kit SQK-LSK109 (Oxford Nanopore Technologies, Oxford, UK) based on the manufacturer's instructions. After purifying using Ampure XP beads and the ABB wash buffer (Oxford Nanopore Technologies), the resulting library was sequenced on R9.4 flow cells using the PromethION DNA sequencer (Oxford Nanopore Technologies). Raw signal data in fast5 format was subsequently base called using Guppy v.2.3.5 (Oxford Nanopore Technologies) with default parameters, and the reads with the mean_qscore_template <7 were filtered.

For chromatin conformation capture (Hi-C) sequencing, fresh leaves from the same 'Changxianggeng 1813' plant that were used for Illumina and Nanopore sequencing were collected. A Hi-C library was created in a similar manner to that described by Lieberman-Aiden et al. [53]. Briefly, chromatin was first fixed in 1% final concentration of formaldehyde, and the extracted fixed chromatin was digested using the restriction enzyme DpnII. The 5′ overhangs were then filled in with biotinylated nucleotides, and free blunt ends were ligated. After ligation, cross-links were reversed, and the DNA was purified from the protein. Purified DNA was further filtered to remove unligated but biotin-labeled fragments and subjected to selection for fragments with lengths between 300 and 700 bp. The quality of the purified library was evaluated with an Agilent 2100 instrument, a Qubit fluorometer (Thermo Fisher Scientific, Waltham, MA, USA), and quantitative PCR (qPCR). Finally, the qualified library was sequenced on an Illumina HiSeq X Ten platform with the layout of pair-ended 150 bp reads.

For transcriptome sequencing (RNA-Seq), the best-quality RNA samples of each tissue (root, branch, leaf, and panicle) were mixed together to build a Nanopore sequencing library using the Ligation Sequencing Kit (SQK-LSK109, Oxford Nanopore Technologies) by following the manufacturer's protocol. The cDNA library was added to FLO-MIN109 flow cells and sequenced on the Nanopore PromethION platform. Raw reads were filtered with the following settings: minimum average read quality score 7 and minimum read length 500 bp. Ribosomal RNA was discarded by searching against the Silva rRNA database (https://www.arb-silva.de, accessed on 27 August 2020). The RNA-Seq data were used to improve the annotation of 'Changxianggeng 1813'. All library construction, sequencing, and data filtering were conducted in Wuhan Benagen Tech Solutions Company Limited, Wuhan, China.

3.3. Genome Size Estimation and Genome Assembly

All Illumina clean reads were used for the estimation of genome size and heterozygosity with *k*-mer analysis. The data were run through Jellyfish v.2.3.0 [54] to generate *k*-mer frequency distribution, with a *k*-mer size of 19. Genome size was estimated by the commonly used formula: genome size = *k*-mer_number/*k*-mer_depth, where *k*-mer_number is the total number of *k*-mers, and *k*-mer_depth is the main peak of *k*-mer frequency.

NextDenovo v.2.4.0 (https://github.com/Nextomics/NextDenovo, accessed on 27 December 2020) was applied to de novo assembly of 'Changxianggeng 1813' genome using nanopore long reads. Briefly, the NextCorrect module was employed to correct raw reads and extract consensus sequences, and then the NextGraph module was used to assemble the draft genome. To improve the accuracy of the draft genome, we used Racon v.1.4.11 [55] and Pilon v.1.23 [56] to polish the assembly for two rounds, respectively, based on the corrected nanopore long reads and the cleaned Illumina short reads. After these two-step polishing strategies, the scaffold-level genome assembly was generated. To further anchor the genome assembly to the chromosome level, HiCUP v.0.6.17 [57] was used to produce cleaned mapped data accompanied with QC reports. Only uniquely aligned read pairs with mapping quality >20 were retained and utilized to cluster, order, and orient the assembly scaffolds onto chromosomes by LACHESIS software [58].

3.4. Quality Assessment of Genome Assembly

To evaluate the accuracy and completeness of the genome assembly, Illumina reads were mapped back to the reference genome using BWA-MEM v.0.7.17 [59] and assessed by their depth of coverage. Furthermore, BUSCO (Benchmarking Universal Single-Copy Orthologs) v.4.1.4 [60], with the database embryophyta_odb10, was employed to assess the completeness of the genome assembly.

3.5. Genome Annotation

The genome of 'Changxianggeng 1813' was annotated at three independent dimensions: (i) repetitive elements, (ii) protein-coding genes, and (iii) noncoding RNAs. For repetitive element annotation, transposable elements (TEs) in the 'Changxianggeng 1813' genome were identified using a hybrid strategy combining homology-based searching in known repeat database and de novo prediction. RepeatMasker v.4.0.6 [61] was used to identify TEs against both the RepBase database of known TEs [62], and a de novo repeat library constructed by RepeatModeler v.1.0.11 (http://www.repeatmasker.org/RepeatModeler/, accessed on 21 October 2017). TEs identified from both homology-based and de novo approaches were further filtered for redundant sequences and merged into a non-redundant repeat library by CD-HIT [63]. In addition, tandem repeats including microsatellites (SSRs) were identified in the reference genome of 'Changxianggeng 1813' using Tandem Repeat Finder (TRF) v.4.0.9 [64].

For protein-coding genes prediction, three different methods, including homology-based, de novo, and transcriptome-based methods were unitedly conducted. Frist, Exonerate v.2.4.0 [65] was used for the homology-based prediction, based on protein sequences of *O. brachyantha*, *O. sativa japonica* Nipponbare, *Aegilops tauschii*, and *Panicum hallii* retrieved from NCBI (http://www.ncbi.nlm.nih.gov, accessed on 28 June 2021). Then, Augustus v.3.3.2 [66] and GlimmerHMM v.3.0.4 [67] were applied for the de novo prediction, with default parameters. Next, TransDecoder v.5.1.0 (https://github.com/TransDecoder/TransDecoder/wiki, accessed on 28 March 2018) was employed to identify the potential coding regions, based on the assembled transcripts using Stringtie v.2.1.1 [68]. Finally, EvidenceModeler v.1.1.1 [69] was used to integrate the prediction results obtained through the above three methods to generate the final gene set of 'Changxianggeng 1813'. Functional gene annotation was performed by aligning the protein sequences against Uniprot, Pfam, GO (Gene Ontology), KEGG (Kyoto Encyclopedia of Genes and Genomes), and NR (Non-redundant) databases, with an E-value threshold of 1×10^{-5}. Furthermore, InterProScan v.5.33 [70] was used to annotate the motifs and domains by searching against the InterPro and Pfam databases. These results were further integrated to produce the final genes set.

For noncoding RNA prediction, transfer RNAs (tRNAs) and ribosomal RNAs (rRNAs) were predicted using tRNAscan-SE v.1.23 [71] and RNAmmer v.1.2 [72], respectively. Other types of noncoding RNAs, including small nuclear RNAs (snRNAs) and microRNAs (miRNAs) were identified by using Infernal v.1.1.2 [73] based on the Rfam database [74].

3.6. Characterization and Evolutionary Analysis of BADH2 Gene

The *BADH2* gene sequence for 'Changxianggeng 1813' was extracted from its genome sequence according to annotation files and then compared to that of the non-fragrant cultivar Nipponbare using the MAFFT multiple sequence alignment program [75], to verify the presence/absence of the mutations associated with fragrance in 'Changxianggeng 1813'. The *BADH2* protein-coding sequence for 'Changxianggeng 1813', was further combined with previously published 38 haplotypes in the *BADH2* coding region for phylogenetic analysis [2]. All 39 *BADH2* coding sequences were aligned using ClustalW [76], and the resulting alignment was used for Neighbor-Joining (NJ) phylogenetic tree construction using MEGA v.11.0.11 [77], with 1000 bootstrap replicates.

3.7. Genome Synteny and Collinearity Analysis

To identify chromosome structural changes between 'Changxianggeng 1813' and its two close relatives, i.e., the non-fragrant *japonica* rice cultivar Nipponbare and the common wild rice *O. rufipogon*, genome syntenic blocks were identified using the Python version of MCscan incorporated in jcvi (https://github.com/tanghaibao/jcvi/wiki/MCscan-(Python-version), accessed on 16 June 2020), with default parameters. In brief, all-against-all LAST [78] was performed, and the LAST hits with a distance cutoff of ten genes and at least five syntenic genes per block were chained. Dot plots for pairwise synteny, and macrosyntenic patterns were generated using the commands 'python-m jcvi.graphics.dotplot' and 'python-m jcvi.graphics.karyotype', respectively.

3.8. Gene Family and Phylogenetic Analysis

OrthoMCL v.2.0.9 [79] was used to identify gene family clusters in the genomes of 'Changxianggeng 1813' and five other members of *Oryza*, including the non-fragrant *japonica* cultivar Nipponbare, the *indica* subspecies, and three wild species (*O. rufipogon*, *O. nivara*, and *O. barthii*). Low-quality protein sequences from these six *Oryza* genomes were firstly filtered, based on default parameters in OrthoMCL. Then, an all-versus-all BLASTP search was conducted for all remaining proteins with an E-value threshold of 1×10^{-5}. Finally, protein sequences were clustered into paralogous and orthologous genes using the program OrthoMCL, with a default inflation parameter for the Markov cluster algorithm.

To resolve the phylogenetic position of 'Changxianggeng 1813', the single-copy orthologous genes extracted from the above six *Oryza* genomes were aligned using MUSCLE v.3.8.3 [80] and then concatenated into a super-gene alignment matrix. Phylogenetic analysis was conducted using RAxML-HPC v.8.2.8 [81] with 1000 bootstrap replicates. The best model and parameter settings were chosen according to the Akaike Information Criterion (AIC) using jModelTest v.2.1.4 [82]. Divergence times between these six *Oryza* species/subspecies/cultivars were estimated by the program MCMCTree in PAML v.4.7 [83]. The following four divergence times obtained from the Timetree database (http://www.timetree.org/, accessed on 7 February 2019) were used for calibrations (in million years ago, Ma): (i) *O. barthii* and *O. sativa* (0.95–2.42 Ma), (ii) *O. nivara* and *O. sativa* (0.603–1.089 Ma), (iii) *O. rufipogon* and *O. nivara* (0.603–1.089 Ma), and (iv) *O. rufipogon* and *O. sativa* (0.598–1.255 Ma). To gain more insights into the evolutionary dynamics of the genes, the expansion and contraction of orthologous gene families were determined in these six members of *Oryza* with CAFÉ [84] and then subjected to GO functional annotation.

3.9. Analysis of Genomic Variations

MUMmer v.3.23 [85] was used to align the 'Changxianggeng 1813' (fragrant *japonica* cultivar) genome against the Nipponbare (non-fragrant *japonica* cultivar) genome by the nucmer utility under the parameters-mum. The delta-filter utility was subsequently used to filter repeats and determine the one-to-one alignment blocks in conjunction with parameters -1 -r -q. Single nucleotide polymorphisms (SNPs) and small insertion-deletion polymorphisms (InDels) were called from the filtered data using the show-snps function under the parameters -Clr TH.

Structural variants (SVs) were detected from the genome alignment between 'Changxianggeng 1813' and Nipponbare by using the show-diff function in MUMmer, and six SV types were obtained, including gap between two mutually consistent alignments (GAP), inserted duplication (DUP), other inserted sequence (BRK), rearrangement (JMP), rearrangement with inversion (INV), and rearrangement with another sequence (SEQ). The SVs with a minimum size of 1000 bp in length were retained in this study.

4. Conclusions

Here, we presented a high-quality reference genome sequence of a new fragrant rice cultivar 'Changxianggeng 1813', using a combination of Nanopore long reads, Illumina short reads, and Hi-C data. To our knowledge, this is the first de novo chromosome-level genome assembly for fragrant *japonica* rice. The 'Changxianggeng 1813' genome has a total length of ~378.78 Mb and comprises 31,671 high-quality protein-coding genes. Based on this annotated genome sequence, we demonstrated that it was the *badh2-E2* type of deletion (a 7 bp deletion in the second exon) that caused fragrance in this *japonica* rice cultivar. Through pairwise genome comparison between 'Changxianggeng 1813' and the non-fragrant *japonica* cultivar Nipponbare, a total of 289,970 SNPs, 96,093 InDels, and 8690 large SVs were identified. Undoubtedly, these genomic resources will promote the genic and genomic studies of rice and be beneficial for cultivar improvement of fragrant *japonica* rice. However, it should also be noted that our study has two notable limitations. First, we sequenced only a single individual, which was insufficient for investigating population genomic diversity, population structure, and cultivar origins of fragrant *japonica* rice. Second, our study still leaves a gap in our knowledge of genomic variations between *japonica*, *indica*, and *aus* type fragrant rice. Hence, we anticipate that further population-scale, long-read sequencing datasets, as well as improvements in genome comparison algorithms, will help overcome these limitations.

Supplementary Materials: The following supporting information can be downloaded at: https://www.mdpi.com/article/10.3390/ijms23179705/s1.

Author Contributions: Conceptualization, X.S.; methodology, R.L.; software, R.L. and J.L.; validation, X.W., X.J., N.L. and X.S.; resources, G.M.; data curation, R.L.; writing—original draft preparation, R.L.; writing—review and editing, Z.S., N.L. and X.S.; funding acquisition, R.L., X.W. and X.S. All authors have read and agreed to the published version of the manuscript.

Funding: This research was funded by the Changshu Agricultural Production and Public Service Project, the Jiangsu Key Laboratory of Plant Resources Research and Utilization grant (JSPKLB201921), and the Jiangsu Innovative and Entrepreneurial Talent Programme (JSSCBS20211311).

Institutional Review Board Statement: Not applicable.

Informed Consent Statement: Not applicable.

Data Availability Statement: The assembled genome of 'Changxianggeng 1813' and all raw sequencing data have been deposited under NCBI BioProject PRJNA856027 with accession nos. SRR20046019–SRR20046022.

Conflicts of Interest: The authors declare no conflict of interests.

References

1. Griglione, A.; Liberto, E.; Cordero, C.; Bressanello, D.; Cagliero, C.; Rubiolo, P.; Bicchi, C.; Sgorbini, B. High-quality Italian rice cultivars: Chemical indices of ageing and aroma quality. *Food Chem.* **2015**, *172*, 305–313. [CrossRef] [PubMed]
2. Phitaktansakul, R.; Kim, K.W.; Aung, K.M.; Maung, T.Z.; Min, M.H.; Somsri, A.; Lee, W.; Lee, S.B.; Nam, J.; Kim, S.H.; et al. Multi-omics analysis reveals the genetic basis of rice fragrance mediated by betaine aldehyde dehydrogenase 2. *J. Adv. Res.* 2021, *in press*. [CrossRef]
3. Sheng, Z.; Fiaz, S.; Li, Q.; Chen, W.; Wei, X.; Xie, L.; Jiao, G.; Shao, G.; Tang, S.; Wang, J.; et al. Molecular breeding of fragrant early-season hybrid rice using the *BADH2* gene. *Pak. J. Bot.* **2019**, *51*, 2089–2095. [CrossRef]
4. Ahmed, F.; Abro, T.F.; Kabir, M.; Latif, M.A. Rice quality: Biochemical composition, eating quality, and cooking quality. In *The Future of Rice Demand: Quality beyond Productivity*; de Oliveira, A.C., Pegoraro, C., Viana, V.E., Eds.; Springer: Cham, Switzerland, 2020; pp. 3–24.
5. Zhao, Q.; Xi, J.; Xu, X.; Yin, Y.; Xu, D.; Jin, Y.; Tong, Q.; Dong, L.; Wu, F. Volatile fingerprints and biomarkers of Chinese fragrant and non-fragrant *japonica* rice before and after cooking obtained by untargeted GC/MS-based metabolomics. *Food Biosci.* **2022**, *47*, 101764. [CrossRef]
6. Shao, G.N.; Tang, A.; Tang, S.Q.; Luo, J.; Jiao, G.A.; Wu, J.L.; Hu, P.S. A new deletion mutation of fragrant gene and the de-velopment of three molecular markers for fragrance in rice. *Plant Breed.* **2011**, *130*, 172–176. [CrossRef]

7. Kovach, M.J.; Calingacion, M.N.; Fitzgerald, M.A.; McCouch, S.R. The origin and evolution of fragrance in rice (*Oryza sativa* L.). *Proc. Natl. Acad. Sci. USA* **2009**, *106*, 14444–14449. [CrossRef]
8. Shao, G.; Tang, S.; Chen, M.; Wei, X.; He, J.; Luo, J.; Jiao, G.; Hu, Y.; Xie, L.; Hu, P. Haplotype variation at *Badh2*, the gene determining fragrance in rice. *Genomics* **2013**, *101*, 157–162. [CrossRef]
9. Chen, S.; Yang, Y.; Shi, W.; Ji, Q.; He, F.; Zhang, Z.; Cheng, Z.; Liu, X.; Xu, M. *Badh2*, encoding betaine aldehyde dehydrogenase, inhibits the biosynthesis of 2-Acetyl-1-Pyrroline, a major component in rice fragrance. *Plant Cell* **2008**, *20*, 1850–1861. [CrossRef]
10. Li, W.; Zeng, X.; Li, S.; Chen, F.; Gao, J. Development and application of two novel functional molecular markers of *BADH2* in rice. *Electron. J. Biotechnol.* **2020**, *46*, 1–7. [CrossRef]
11. Sansenya, S.; Hua, Y.; Chumanee, S.; Phasai, K.; Sricheewin, C. Effect of gamma irradiation on 2-Acetyl-1-pyrroline content, GABA content and volatile compounds of germinated rice (Thai upland rice). *Plants* **2017**, *6*, 18. [CrossRef]
12. Bradbury, L.M.T.; Fitzgerald, T.L.; Henry, R.J.; Jin, Q.; Waters, D.L.E. The gene for fragrance in rice. *Plant Biotechnol. J.* **2005**, *3*, 363–370. [CrossRef]
13. Shi, W.; Yang, Y.; Chen, S.; Xu, M. Discovery of a new fragrance allele and the development of functional markers for the breeding of fragrant rice varieties. *Mol. Breed.* **2008**, *22*, 185–192. [CrossRef]
14. Chan-In, P.; Jamjod, S.; Yimyam, N.; Rerkasem, B.; Pusadee, T. Grain quality and allelic variation of the *Badh2* gene in Thai fragrant rice landraces. *Agronomy* **2020**, *10*, 779. [CrossRef]
15. Lan, G.; Lu, Y.; Ke, Y.; Tao, J.; Ma, G.; Pan, B.; Tang, Y.; Yu, L.; Wang, X.; Wang, X.; et al. Breeding of high-quality *japonica* rice variety Changxianggeng 1813. *China Seed Ind.* **2020**, *12*, 87–88. (in Chinese).
16. Choi, J.Y.; Lye, Z.N.; Groen, S.C.; Dai, X.; Rughani, P.; Zaaijer, S.; Harrington, E.D.; Juul, S.; Purugganan, M.D. Nanopore sequencing-based genome assembly and evolutionary genomics of circum-basmati rice. *Genome Biol.* **2020**, *21*, 21. [CrossRef]
17. Yu, J.; Hu, S.; Wang, J.; Wong, G.K.S.; Li, S.; Liu, B.; Deng, Y.; Dai, L.; Zhou, Y.; Zhang, X.; et al. A draft sequence of the rice genome (*Oryza sativa* L. ssp. *indica*). *Science* **2002**, *296*, 79–92. [CrossRef]
18. Goff, S.A.; Ricke, D.; Lan, T.H.; Presting, G.; Wang, R.; Dunn, M.; Glazebrook, J.; Sessions, A.; Oeller, P.; Varma, H.; et al. A draft sequence of the rice genome (*Oryza sativa* L. ssp. *japonica*). *Science* **2002**, *296*, 92–100. [CrossRef]
19. Chen, J.; Huang, Q.; Gao, D.; Wang, J.; Lang, Y.; Liu, T.; Li, B.; Bai, Z.; Goicoechea, J.L.; Liang, C.; et al. Whole-genome se-quencing of *Oryza brachyantha* reveals mechanisms underlying *Oryza* genome evolution. *Nat. Commun.* **2013**, *4*, 1595. [CrossRef]
20. Wang, M.; Yu, Y.; Haberer, G.; Marri, P.R.; Fan, C.; Goicoechea, J.L.; Zuccolo, A.; Song, X.; Kudrna, D.; Ammiraju, J.S.S.; et al. The genome sequence of African rice (*Oryza glaberrima*) and evidence for independent domestication. *Nat. Genet.* **2014**, *46*, 982–988. [CrossRef]
21. Zhang, Q.J.; Zhu, T.; Xia, E.H.; Shi, C.; Liu, Y.L.; Zhang, Y.; Liu, Y.; Jiang, W.K.; Zhao, Y.J.; Mao, S.Y.; et al. Rapid diversification of five *Oryza* AA genomes associated with rice adaptation. *Proc. Natl. Acad. Sci. USA* **2014**, *111*, E4954–E4962. [CrossRef]
22. Mondal, T.K.; Rawal, H.C.; Chowrasia, S.; Varshney, D.; Panda, A.K.; Mazumdar, A.; Kaur, H.; Gaikwad, K.; Sharma, T.R.; Singh, N.K. Draft genome sequence of first monocot-halophytic species *Oryza coarctata* reveals stress-specific genes. *Sci. Rep.* **2018**, *8*, 13698. [CrossRef] [PubMed]
23. Wang, W.; Mauleon, R.; Hu, Z.; Chebotarov, D.; Tai, S.; Wu, Z.; Li, M.; Zheng, T.; Fuentes, R.R.; Zhang, F.; et al. Genomic variation in 3,010 diverse accessions of Asian cultivated rice. *Nature* **2018**, *557*, 43–49. [CrossRef] [PubMed]
24. Wu, Z.; Fang, D.; Yang, R.; Gao, F.; An, X.; Zhuo, X.; Li, Y.; Yi, C.; Zhang, T.; Liang, C.; et al. De novo genome assembly of *Oryza granulata* reveals rapid genome expansion and adaptive evolution. *Commun. Biol.* **2018**, *1*, 84. [CrossRef] [PubMed]
25. Stein, J.C.; Yu, Y.; Copetti, D.; Zwickl, D.J.; Zhang, L.; Zhang, C.; Chougule, K.; Gao, D.; Iwata, A.; Goicoechea, J.L.; et al. Genomes of 13 domesticated and wild rice relatives highlight genetic conservation, turnover and innovation across the genus *Oryza*. *Nat. Genet.* **2018**, *50*, 285–296. [CrossRef]
26. Zhao, Q.; Feng, Q.; Lu, H.; Li, Y.; Wang, A.; Tian, Q.; Zhan, Q.; Lu, Y.; Zhang, L.; Huang, T.; et al. Pan-genome analysis highlights the extent of genomic variation in cultivated and wild rice. *Nat. Genet.* **2018**, *50*, 278–284. [CrossRef]
27. Li, W.; Zhang, Q.J.; Zhu, T.; Tong, Y.; Li, K.; Shi, C.; Zhang, Y.; Liu, Y.L.; Jiang, J.J.; Liu, Y.; et al. Draft genomes of two out-crossing wild rice, *Oryza rufipogon* and *O. longistaminata*, reveal genomic features associated with mating-system evolution. *Plant Direct* **2020**, *4*, e00232. [CrossRef]
28. Shenton, M.; Kobayashi, M.; Terashima, S.; Ohyanagi, H.; Copetti, D.; Hernández-Hernández, T.; Zhang, J.; Ohmido, N.; Fujita, M.; Toyoda, A.; et al. Evolution and diversity of the wild rice *Oryza officinalis* complex, across continents genome types, and ploidy levels. *Genome Biol. Evol.* **2020**, *12*, 413–428. [CrossRef]
29. Yu, H.; Lin, T.; Meng, X.; Du, H.; Zhang, J.; Liu, G.; Chen, M.; Jing, Y.; Kou, L.; Li, X.; et al. A route to de novo domestication of wild allotetraploid rice. *Cell* **2021**, *184*, 1156–1170.e14. [CrossRef]
30. Zhang, J.; Chen, L.L.; Xing, F.; Kudrna, D.A.; Yao, W.; Copetti, D.; Mu, T.; Li, W.; Song, J.M.; Xie, W.; et al. Extensive sequence divergence between the reference genomes of two elite *indica* rice varieties Zhenshan 97 and Minghui 63. *Proc. Natl. Acad. Sci. USA* **2016**, *113*, E5163–E5171. [CrossRef]
31. Du, H.; Yu, Y.; Ma, Y.; Gao, Q.; Cao, Y.; Chen, Z.; Ma, B.; Qi, M.; Li, Y.; Zhao, X.; et al. Sequencing and de novo assembly of a near complete *indica* rice genome. *Nat. Commun.* **2017**, *8*, 15324. [CrossRef]
32. Jain, R.; Jenkins, J.; Shu, S.; Chern, M.; Martin, J.A.; Copetti, D.; Duong, P.Q.; Pham, N.T.; Kudrna, D.A.; Talag, J.; et al. Genome sequence of the model rice variety KitaakeX. *BMC Genom.* **2019**, *20*, 905. [CrossRef]

33. Zhang, H.; Wang, Y.; Deng, C.; Zhao, S.; Zhang, P.; Feng, J.; Huang, W.; Kang, S.; Qian, Q.; Xiong, G.; et al. High-quality genome assembly of Huazhan and Tianfeng, the parents of an elite rice hybrid Tian-you-hua-zhan. *Sci. China Life Sci.* **2021**, *65*, 398–411. [CrossRef]
34. Belser, C.; Istace, B.; Denis, E.; Dubarry, M.; Baurens, F.C.; Falentin, C.; Genete, M.; Berrabah, W.; Chèvre, A.M.; Delourme, R.; et al. Chromosome-scale assemblies of plant genomes using nanopore long reads and optical maps. *Nat. Plants* **2018**, *4*, 879–887. [CrossRef]
35. Howe, K.; Wood, J.M.D. Using optical mapping data for the improvement of vertebrate genome assemblies. *GigaScience* **2015**, *4*, 10. [CrossRef]
36. Udall, J.A.; Dawe, R.K. Is it ordered correctly? Validating genome assemblies by optical mapping. *Plant Cell* **2017**, *30*, 7–14. [CrossRef]
37. Schatz, M.C.; Maron, L.G.; Stein, J.C.; Wences, A.H.; Gurtowski, J.; Biggers, E.; Lee, H.; Kramer, M.; Antoniou, E.; Ghiban, E.; et al. Whole genome de novo assemblies of three divergent strains of rice, *Oryza sativa*, document novel gene space of *aus* and *indica*. *Genome Biol.* **2014**, *15*, 506. [CrossRef]
38. Tanaka, T.; Nishijima, R.; Teramoto, S.; Kitomi, Y.; Hayashi, T.; Uga, Y.; Kawakatsu, T. De novo genome assembly of the *indica* rice variety IR64 using linked-read sequencing and Nanopore sequencing. *G3 Genes Genomes Genet.* **2020**, *10*, 1495–1501. [CrossRef]
39. Vitte, C.; Panaud, O. LTR retrotransposons and flowering plant genome size: Emergence of the increase/decrease model. *Cytogenet. Genome Res.* **2005**, *110*, 91–107. [CrossRef]
40. Yuan, Z.; Fang, Y.; Zhang, T.; Fei, Z.; Han, F.; Liu, C.; Liu, M.; Xiao, W.; Zhang, W.; Wu, S.; et al. The pomegranate (*Punica granatum* L.) genome provides insights into fruit quality and ovule developmental biology. *Plant Biotechnol. J.* **2017**, *16*, 1363–1374. [CrossRef] [PubMed]
41. Lee, S.I.; Kim, N.S. Transposable elements and genome size variations in plants. *Genom. Inform.* **2014**, *12*, 87–97. [CrossRef]
42. Zuccolo, A.; Sebastian, A.; Talag, J.; Yu, Y.; Kim, H.; Collura, K.; Kudrna, D.; Wing, R.A. Transposable element distribution, abundance and role in genome size variation in the genus *Oryza*. *BMC Evol. Biol.* **2007**, *7*, 152. [CrossRef] [PubMed]
43. Liu, D.; Hunt, M.; Tsai, I.J. Inferring synteny between genome assemblies: A systematic evaluation. *BMC Bioinform.* **2018**, *19*, 26. [CrossRef] [PubMed]
44. Kovach, M.J.; Sweeney, M.T.; McCouch, S.R. New insights into the history of rice domestication. *Trends Genet.* **2007**, *23*, 578–587. [CrossRef]
45. Harris, R.M.; Hofmann, H.A. Seeing is believing: Dynamic evolution of gene families. *Proc. Natl. Acad. Sci. USA* **2015**, *112*, 1252–1253. [CrossRef]
46. Liu, J.; Liu, S.; Zheng, K.; Tang, M.; Gu, L.; Young, J.; Wang, Z.; Qiu, Y.; Dong, J.; Gu, S.; et al. Chromosome-level genome assembly of the Chinese three-keeled pond turtle (*Mauremys reevesii*) provides insights into freshwater adaptation. *Mol. Ecol. Resour.* **2021**, *22*, 1596–1605. [CrossRef]
47. Sharma, P.; Dubey, R.S. Drought induces oxidative stress and enhances the activities of antioxidant enzymes in growing rice seedlings. *Plant Growth Regul.* **2005**, *46*, 209–221. [CrossRef]
48. Fitzgerald, T.L.; Waters, D.L.E.; Brooks, L.O.; Henry, R.J. Fragrance in rice (*Oryza sativa*) is associated with reduced yield under salt treatment. *Environ. Exp. Bot.* **2010**, *68*, 292–300. [CrossRef]
49. Qin, P.; Lu, H.; Du, H.; Wang, H.; Chen, W.; Chen, Z.; He, Q.; Ou, S.; Zhang, H.; Li, X.; et al. Pan-genome analysis of 33 ge-netically diverse rice accessions reveals hidden genomic variations. *Cell* **2021**, *184*, 3542–3558. [CrossRef]
50. Saxena, R.K.; Edwards, D.; Varshney, R.K. Structural variations in plant genomes. *Briefings Funct. Genom.* **2014**, *13*, 296–307. [CrossRef]
51. Alonge, M.; Wang, X.; Benoit, M.; Soyk, S.; Pereira, L.; Zhang, L.; Suresh, H.; Ramakrishnan, S.; Maumus, F.; Ciren, D.; et al. Major impacts of widespread structural variation on gene expression and crop improvement in tomato. *Cell* **2020**, *182*, 145–161. [CrossRef]
52. Yalcin, B.; Wong, K.; Agam, A.; Goodson, M.; Keane, T.; Gan, X.; Nellåker, C.; Goodstadt, L.; Nicod, J.; Bhomra, A.; et al. Sequence-based characterization of structural variation in the mouse genome. *Nature* **2011**, *477*, 326–329. [CrossRef]
53. Lieberman-Aiden, E.; Van Berkum, N.L.; Williams, L.; Imakaev, M.; Ragoczy, T.; Telling, A.; Amit, I.; Lajoie, B.R.; Sabo, P.J.; Dorschner, M.O.; et al. Comprehensive mapping of long-range interactions reveals folding principles of the human genome. *Science* **2009**, *326*, 289–293. [CrossRef]
54. Marçais, G.; Kingsford, C. A fast, lock-free approach for efficient parallel counting of occurrences of *k*-mers. *Bioinformatics* **2011**, *27*, 764–770. [CrossRef]
55. Vaser, R.; Sović, I.; Nagarajan, N.; Šikić, M. Fast and accurate de novo genome assembly from long uncorrected reads. *Genome Res.* **2017**, *27*, 737–746. [CrossRef] [PubMed]
56. Walker, B.J.; Abeel, T.; Shea, T.; Priest, M.; Abouelliel, A.; Sakthikumar, S.; Cuomo, C.A.; Zeng, Q.; Wortman, J.; Young, S.K.; et al. Pilon: An integrated tool for comprehensive microbial variant detection and genome assembly improvement. *PLoS ONE* **2014**, *9*, e112963. [CrossRef]
57. Wingett, S.W.; Ewels, P.; Furlan-Magaril, M.; Nagano, T.; Schoenfelder, S.; Fraser, P.; Andrews, S. HiCUP: Pipeline for mapping and processing Hi-C data. *F1000Research* **2015**, *4*, 1310. [CrossRef]
58. Burton, J.N.; Adey, A.; Patwardhan, R.P.; Qiu, R.; Kitzman, J.O.; Shendure, J. Chromosome-scale scaffolding of de novo genome assemblies based on chromatin interactions. *Nat. Biotechnol.* **2013**, *31*, 1119–1125. [CrossRef]

59. Li, H.; Durbin, R. Fast and accurate short read alignment with Burrows—Wheeler transform. *Bioinformatics* **2009**, *25*, 1754–1760. [CrossRef]
60. Simão, F.A.; Waterhouse, R.M.; Ioannidis, P.; Kriventseva, E.V.; Zdobnov, E.M. BUSCO: Assessing genome assembly and an-notation completeness with single-copy orthologs. *Bioinformatics* **2015**, *31*, 3210–3212. [CrossRef]
61. Chen, N. Using Repeat Masker to identify repetitive elements in genomic sequences. *Curr. Protoc. Bioinform.* **2004**, *5*, 4–10. [CrossRef]
62. Bao, W.; Kojima, K.K.; Kohany, O. Repbase Update, a database of repetitive elements in eukaryotic genomes. *Mob. DNA* **2015**, *6*, 11. [CrossRef] [PubMed]
63. Huang, Y.; Niu, B.; Gao, Y.; Fu, L.; Li, W. CD-HIT Suite: A web server for clustering and comparing biological sequences. *Bioinformatics* **2010**, *26*, 680–682. [CrossRef] [PubMed]
64. Benson, G. Tandem repeats finder: A program to analyze DNA sequences. *Nucleic Acids Res.* **1999**, *27*, 573–580. [CrossRef] [PubMed]
65. Slater, G.S.C.; Birney, E. Automated generation of heuristics for biological sequence comparison. *BMC Bioinform.* **2005**, *6*, 31. [CrossRef]
66. Stanke, M.; Keller, O.; Gunduz, I.; Hayes, A.; Waack, S.; Morgenstern, B. Augustus: Ab initio prediction of alternative transcripts. *Nucleic Acids Res.* **2006**, *34*, 435–439. [CrossRef]
67. Majoros, W.H.; Pertea, M.; Salzberg, S. TigrScan and GlimmerHMM: Two open source ab initio eukaryotic gene-finders. *Bioinformatics* **2004**, *20*, 2878–2879. [CrossRef]
68. Pertea, M.; Pertea, G.M.; Antonescu, C.M.; Chang, T.C.; Mendell, J.T.; Salzberg, S.L. StringTie enables improved reconstruction of a transcriptome from RNA-seq reads. *Nat. Biotechnol.* **2015**, *33*, 290–295. [CrossRef]
69. Haas, B.J.; Salzberg, S.L.; Zhu, W.; Pertea, M.; Allen, J.E.; Orvis, J.; White, O.; Buell, C.R.; Wortman, J.R. Automated eukaryotic gene structure annotation using EVidenceModeler and the Program to Assemble Spliced Alignments. *Genome Biol.* **2008**, *9*, R7. [CrossRef]
70. Jones, P.; Binns, D.; Chang, H.Y.; Fraser, M.; Li, W.; McAnulla, C.; McWilliam, H.; Maslen, J.; Mitchell, A.; Nuka, G.; et al. InterProScan 5: Genome-scale protein function classification. *Bioinformatics* **2014**, *30*, 1236–1240. [CrossRef]
71. Lowe, T.M.; Eddy, S.R. tRNAscan-SE: A program for improved detection of transfer RNA genes in genomic sequence. *Nucleic Acids Res.* **1997**, *25*, 955–964. [CrossRef]
72. Lagesen, K.; Hallin, P.; Rødland, E.A.; Staerfeldt, H.H.; Rognes, T.; Ussery, D.W. RNAmmer: Consistent and rapid annotation of ribosomal RNA genes. *Nucleic Acids Res.* **2007**, *35*, 3100–3108. [CrossRef]
73. Nawrocki, E.P.; Eddy, S.R. Infernal 1.1: 100-fold faster RNA homology searches. *Bioinformatics* **2013**, *29*, 2933–2935. [CrossRef]
74. Kalvari, I.; Argasinska, J.; Quinones-Olvera, N.; Nawrocki, E.P.; Rivas, E.; Eddy, S.R.; Bateman, A.; Finn, R.D.; Petrov, A.I. Rfam 13.0: Shifting to a genome-centric resource for non-coding RNA families. *Nucleic Acids Res.* **2017**, *46*, D335–D342. [CrossRef]
75. Katoh, K.; Standley, D.M. MAFFT multiple sequence alignment software version 7: Improvements in performance and usability. *Mol. Biol. Evol.* **2013**, *30*, 772–780. [CrossRef]
76. Thompson, J.D.; Higgins, D.G.; Gibson, T.J. CLUSTAL W: Improving the sensitivity of progressive multiple sequence alignment through sequence weighting, position-specific gap penalties and weight matrix choice. *Nucleic Acids Res.* **1994**, *22*, 4673–4680. [CrossRef]
77. Tamura, K.; Stecher, G.; Kumar, S. MEGA11: Molecular evolutionary genetics analysis version 11. *Mol. Biol. Evol.* **2021**, *38*, 3022–3027. [CrossRef]
78. Kiełbasa, S.M.; Wan, R.; Sato, K.; Horton, P.; Frith, M.C. Adaptive seeds tame genomic sequence comparison. *Genome Res.* **2011**, *21*, 487–493. [CrossRef]
79. Li, L.; Stoeckert, C.J., Jr.; Roos, D.S. OrthoMCL: Identification of ortholog groups for eukaryotic genomes. *Genome Res.* **2003**, *13*, 2178–2189. [CrossRef]
80. Edgar, R.C. MUSCLE: Multiple sequence alignment with high accuracy and high throughput. *Nucleic Acids Res.* **2004**, *32*, 1792–1797. [CrossRef]
81. Stamatakis, A. RAxML version 8: A tool for phylogenetic analysis and post-analysis of large phylogenies. *Bioinformatics* **2014**, *30*, 1312–1313. [CrossRef]
82. Darriba, D.; Taboada, G.L.; Doallo, R.; Posada, D. jModelTest 2: More models, new heuristics and parallel computing. *Nat. Methods* **2012**, *9*, 772. [CrossRef] [PubMed]
83. Yang, Z. PAML 4: Phylogenetic analysis by maximum likelihood. *Mol. Biol. Evol.* **2007**, *24*, 1586–1591. [CrossRef] [PubMed]
84. De Bie, T.; Cristianini, N.; DeMuth, J.P.; Hahn, M.W. CAFE: A computational tool for the study of gene family evolution. *Bioinformatics* **2006**, *22*, 1269–1271. [CrossRef] [PubMed]
85. Kurtz, S.; Phillippy, A.; Delcher, A.L.; Smoot, M.; Shumway, M.; Antonescu, C.; Salzberg, S.L. Versatile and open software for comparing large genomes. *Genome Biol.* **2004**, *5*, R12. [CrossRef]

International Journal of *Molecular Sciences*

Article

Creation of Two-Line Fragrant Glutinous Hybrid Rice by Editing the *Wx* and *OsBADH2* Genes via the CRISPR/Cas9 System

Yahong Tian †, Yin Zhou †, Guanjun Gao, Qinglu Zhang, Yanhua Li, Guangming Lou and Yuqing He *

National Key Laboratory of Crop Genetic Improvement and National Center of Plant Gene Research, Hubei Hongshan Laboratory, Huazhong Agricultural University, Wuhan 430070, China

* Correspondence: yqhe@mail.hzau.edu.cn

† These authors contributed equally to this work.

Abstract: Global food security has benefited from the development and promotion of the two-line hybrid rice system. Excellent eating quality determines the market competitiveness of hybrid rice varieties based on achieving the fundamental requirements of high yield and good adaptability. Developing sterile and restorer lines with improved quality for two-line hybrid breeding by editing quality genes with clustered regularly interspaced short palindromic repeat (CRISPR)/Cas9 is an efficient and practical alternative to the lengthy and laborious process of conventional breeding to improve rice quality. We edited *Wx* and *OsBADH2* using CRISPR/Cas9 technology to produce both homozygous male sterile mutant lines and homozygous restorer mutant lines with Cas9-free. These mutants have a much lower amylose content while having a significantly higher 2-acetyl-1-pyrroline aroma content. Based on this, a fragrant glutinous hybrid rice was developed without too much effect on most agronomic traits. This study demonstrates the use of CRISPR/Cas9 in creating two-line fragrant glutinous hybrid rice by editing the components of the male sterile and the restorative lines.

Keywords: *Wx*; *OsBADH2*; CRISPR/Cas9; two-line hybrid rice; 2-acetyl-1-pyrroline aroma; quality

check for updates

Citation: Tian, Y.; Zhou, Y.; Gao, G.; Zhang, Q.; Li, Y.; Lou, G.; He, Y. Creation of Two-Line Fragrant Glutinous Hybrid Rice by Editing the *Wx* and *OsBADH2* Genes via the CRISPR/Cas9 System. *Int. J. Mol. Sci.* **2023**, *24*, 849. https://doi.org/10.3390/ijms24010849

Academic Editor: Luigi Cattivelli

Received: 19 October 2022
Revised: 15 December 2022
Accepted: 16 December 2022
Published: 3 January 2023

Copyright: © 2023 by the authors. Licensee MDPI, Basel, Switzerland. This article is an open access article distributed under the terms and conditions of the Creative Commons Attribution (CC BY) license (https://creativecommons.org/licenses/by/4.0/).

1. Introduction

The most significant physical and chemical factor that impacts rice quality is the amylose content in rice grains [1–3]. *Wx* is important for the regulation of rice quality (including appearance quality and eating and cooking quality (ECQ)) and encodes granule-bound starch synthase I [4–6]. The abundant natural allelic variations confer the extensive variation of amylose content and rice quality among modern cultivated rice [6]. At least 10 distinct functional alleles of *Wx* have been identified, including Wx^{a}, Wx^{b}, wx, Wx^{in}, Wx^{op}, Wx^{mp}, Wx^{mq}, Wx^{hp}, Wx^{lv}, and Wx^{la}/Wx^{mw} [6–13]. Of these, Wx^{lv} is an ancestral allele derived from wild rice. Wx^{la}/Wx^{mw} is a recently identified *Wx* allele derived from intragenic recombination, giving rice good eating qualities and grain transparency. The clustered regularly interspaced short palindromic repeat (CRISPR)/Cas9 system appears to be a popular trend for editing various components of the *Wx* gene to enhance rice quality. Zhang et al. [14] and Xu et al. [15] edited the exon of the *Wx* gene, while Huang et al. [16] and Zeng et al. [17] edited the promoter and 5′UTR intron of the *Wx* gene, respectively. Thus, several mutants with different quality traits were produced. According to Liu et al. [18], the amylose content was significantly increased in rice seeds by editing the first intron of the *Wx* gene. These findings suggest that CRISPR/Cas9-mediated gene editing of suitable target sites can produce ideal amylose content and quality breeding materials.

The eating quality of rice is significantly influenced by several factors, including aroma, which is mainly controlled by the recessive gene *fgr*/*OsBADH2* [19]. When the *fgr*/*OsBADH2* gene underwent loss of function, it caused the BADH2 protein to lose its ability to catalyze the oxidation of 4-amino-butanal. This led to an accumulation of 4-amino-butanal, which promoted the synthesis of 2-acetyl-1-pyrroline (2-AP), and resulted in the production of fragrance in rice [20]. Shan et al. edited the *fgr/OsBADH2* gene in Nipponbare using transcription activator-like effector nuclease (TALEN) technology, and the 2-AP content in the homozygous T_1 line was significantly increased [21]. Ashokkumar et al. used the CRISPR/Cas9 method to produce novel alleles of *fgr/OsBADH2* to induce aroma into the top non-aromatic rice variety ASD16. The phenotype was stably inherited in the T_1 generation [22]. Any non-aromatic rice variety can be made aromatic by gene editing *fgr/OsBADH2*.

Heterosis refers to the superior performance of hybrids over their parents. A typical example of using heterosis is the creation of two-line hybrid rice systems. One of the fundamental components of two-line hybrid rice breeding is the light- and temperature-sensitive male sterile line. Although the adoption and use of two-line hybrid rice have significantly improved rice yield, the quality of hybrid rice is typically subpar. Three efficient gene editing techniques can be utilized to accurately and rapidly modify crop target traits: TALEN technology [23], zinc finger nuclease technology [24], and CRISPR technology [25]. In this study, we edited *Wx* and *OsBADH2* using the CRISPR/Cas9 system to create homozygous mutants of the Zhinong 1S (ZN1S) sterile and the Zhinong 1307 (ZN1307) restorer lines, resulting in fragrant glutinous hybrid rice with better yield performance than its parents and providing new ideas and insights for quality improvement of hybrid rice.

2. Results

2.1. Creation of Fragrant Glutinous Mutants with Cas9-Free

We developed *wx-fgr* double mutants using CRISPR/Cas9 technology on the genetic background of the sterile line ZN1S and restorer line ZN1307 to produce fragrant and waxy hybrid rice (Figure 1A). PAGE combined with Sanger sequencing analysis revealed that ZN1S and ZN1307, respectively, had heterozygous mutations of 9 and 6 distinct mutation types in T_0 transgenic plants (Supplementary Tables S2 and S3). We PCR-selected Cas9-free plants from transgenic (T_1–T_2) segregating families and identified 5 and 6 homozygous mutant T_3-lines for ZN1S and ZN1307, respectively (Figure 1B). Unlike the translucent endosperm of wild type grains, the endosperm of all the mutants showed milky white similar to that of glutinous rice (Figure 1C). The amylose content of grains from these 11 mutant T_3-lines was measured, and the results revealed that once the *Wx* gene was mutated, the amylose content in every mutant drastically dropped (Figure 1D). The lowest was the *wx-fgr*-S2 mutant line, whose amylose content was 2.2%, which was very close to that of wild glutinous rice. Among them, the amylose content in grains of ZN1S mutants ranged between 2–4%. Amylose content ranged between 2.5–3.5% in ZN1307 mutants, with *wx-fgr*-R2 having the lowest value at 2.6% (Figure 1D). Additionally, we used the potassium hydroxide method to conduct a sensory evaluation of these mutants' aromas. The results revealed that the mutant brown rice from both materials could generate a light aroma of rice (Figure 1E). Next, we assessed the 11 homozygous mutant T_4-lines for ZN1S and ZN1307's quality traits. All mutants' amylose content (for convenience, the abbreviation AC is used below for amylose content) was significantly lower than the corresponding wild type, consistent with the results of T_3 generation, while their 2-AP and GC were significantly higher (Figure 1F–I). Rapid viscosity analysis (RVA) was used to evaluate the starch quality [26]. Compared to WT-S, the viscosity indexes of the remaining three mutants from ZN1S were lower than those of *wx-fgr*-S1 and *wx-fgr*-S4 (Figure 1J). Different from mutants from ZN1S, all mutants from ZN1307 showed similar viscosity indexes, but these were obviously less than for WT-R (Figure 1J).

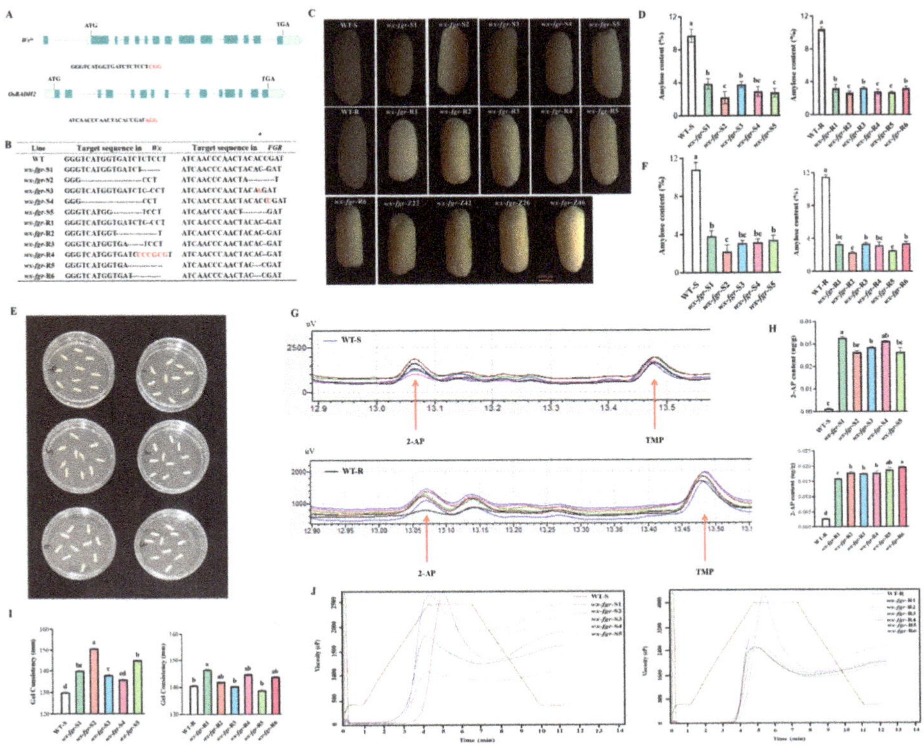

Figure 1. Improvement of rice grain quality by editing of the Wx^{la} and *OsBADH2* using CRISPR/Cas9. (**A**) Schematic diagram of the targeted sites in *Wx* and *OsBADH2*. The protospacer-adjacent motifs (PAMs) are shown in red. (**B**) Mutations in the edited T_3 lines. Inserted bases are marked in red, and missing bases are indicated by short dashed lines. (**C**) Morphology of milled rice from ZN1S, ZN1307 and the F_1 hybrid ZN1S mutants × ZN1307 mutants. WT-S represents the wild type of sterile line ZN1S and WT-R represents the wild type of restorer line ZN1307. Scale bars, 1000 μm. (**D**) The amylose contents determined by iodine colorimetry of ZN1S, ZN1307 and their T_3 generation homozygous mutant lines. (**E**) Sensory evaluation of the aroma of ZN1S, ZN1307 and their mutants by potassium hydroxide method. (**F**) The grain amylose contents determined by iodine colorimetry of ZN1S, ZN1307 and their T_4 generation homozygous mutant lines. (**G**) Total ion chromatograms (TIC) of 2-AP and TMP in the grains of ZN1S, ZN1307 and their T_4 generation homozygous mutant lines. (**H**) 2-AP content in grains of ZN1S, ZN1307 and their T_4 generation homozygous mutant lines. 2, 4, 6-trimethyl pyridine (TMP) was used as the internal standard. (**I**) Gel consistency. (**J**) Rapid visco analysis profiles of grain starches of ZN1S, ZN1307 and their T_4 generation homozygous mutant lines. cP (centi Poise), viscosity unit. Error bars are means ± SD (n = 3). Samples without the same letter show significant difference by Duncan's test ($p < 0.05$).

Furthermore, we looked at the agronomic traits of these homozygous mutant T_4-lines for ZN1S and ZN1307 (Supplementary Table S4). Only specific agronomic traits of some mutants were altered, such as plant height in *wx-fgr*-S5, number of effective tillers in *wx-fgr*-S1 and *wx-fgr*-S3, number of primary branches in *wx-fgr*-S2, and grain number per panicle in *wx-fgr*-S1 and *wx-fgr*-S5, etc. Similar to those of ZN1S mutants, most agronomic traits did not change significantly in ZN1307 mutants; only certain mutants showed changes in specific agronomic traits, such as the number of primary branches in *wx-fgr*-R1 and *wx-fgr*-R2, the setting rate in *wx-fgr*-R3, the 1000-grain weight in *wx-fgr*-R1 and *wx-fgr*-R5. The only noteworthy thing is that all ZN1307 mutants had varying degrees of decreased yield per plant.

2.2. Creation of Fragrant Glutinous Hybrid Rice

wx-fgr-S2 with the lowest amylose content among the ZN1S sterile mutants and *wx-fgr*-S4 with similar agronomic traits to the wild-type were selected as recipients. In contrast, *wx-fgr*-R2 with the lowest amylose content among the ZS1307 restorer mutants and *wx-fgr*-R6 with the least impact on yield per plant were selected as donor parents to obtain four types of hybrid rice, namely *wx-fgr*-Z22 (*wx-fgr*-S2/*wx-fgr*-R2), *wx-fgr*-Z42 (*wx-fgr*-S4/*wx-fgr*-R2), *wx-fgr*-Z26 (*wx-fgr*-S2/*wx-fgr*-R6), and *wx-fgr*-Z46 (*wx-fgr*-S4/*wx-fgr*-R6). We identified the four transgenic hybrid rice's primary quality features. Similar to ZN1S sterile mutants and ZS1307 restorer mutants, the grain endosperms produced by all hybrid combinations had low transparency and milky white appearance, obviously different from that of WT-S and WT-R (Figure 1C). Iodine staining results showed that refined rice grains from hybrid rice had lighter coloration in transverse sections than WT-S and WT-R (Figure 2A). Correspondingly, the four transgenic hybrid rice variants have much less amylose than their two parental strains (Figure 2B). In particular, *wx-fgr*-Z42 had an amylose content of 1.76%, which was on par with wild-type waxy rice. The four transgenic hybrid rice strains had an aroma substance 2-AP level that was significantly higher than that of the sterile line ZN1S while not comparable to that of the restorer line ZN1307 (Figure 2C,D). However, rice flour from the four transgenic hybrid rice variants had a worse RVA curve pattern than that from thetwo parents, WT-S and WT-R, while having a greater gel consistency than the wild type (Figure 2E,F). In addition, we also investigated the main agronomic traits of the transgenic hybrid rice (Supplementary Table S4). In general, hybrid rice showed obvious superparent advantage in plant height and panicle length, but no obvious changes in other traits, except for specific traits of several mutants, such as the grain weight per panicle in *wx-fgr*-Z46, the setting rate in *wx-fgr*-Z26 and *wx-fgr*-Z46 and the yield per plant in *wx-fgr*-Z26.

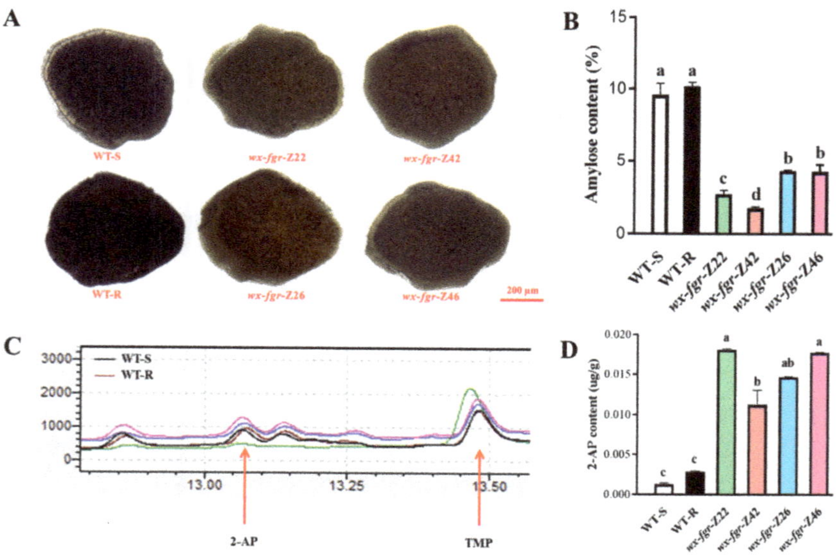

Figure 2. *Cont.*

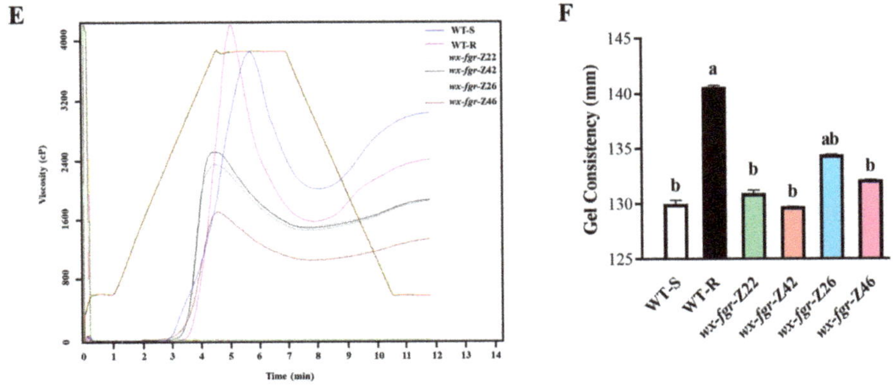

Figure 2. Creation of two-line fragrant glutinous hybrid rice. (**A**) Microscopic observation of iodine-stained endosperm. WT-S represents the wild type of sterile line ZN1S and WT-R represents the wild type of restorer line ZN1307. Scale bars, 200 μm. (**B**) The amylose contents determined by iodine colorimetry of ZN1S, ZN1307 and the F_1 hybrid ZN1S mutants × ZN1307 mutants. (**C**) Total ion chromatograms (TIC) of 2-AP and TMP in the grains of ZN1S, ZN1307 and the F_1 hybrid ZN1S mutants × ZN1307 mutants. 2, 4, 6-trimethyl pyridine (TMP) was used as the internal standard. (**D**) 2-AP content. (**E**) Rapid visco analysis. cP (centi Poise), viscosity unit. (**F**) Gel consistency. Error bars are means ± SD (n = 3). Samples without the same letter show significant difference by Duncan's test ($p < 0.05$).

3. Discussion

Although hybrid rice's cooking and eating quality have somewhat improved recently, they still fall short of high-grade and high-quality conventional rice. Incorporating the *wx* mutation into varieties with low initial AAC levels led to further reductions in AAC; however, these effects had little to no impact on the desired gelatinization traits, amylopectin structure types, or the major agronomic traits [27]. Therefore, introducing the *wx* mutation into rice varieties with low baseline AAC levels is a feasible strategy for increasing the ECQ of rice. In this study, the wild-type sterile line ZN1S and the restorer line ZN1307 had initial amylose contents of about 10% (Figure 1D,F), which were not very high. All of the mutants' amylose contents were reduced to 2–4% levels by Cas9-mediated gene editing, along with some modifications to their gelatinizing properties (Figure 1). Notably, sterile line mutants and restorer line mutants have gel consistency that is significantly higher than that of the corresponding wild type (Figure 1I). However, different from the great changes in ZN1S and ZN1307 mutants, only *wx-fgr*-Z26 had significantly higher gel consistency in hybrid rice, but only between the two parents (Figure 2F). The viscosity curve of the sterile line mutants varied in degree from the wild type, whereas all restorer line mutants and hybrid rice had softer pasting properties (Figure 1J). These results indicate that the heredity of quality traits in hybrid rice may not be determined by simple additive effect. In addition, it may not be enough to manipulate only *Wx*, the main quality gene. Combining with other quality genes, even some quality genes with minor effects, may be needed to finally improve rice quality.

Overall, our study shows how to directly alter high-quality genes of interest in elite sterile and restorer lines to produce enhanced hybrid rice with the potential for commercialization.

4. Materials and Methods

4.1. Plant Materials and Growing Conditions

ZN1S is a photoperiod-sensitive genic male sterile rice variety, whereas ZN1307 is an indica restorer rice variety with high quality and high yield, and both carry a Wx^{la} allele. Plants were grown in the field throughout the regular rice growing season at Huazhong

Agricultural University's experimental station, Wuhan (Hubei), and Lingshui (Hainan). The seeds were sown on May 15 every year, and the seedlings (25 or 30 days old) were transplanted onto the field with a single plant spacing of 16.5 cm and 26.4 cm between rows. The field was managed using customary agricultural practices. At the end of the third stage of panicle differentiation, ZN1S male sterile lines—whose fertility transformation temperature was 24 °C—were generally transferred to a cold water pool (22.5 °C), then after 15 days of treatment from the cold water pool to the field.

4.2. Design of the Wx and OsBADH2 Target Sites and Construction of the CRISPR/Cas9 Double-Targeting Vector

We designed target sites at bases 14–33 of the third exon of the *Wx* gene and bases 122–141 of the fourth exon of the *OsBADH2* gene, respectively, to produce waxy and fragrant transgenic lines. The online tool CRISPR GE was used to design the target sites [28]. The target sequences for the *Wx* gene were GGGTCATGGTGATCTCTCCTCGG and for the *OsBADH2* gene ATCAACCCAACTACACCGATAGG. The U6-sgRNA expression cassette encoding the 20 nucleotides (nt) *Wx* target sequence was amplified by polymerase chain reaction (PCR) and ligated into the pCXUN vector, which *Kpn I* digested to create the pCXUN-U6-sgRNA intermediate vector. The 20 nt *OsBADH2* target sequence from the U3-sgRNA expression cassette was then amplified by PCR and ligated into the pCXUN-U6-sgRNA intermediate vector, digested by *Sac I* to create the final vector pCXUN-U6-U3-sgRNA. The constructed CRISPR/Cas9 final vector was introduced into the ZN1S and the ZN1307 receptor using *Agrobacterium tumfaciens*-mediated genetic transformation [29]. The primer sequences used to construct the vector are listed in Supplementary Table S1.

4.3. Molecular Characterization of the Mutant Plants

Genomic DNA was extracted from seedling leaves using the sodium dodecyl sulfate method [30]. PCR amplification was performed using primer pairs that produce amplicons containing the target sites. The amplified products were then sequenced using the Sanger sequencing method and assembled using the software SeqMan from the Lasergene package. We compared the amplicon sequences generated from the corresponding wild-type and real transgenic templates to identify mutations. Polyacrylamide gel electrophoresis (PAGE) was used to determine the homozygosity/heterozygosity for a mutant individual. Indel primer pairs that produce an amplicon, including the target sites, were created. Agarose gel electrophoresis was used to identify Cas9 in each individual. The relevant PCR primers for these steps are listed in Supplementary Table S1.

4.4. Evaluation of Rice Quality

Amylose content and gel consistency were measured using the previously described method [31]. Ten full grains of brown rice were placed in a Petri dish for the sensory evaluation of rice fragrance. Following this, 10 mL of 1.7% potassium hydroxide solution was added, and the dish was covered and left at room temperature for 10 min. The Petri dishes were then opened one by one, and at least three subjects with a normal sense of smell were requested to assess and average each sample. Gas chromatography-mass spectrometry (GC-MS) was used to determine the content of flavor ingredient 2-AP. The extract was added after the brown rice was ground into a powder. The final sample was injected into the Agilent6890 GC connected to the HP5973 MS detector (Agilent Technologies, Palo Alto, CA, USA) for detection. MS database and retention time were used to identify the volatile compounds. Internal standard methylene chloride. 2, 4, 6-trimethyl pyridine was injected at a concentration of 0.4585 ng/μL, and the injection volume was 1 μL. A previous study has outlined the detailed steps [32].

4.5. Investigation of Agronomic Characters

In the paddy field, plant height, grains per panicle, panicle length, effective panicles, primary branches, and secondary branches were measured. Grain number per panicle, grain weight per panicle, setting rate, 1000-grain weight and yield per plant were investigated indoors. Seeds were collected for each plant and dried at 37 °C for two weeks to determine the yield per plant. Using the SC-A grain analysis system (Wseen company, Hangzhou, China), 1000-grain weight was measured. Three similar-sized panicles from each plant were selected, dried at 37 °C for two weeks, and the average value was then calculated to determine the number of grains and grain weight per panicle.

4.6. Performance of the F_1 Hybrids Obtained from ZN1S × ZN1307

ZN1S mutant lines (*wx-fgr*-S2 and *wx-fgr*-S4) and ZN1307 restorer lines (*wx-fgr*-R2 and *wx-fgr*-R6) were crossed in pairs. To avoid unforeseen pollination, the panicles were covered with brown paper bags. The F_1 seeds were harvested 30 days after the grains matured, and the F_1 hybrids were sown in the field plots. The experiment used a replicative complete block design with 36 plants in each plot that was replicated twice.

4.7. Microscopy

In the iodine staining experiment, the mature seeds were first processed into milled rice, and the milled rice was cut with a sharp blade after being stained with iodine solution, and then observed and photographed with a stereo fluorescence microscope (SMZ25, Nikon, Tokyo, Japan). The appearance of milled rice was directly observed and photographed by a stereo fluorescence microscope (SMZ25, Nikon, Tokyo, Japan).

4.8. Statistical Analysis

The various diagrams in this study were drawn using GraphPad Prism 8. One-way analysis of variance and Duncan multiple comparisons were used to analyze differences between different groups using IBM Statistical Package for Social Sciences v16.0 software. Microsoft Excel 2016 was used to perform the preliminary processing and analysis of phenotypic data.

Supplementary Materials: The following supporting information can be downloaded at: https://www.mdpi.com/article/10.3390/ijms24010849/s1.

Author Contributions: G.L. wrote the manuscript. Y.T. and G.L. performed most of the experiments. Y.Z. constructed the CRISPR/Cas9 double-targeting vector and completed the preliminary genotype detection work. Y.L. provided guidance for determination of some quality traits. G.G. and Q.Z. participated in part of the field experiments. Y.H. designed experiments. All authors have read and agreed to the published version of the manuscript.

Funding: This work was supported by grants from the Ministry of Science and Technology (2021YFF1000200), National Natural Science Foundation of China (U21A20211, 91935303), the Science and Technology Major Program of Hubei Province (2021ABA011), and China Agriculture Research System (CARS-01-03).

Institutional Review Board Statement: Not applicable.

Informed Consent Statement: Not applicable.

Data Availability Statement: For materials, please contact the corresponding author's email address.

Conflicts of Interest: The authors declare that there are no conflict of interest.

References

1. Tian, Z.; Qian, Q.; Liu, Q.; Yan, M.; Liu, X.; Yan, C.; Liu, G.; Gao, Z.; Tang, S.; Zeng, D.; et al. Allelic diversities in rice starch biosynthesis lead to a diverse array of rice eating and cooking qualities. *Proc. Natl. Acad. Sci. USA* **2009**, *106*, 21760–21765. [CrossRef] [PubMed]
2. Fitzgerald, M.A.; McCouch, S.R.; Hall, R.D. Not just a grain of rice: The quest for quality. *Trends Plant Sci.* **2009**, *14*, 133–139. [CrossRef] [PubMed]
3. Li, H.; Prakash, S.; Nicholson, T.M.; Fitzgerald, M.A.; Gilbert, R.G. The importance of amylose and amylopectin fine structure for textural properties of cooked rice grains. *Food Chem.* **2016**, *196*, 702–711. [CrossRef] [PubMed]

4. Sano, Y. Differential regulation of waxy gene expression in rice endosperm. *Theor. Appl. Genet.* **1984**, *68*, 467–473. [CrossRef] [PubMed]
5. Wang, Z.-y.; Wu, Z.-l.; Xing, Y.-y.; Zheng, F.-g.; Guo, X.-l.; Zhang, W.-g.; Hong, M.-m. Nucleotide sequence of rice waxy gene. *Nucleic Acids Res.* **1990**, *18*, 5898. [CrossRef]
6. Zhang, C.; Zhu, J.; Chen, S.; Fan, X.; Li, Q.; Lu, Y.; Wang, M.; Yu, H.; Yi, C.; Tang, S.; et al. Wxlv, the Ancestral Allele of Rice Waxy Gene. *Mol. Plant* **2019**, *12*, 1157–1166. [CrossRef]
7. Ayres, N.M.; McClung, A.M.; Larkin, P.D.; Bligh, H.F.J.; Jones, C.A.; Park, W.D. Microsatellites and a single-nucleotide polymorphism differentiate apparentamylose classes in an extended pedigree of US rice germ plasm. *Theor. Appl. Genet.* **1997**, *94*, 773–781. [CrossRef]
8. Wanchana, S.; Toojinda, T.; Tragoonrung, S.; Vanavichit, A. Duplicated coding sequence in the waxy allele of tropical glutinous rice (*Oryza sativa* L.). *Plant Sci.* **2003**, *165*, 1193–1199. [CrossRef]
9. Mikami, I.; Uwatoko, N.; Ikeda, Y.; Yamaguchi, J.; Hirano, H.Y.; Suzuki, Y.; Sano, Y. Allelic diversification at the wx locus in landraces of Asian rice. *Theor. Appl. Genet.* **2008**, *116*, 979–989. [CrossRef]
10. Liu, L.; Ma, X.; Liu, S.; Zhu, C.; Jiang, L.; Wang, Y.; Shen, Y.; Ren, Y.; Dong, H.; Chen, L.; et al. Identification and characterization of a novel Waxy allele from a Yunnan rice landrace. *Plant Mol. Biol.* **2009**, *71*, 609–626. [CrossRef]
11. Yang, J.; Wang, J.; Fan, F.-J.; Zhu, J.-Y.; Chen, T.; Wang, C.-L.; Zheng, T.-Q.; Zhang, J.; Zhong, W.-G.; Xu, J.-L. Development of AS-PCR marker based on a key mutation confirmed by resequencing of Wx-mp in Milky Princess and its application in japonica soft rice (*Oryza sativa* L.) breeding. *Plant Breed.* **2013**, *132*, 595–603. [CrossRef]
12. Zhou, H.; Xia, D.; Zhao, D.; Li, Y.; Li, P.; Wu, B.; Gao, G.; Zhang, Q.; Wang, G.; Xiao, J.; et al. The origin of Wxla provides new insights into the improvement of grain quality in rice. *J. Integr. Plant Biol.* **2021**, *63*, 878–888. [CrossRef]
13. Zhang, C.; Yang, Y.; Chen, S.; Liu, X.; Zhu, J.; Zhou, L.; Lu, Y.; Li, Q.; Fan, X.; Tang, S.; et al. A rare Waxy allele coordinately improves rice eating and cooking quality and grain transparency. *J. Integr. Plant Biol.* **2021**, *63*, 889–901. [CrossRef]
14. Zhang, J.; Zhang, H.; Botella, J.R.; Zhu, J.-K. Generation of new glutinous rice by CRISPR/Cas9-targeted mutagenesis of the Waxy gene in elite rice varieties. *J. Integr. Plant Biol.* **2018**, *60*, 369–375. [CrossRef]
15. Xu, Y.; Lin, Q.; Li, X.; Wang, F.; Chen, Z.; Wang, J.; Li, W.; Fan, F.; Tao, Y.; Jiang, Y.; et al. Fine-tuning the amylose content of rice by precise base editing of the Wx gene. *Plant Biotechnol. J.* **2021**, *19*, 11–13. [CrossRef]
16. Huang, L.; Li, Q.; Zhang, C.; Chu, R.; Gu, Z.; Tan, H.; Zhao, D.; Fan, X.; Liu, Q. Creating novel Wx alleles with fine-tuned amylose levels and improved grain quality in rice by promoter editing using CRISPR/Cas9 system. *Plant Biotechnol. J.* **2020**, *18*, 2164–2166. [CrossRef]
17. Zeng, D.; Liu, T.; Ma, X.; Wang, B.; Zheng, Z.; Zhang, Y.; Xie, X.; Yang, B.; Zhao, Z.; Zhu, Q.; et al. Quantitative regulation of Waxy expression by CRISPR/Cas9-based promoter and 5′UTR-intron editing improves grain quality in rice. *Plant Biotechnol. J.* **2020**, *18*, 2385–2387. [CrossRef]
18. Liu, X.; Ding, Q.; Wang, W.; Pan, Y.; Tan, C.; Qiu, Y.; Chen, Y.; Li, H.; Li, Y.; Ye, N.; et al. Targeted Deletion of the First Intron of the Wxb Allele via CRISPR/Cas9 Significantly Increases Grain Amylose Content in Rice. *Rice* **2022**, *15*, 1. [CrossRef]
19. Grimm, C.C.; Bergman, C.; Delgado, J.T.; Bryant, R. Screening for 2-Acetyl-1-pyrroline in the Headspace of Rice Using SPME/GC-MS. *J. Agric. Food Chem.* **2001**, *49*, 245–249. [CrossRef]
20. Bradbury, L.M.T.; Fitzgerald, T.L.; Henry, R.J.; Jin, Q.; Waters, D.L.E. The gene for fragrance in rice. *Plant Biotechnol. J.* **2005**, *3*, 363–370. [CrossRef]
21. Shan, Q.; Zhang, Y.; Chen, K.; Zhang, K.; Gao, C. Creation of fragrant rice by targeted knockout of the OsBADH2 gene using TALEN technology. *Plant Biotechnol. J.* **2015**, *13*, 791–800. [CrossRef] [PubMed]
22. Ashokkumar, S.; Jaganathan, D.; Ramanathan, V.; Rahman, H.; Palaniswamy, R.; Kambale, R.; Muthurajan, R. Creation of novel alleles of fragrance gene OsBADH2 in rice through CRISPR/Cas9 mediated gene editing. *PLoS ONE* **2020**, *15*, e0237018. [CrossRef] [PubMed]
23. Shan, Q.; Wang, Y.; Chen, K.; Liang, Z.; Li, J.; Zhang, Y.; Zhang, K.; Liu, J.; Voytas, D.F.; Zheng, X.; et al. Rapid and Efficient Gene Modification in Rice and Brachypodium Using TALENs. *Mol. Plant* **2013**, *6*, 1365–1368. [CrossRef] [PubMed]
24. Cantos, C.; Francisco, P.; Trijatmiko, K.R.; Slamet-Loedin, I.; Chadha-Mohanty, P.K. Identification of "safe harbor" loci in indica rice genome by harnessing the property of zinc-finger nucleases to induce DNA damage and repair. *Front. Plant Sci.* **2014**, *5*, 302. [CrossRef] [PubMed]
25. Liang, Z.; Zhang, K.; Chen, K.; Gao, C. Targeted Mutagenesis in Zea mays Using TALENs and the CRISPR/Cas System. *J. Genet. Genom.* **2014**, *41*, 63–68. [CrossRef]
26. Fitzgerald, M.A.; Martin, M.; Ward, R.M.; Park, W.D.; Shead, H.J. Viscosity of Rice Flour: A Rheological and Biological Study. *J. Agric. Food Chem.* **2003**, *51*, 2295–2299. [CrossRef]
27. Fu, Y.; Luo, T.; Hua, Y.; Yan, X.; Liu, X.; Liu, Y.; Liu, Y.; Zhang, B.; Liu, R.; Zhu, Z.; et al. Assessment of the Characteristics of Waxy Rice Mutants Generated by CRISPR/Cas9. *Front. Plant Sci.* **2022**, *13*, 881964. [CrossRef]
28. Xie, X.; Ma, X.; Zhu, Q.; Zeng, D.; Li, G.; Liu, Y.-G. CRISPR-GE: A Convenient Software Toolkit for CRISPR-Based Genome Editing. *Mol. Plant* **2017**, *10*, 1246–1249. [CrossRef]
29. Hiei, Y.; Komari, T.; Kubo, T. Transformation of rice mediated by Agrobacterium tumefaciens. *Plant Mol. Biol.* **1997**, *35*, 205–218. [CrossRef]
30. Dellaporta, S.L.; Wood, J.; Hicks, J.B. A plant DNA minipreparation: Version II. *Plant Mol. Biol. Rep.* **1983**, *1*, 19–21. [CrossRef]

31. Qiu, X.J.; Peng, B.; Yu, S.B. Rice Grain Quality Analysis. *Bio-Protocol* **2018**, e1010163. [CrossRef]
32. Huang, T.-C.; Teng, C.-S.; Chang, J.-L.; Chuang, H.-S.; Ho, C.-T.; Wu, M.-L. Biosynthetic Mechanism of 2-Acetyl-1-pyrroline and Its Relationship with Δ1-Pyrroline-5-carboxylic Acid and Methylglyoxal in Aromatic Rice (*Oryza sativa* L.) Callus. *J. Agric. Food Chem.* **2008**, *56*, 7399–7404. [CrossRef]

Disclaimer/Publisher's Note: The statements, opinions and data contained in all publications are solely those of the individual author(s) and contributor(s) and not of MDPI and/or the editor(s). MDPI and/or the editor(s) disclaim responsibility for any injury to people or property resulting from any ideas, methods, instructions or products referred to in the content.

International Journal of *Molecular Sciences*

MDPI

Article

Effect of Heading Date on the Starch Structure and Grain Yield of Rice Lines with Low Gelatinization Temperature

Naoko Crofts, Kaito Hareyama, Satoko Miura, Yuko Hosaka, Naoko F. Oitome and Naoko Fujita *

Department of Biological Production, Akita Prefectural University, 241-438 Kaidobata-Nishi, Shimoshinjo-Nakano, Akita City 010-0195, Japan
* Correspondence: naokof@akita-pu.ac.jp

Abstract: Early flowering trait is essential for rice cultivars grown at high latitude since delayed flowering leads to seed development at low temperature, which decreases yield. However, early flowering at high temperature promotes the formation of chalky seeds with low apparent amylose content and high starch gelatinization temperature, thus affecting grain quality. Deletion of starch synthase IIa (SSIIa) shows inverse effects of high temperature, and the *ss2a* mutant shows higher apparent amylose content and lower gelatinization temperature. *Heading date 1* (*Hd1*) is the major regulator of flowering time, and a nonfunctional *hd1* allele is required for early flowering. To understand the relationship among heading date, starch properties, and yield, we generated and characterized near-isogenic rice lines with *ss2a Hd1*, *ss2a Hd1 hd1*, and *ss2a hd1* genotypes. The *ss2a Hd1* line showed the highest plant biomass; however, its grain yield varied by year. The *ss2a Hd1 hd1* showed higher total grain weight than *ss2a hd1*. The *ss2a hd1* line produced the lowest number of premature seeds and showed higher gelatinization temperature and lower apparent amylose content than *ss2a Hd1*. These results highlight *Hd1* as the candidate gene for developing high-yielding rice cultivars with the desired starch structure.

Keywords: rice; heading date; starch; yield; low gelatinization temperature; starch synthase IIa

check for updates

Citation: Crofts, N.; Hareyama, K.; Miura, S.; Hosaka, Y.; Oitome, N.F.; Fujita, N. Effect of Heading Date on the Starch Structure and Grain Yield of Rice Lines with Low Gelatinization Temperature. *Int. J. Mol. Sci.* **2022**, *23*, 10783. https://doi.org/10.3390/ijms231810783

Academic Editors: Jinsong Bao and Jianhong Xu

Received: 25 August 2022
Accepted: 13 September 2022
Published: 15 September 2022

Publisher's Note: MDPI stays neutral with regard to jurisdictional claims in published maps and institutional affiliations.

Copyright: © 2022 by the authors. Licensee MDPI, Basel, Switzerland. This article is an open access article distributed under the terms and conditions of the Creative Commons Attribution (CC BY) license (https://creativecommons.org/licenses/by/4.0/).

1. Introduction

Heading date is one of the most important agricultural traits, particularly for rice (*Oryza sativa* L.) cultivars cultivated in high-latitude areas, because early flowering ensures seed development at optimum temperature during the short summer, thus maximizing yield [1]. Cultivars with different heading dates have been selected at different latitudes through natural and artificial means [1–3]. Several genes governing heading date have been identified in rice [4–15] and are shown in Figure 1. *Heading date 1* (*Hd1*) encodes a zinc-finger protein and is the major determinant of heading date [4]. *Hd1* represses the expression of florigen, *Hd3a*, under a long-day photoperiod but promotes its expression under short days [5,14,16]. Once heading is initiated, flowering generally occurs within a couple of days. Therefore, the nonfunctional *hd1* allele is required for early flowering under long-day conditions. Different rice cultivars have acquired several single nucleotide polymorphisms (SNPs) in *Hd1* during the process of domestication [2,3,17,18].

Starch, the major component of rice grain, is composed of glucose polymers of essentially linear amylose and precisely, but highly, branched amylopectin [19,20]. The ratio of amylose to amylopectin as well as the length and frequency of amylopectin branches affect the physicochemical properties of starch and transparency of grains, thus affecting the quality of rice [21–23]. The amylose found in rice endosperm is exclusively synthesized by granule-bound starch synthase I (GBSSI); thus, the expression level of *GBSSI* determines the amylose content of rice grains. Polymorphisms at the last nucleotide of the first intron of the *GBSSI* gene are commonly seen in japonica rice (*O. sativa* L. ssp. *japonica*) [24–27] and are known to reduce the splicing efficiency of *GBSSI* mRNA, especially under high temperature during seed development, which decreases GBSSI protein production and consequently amylose content [28–30].

Amylopectin is synthesized in the rice endosperm by the synergistic and balanced actions of multiple isozymes of starch synthases (SSs), branching enzymes (BEs), and debranching enzymes by forming multiprotein complexes [31,32]. Chromosomal locations of genes encoding these isozymes are summarized in Figure 1. According to the current understanding of amylopectin biosynthesis, SSIIIa synthesizes long glucan chains (amylopectin backbone) with degree of polymerization (DP) > 30, and BEI generates long branches. BEIIb generates short amylopectin branches with DP 6–7, and SSI elongates these short branches to DP 8–12. SSIIa further elongates these branches to DP 12–24 in most indica rice (*O. sativa* L. ssp. *indica*) cultivars, but the SSIIa isozyme of japonica rice is less active than that of indica rice and produces fewer intermediate chains with DP 12–24 [33]. Unnecessary branches are trimmed off by debranching enzymes such as isoamylase 1 [31,34]. Rice lines lacking BEIIb exhibit fewer short amylopectin chains, greater long amylopectin chains, and consequently higher gelatinization temperature than the wild type [35–38]. High temperature during seed development also impacts the expression level and activity of BEIIb [39–42], which increases the long amylopectin branch chains, gelatinization temperature, and chalky seed frequency and decreases the palatability of cooked rice [39,42].

The effects of SSIIa loss on amylopectin structure are opposite to those of BEIIb loss, although loss of either one of these enzymes has the same effect on amylose content. A *ss2a* null mutant rice line, EM204, was previously isolated from the N-nitroso-N-methylurea (NMU)-treated mutant panel of the japonica rice cultivar Kinmaze [43]. EM204 harbors a point mutation at the last nucleotide of the intron 5 of *SSIIa*, resulting in the loss of exon 6 and no detectable SSIIa activity in developing seeds [43]. Loss of SSIIa activity increased short amylopectin branches with DP < 11 and lowered the gelatinization temperature by 5 °C compared with the parental line (Kinmaze), although Kinmaze and other typical japonica rice cultivars exhibit lower SSIIa activity than typical indica rice varieties [33,43]. In addition, loss of SSIIa activity increased the apparent amylose content to 24%, which was considerably higher than that of Kinmaze (20%) [43]. Both Kinmaze and EM204 flower in early September in Akita, Japan (39.7° N, 140.1° E). Although the starch of EM204 shows great potential as an anti-retrogradation agent, the agricultural traits of this mutant line, such as heading date and yield, need further improvement since nighttime temperature sharply declines in September, which drastically reduces grain yield, depending on the harvest year. Thus, EM204 was backcrossed twice with a high-yielding elite rice cultivar, Akita 63 [44], which flowers in early August. Although more than half of the backcrossed lines flowered in early August, some flowered in September because the *SSIIa* and *Hd1* genes are located in close proximity to each other on chromosome 6 (Figure 1).

A previous study analyzed the effects of different *Hd1* alleles on agronomic traits and amylose content using multiple genetic backgrounds, such as glutinous rice, japonica rice, and indica rice. However, because these rice genotypes harbor different alleles of *SSIIa* and *GBSSI*, in addition to the genes responsible for plant biomass and yield components [18], the effects of *Hd1* alleles on starch properties could not be evaluated properly. Therefore, in this study, we used Kinmaze (the parental line of EM204) and Akita 63, both of which have $ss2a^L$, to identify the allele(s) responsible for the differences in heading dates. In addition, to accurately evaluate the effects of different heading dates on starch properties and agricultural traits in the absence of SSIIa, we backcrossed EM204 (late-heading *ss2a* mutant) with Akita 63 (early-heading elite rice cultivar) and generated near-isogenic lines (NILs) with three different combinations: *ss2a ss2a Hd1 Hd1* (*ss2a Hd1*), *ss2a ss2a Hd1 hd1* (*ss2a Hd1 hd1*), and *ss2a ss2a hd1 hd1* (*ss2a hd1*). The effects of three *Hd1* genotypes on agricultural traits, apparent amylose content, amylopectin structure, and starch gelatinization temperature, in the absence of SSIIa, are discussed.

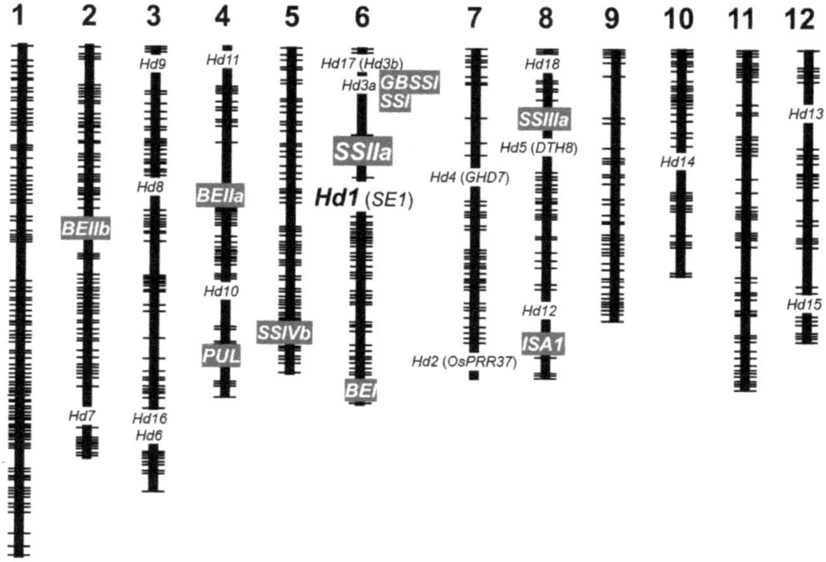

Figure 1. Chromosomal locations of genes responsible for the regulation of heading date and endosperm starch biosynthesis in rice. Genes controlling heading date are written in black writing (*Hd1* is enlarged) and those involved in starch biosynthesis in the rice endosperm are highlighted in gray. Note that *Hd1* and *SSIIa* are located in close proximity of each other on chromosome 6.

2. Results

2.1. Nucleotide Sequence of Hd1 in Kinmaze, Akita 63, and Akitakomachi

The *SSIIa* and *Hd1* genotypes and heading dates of different rice accessions are summarized in Table 1. Genomic DNA sequence of *Hd1* was amplified from Kinmaze, Akita 63, Akitakomachi, and Nipponbare using primers #5 and #12 (Table S1 and Figure S1) and compared (Figure 2b–d). Kinmaze is the parental line of EM204, which flowers in early September; Akita 63 is the high-yielding elite rice cultivar used for backcrossing and flowers in early August; Akitakomachi is commonly grown in Akita, Japan, and flowers in late July (1 week before Akita 63); and Nipponbare is the model japonica rice cultivar that flowers in late August in Akita, Japan (Table 1). Locations of SNPs found in *Hd1* sequences and the resulting amino acid substitutions are summarized in Figure 2a. The results showed that the *Hd1* sequence of Akita 63 was identical to that of Akitakomachi but different from the *Hd1* sequences of Nipponbare and Kinmaze (Figure 2b–d). The *Hd1* sequence of Kinmaze was also different from that of Nipponbare. In addition, the *Hd1* sequences of Akita 63 and Akitakomachi were 4807 bp in length, while that of Kinmaze was 4850 bp. The *Hd1* of Nipponbare was 4814 bp in length and contained two exons (1325–2152 bp and 2790–3149 bp) (Figure 2a) [4]. The *Hd1* of Akita 63, Akitakomachi, and Kinmaze carried a cytosine to thymine polymorphism at the 1640th nucleotide relative to the *Hd1* of Nipponbare, resulting in a histidine to tyrosine substitution (Figure 2a,b,d). In addition, the *Hd1* of Akita 63, Akitakomachi, and Kinmaze harbored 36 nucleotide insertions between the 1657th and 1658th nucleotides, resulting in 12 amino acid insertions between the 110th and 111th amino acid residues, compared with Nipponbare (Figure 2a,b,d). The remaining *Hd1* sequence in Kinmaze was the same as that in Nipponbare. Therefore, Kinmaze was predicted to produce a functional Hd1 protein (Figure 2b–d). On the contrary, the *Hd1* of Akita 63 and Akitakomachi contained 43 additional nucleotide deletions between the 2032nd and 2074th nucleotides compared with the *Hd1* of Nipponbare, resulting in a frame shift at the 236th amino acid and a premature stop codon at the end of exon 1 (Figure 2a–d). Akita 63 and Akitakomachi theoretically produced only 259 of the 407 amino acids of the

Hd1 protein (Figure 2d), although the truncated protein could be degraded. Therefore, Akita 63 and Akitakomachi were speculated to produce a nonfunctional hd1 protein.

Table 1. Summary of the *SSIIa* and *Hd1* genotypes and typical heading dates of rice accessions.

Rice Accession	*SSIIa* Genotype [1]	*Hd1* Genotype	Heading Date [2]
Nipponbare	$ss2a^L$	*Hd1*	Late August
Kasalath	*SS2a*	$hd1^{Kas}$	Early August
Akitakomachi	$ss2a^L$	*hd1*	Late July
Akita 63	$ss2a^L$	*hd1*	Early August
Kinmaze	$ss2a^L$	*Hd1*	Early September
EM204	*ss2a*	*Hd1*	Early September
ss2a Hd1	*ss2a*	*Hd1*	Early September
ss2a Hd1 hd1	*ss2a*	*Hd1 hd1*	Late August
ss2a hd1	*ss2a*	*hd1*	Early August

[1] Superscript L denotes leaky mutation present in the *SS2a* allele of wild-type japonica rice. [2] Typical heading dates from 2017 to 2021 in Akita, Japan. *hd1* allele of Kasalath is shown as $hd1^{Kas}$ to distinguish from that of Akitakomachi, Akita 63, and *ss2a hd1*.

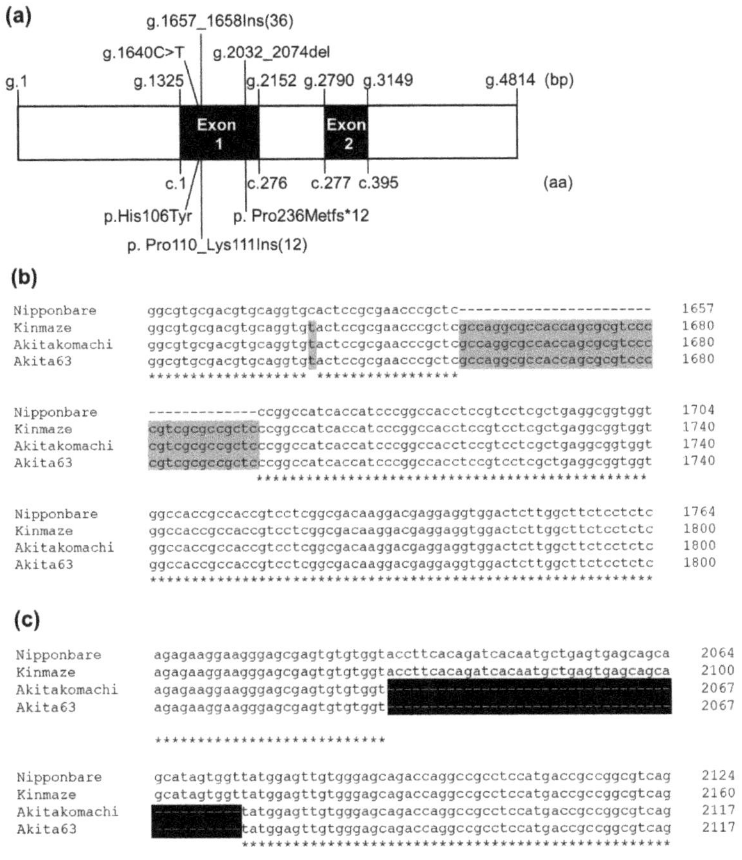

Figure 2. *Cont.*

(d)

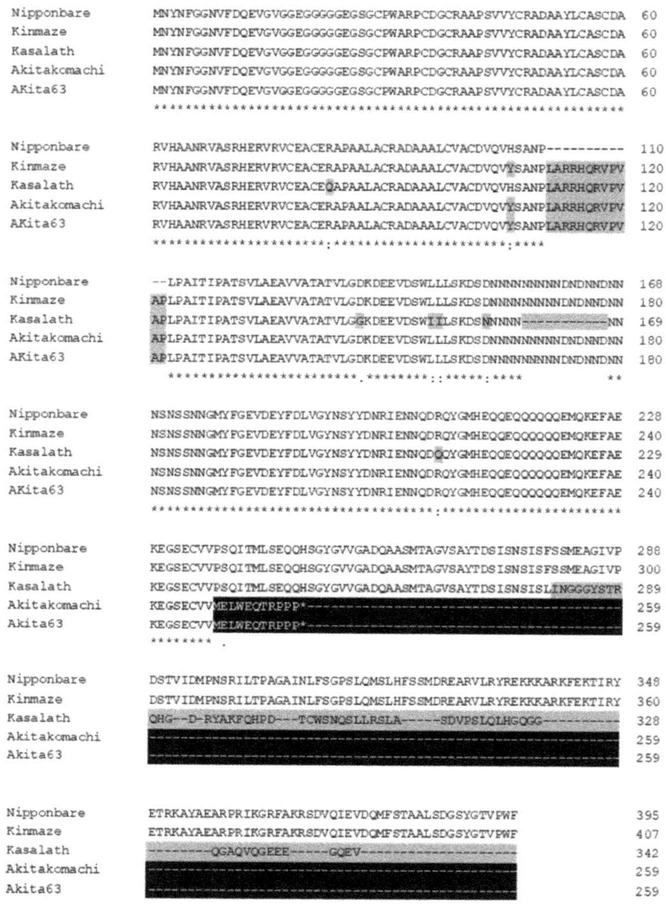

Figure 2. Comparisons of *Hd1* DNA sequences and deduced amino acid sequences in various rice lines. (**a**) Schematic representation of the *Hd1* gene structure in Akita 63. The positions of SNPs and resulting amino acid substitutions relative to Nipponbare are indicated. Ins, insertion; del, deletion; fs*12, frame shift-generated stop codon after 12 amino acids. The letter 'g' followed by a number indicates the nucleotide position in genomic DNA. Similarly, letters 'c' and 'p' followed by numbers represent the nucleotide position in cDNA and amino acid position in protein, respectively. Numbers in brackets indicate the number of nucleotide or amino acid insertions. (**b**,**c**) DNA sequence alignments of *Hd1* from 1598 to 1764 bp (**b**) and from 2005 to 2117 bp (**c**). The nucleotide positions correspond to the *Hd1* sequence of Nipponbare. (**d**) Full-length amino acid sequence alignment of Hd1. DNA and protein sequences different from Nipponbare are indicated with gray boxes, and regions missing in Akita 63 and Akitakomachi are indicated by black boxes. Sequences used to create the alignments are as follows: Nipponbare (AB041838), Kinmaze (MK449352), Kasalath (AB041839), Akitakomachi (MK449350), and Akita 63 (MK449351). Asterisks indicate identical nucleotides (**b**,**c**) and amino acid residues (**d**).

2.2. Genotyping and Western Blotting of Rice Accessions with Different Hd1 and SSIIa Alleles

PCR markers for *Hd1* have been generated for the selection of the early-flowering trait in rice cultivars such as KantoHD1 [45] and Milky Summer [46], which were generated via the introduction of the nonfunctional $hd1^{Kas}$ allele from Kasalath. It is important to note that although both Kasalath and Akita 63 flower at the same time (early August in Akita, Japan), the *hd1*^{Kas} allele of Kasalath is different from that of Akita 63 (Table 1, Figure 2d). Therefore, such selection markers would not be applicable to Akita 63 (Figure 2d). To distinguish the *hd1*, *Hd1 hd1*, and *Hd1* seedlings from the NILs generated by crossing EM204 and Akita 63, a new molecular marker was generated (Figure 3a, Table S1 and Figure S1). Early-flowering lines with the *hd1* allele (such as Akita 63, Akitakomachi, and *ss2a hd1*) generated 130 bp PCR products, whereas late-flowering lines with the *Hd1* allele (such as Nipponbare, Kinmaze, EM204, and *ss2a Hd1*) generated 173 bp PCR products (Figure 3a). Both 173 and 130 bp PCR products were detected in the heterozygous (*Hd1 hd1*) line (Figure 3a). The PCR products exhibited clear differences in migration patterns, thus enabling the distinction among the *Hd1*, *Hd1 hd1*, and *hd1* lines (Figure 3a). Presence of the *ss2a* allele was confirmed via the derived cleaved amplified polymorphic sequence (dCAPS) marker (Figure 3b, [43]); the 141 bp PCR product amplified from Akita 63 was not digested by *Bgl*II, while that amplified from *ss2a Hd1*, *ss2a Hd1 hd1*, *ss2a hd1*, and EM204 was digested into 111- and 30-bp products by *Bgl*II.

Western blotting of the total protein extracted from mature seeds using anti-SSIIa antibody confirmed the absence of SSIIa in *ss2a Hd1*, *ss2a Hd1 hd1*, *ss2a hd1*, and EM204 and the presence of SSIIa in Kinmaze and Akita 63 (Figure 4). The *Hd1* and *SSIIa* genotypes of rice accessions used in this study are summarized in Table 1. Differences in protein levels of SSI, GBSSI, and BEIIb are explained below (Sections 2.4 and 2.5).

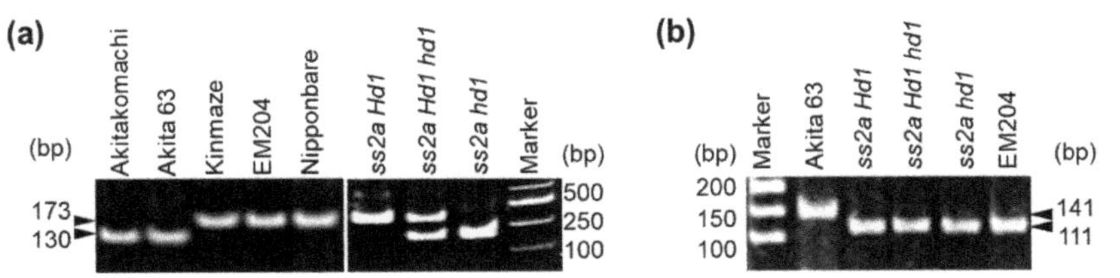

Figure 3. PCR-based screening of rice accessions with variable *Hd1* and *SSIIa* genotypes. (**a**) Screening of *Hd1* alleles using #5 and #12 PCR primers listed in Table S1. Lines with the *hd1* allele generated 130 bp PCR products, while those with the *Hd1* allele generated 173 bp PCR products; the heterozygous (*Hd1 hd1*) line produced both PCR products. (**b**) Screening of the *ss2a* allele. Lines carrying the *ss2a* allele showed 111 bp fragment after digesting with *Bgl*II, whereas Akita 63 (harboring the *ss2a*^{L} allele) showed a 141 bp band (undigested by *Bgl*II).

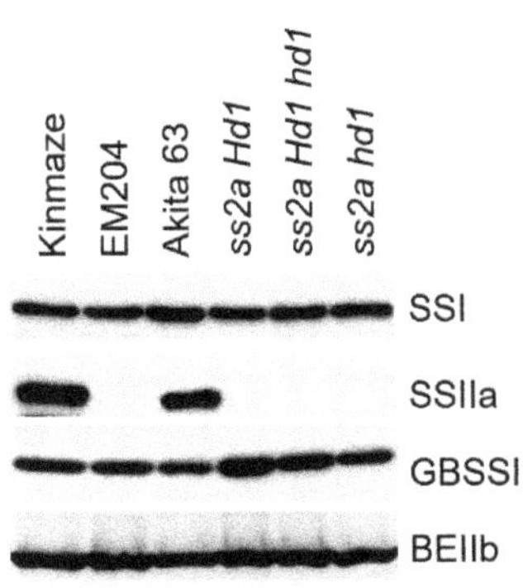

Figure 4. Western blotting analysis of total protein extract prepared from the mature seeds of different rice accessions. Starch biosynthetic enzymes were detected using the corresponding antibodies.

2.3. Effect of Hd1 Alleles on the Agricultural Traits of NILs

The agricultural traits of *ss2a Hd1*, *ss2a Hd1 hd1*, and *ss2a hd1* NILs were examined over 2 years (Figures 5 and 6, Tables S2 and S3). The three NILs were germinated or transplanted on the same respective dates and grown in the same paddy field under the same growth conditions. The heading dates of *ss2a Hd1*, *ss2a Hd1 hd1*, and *ss2a hd1* lines were remarkably different (Figure 5a,b). Although the actual heading dates of NILs slightly differed between the two years, they showed the same trend (Figure 5b, Table 1). The *ss2a hd1* line showed the earliest heading date (early August; August 4 or 7), followed by *ss2a Hd1 hd1* (late August; August 21 or 26) and *ss2a Hd1* (early September; September 2 or 13). The flowering period of individual plants of the same genotype was well synchronized; plants of the same genotype flowered within 2–3 days. In addition, the seed development and maturation period showed the order *ss2a hd1* < *ss2a Hd1 hd1* < *ss2a Hd1*, and the *ss2a hd1*, *ss2a Hd1 hd1*, and *ss2a Hd1* lines took 40, 44–47, and 48–54 days, respectively, to reach maturity after heading (Figure 5b). Only the *ss2a Hd1* line was prematurely harvested on 1 November 2021, since no further seed development was expected because of the arrival of winter (Figure 5c). The vegetative phase of lines *ss2a Hd1*, *ss2a Hd1 hd1*, and *ss2a hd1* was 104–112, 92–94, and 72–78 days, respectively (Figure 5b), and correlated well with the whole-plant dry weight, dry straw weight, plant height, and culm length (Figure 6a–d, Tables S2 and S3). The longer the vegetative period, the longer the culm and the heavier the straw weights. However, the duration of the vegetative phase did not influence the length and number of ears (Figure 6e,f, Tables S2 and S3).

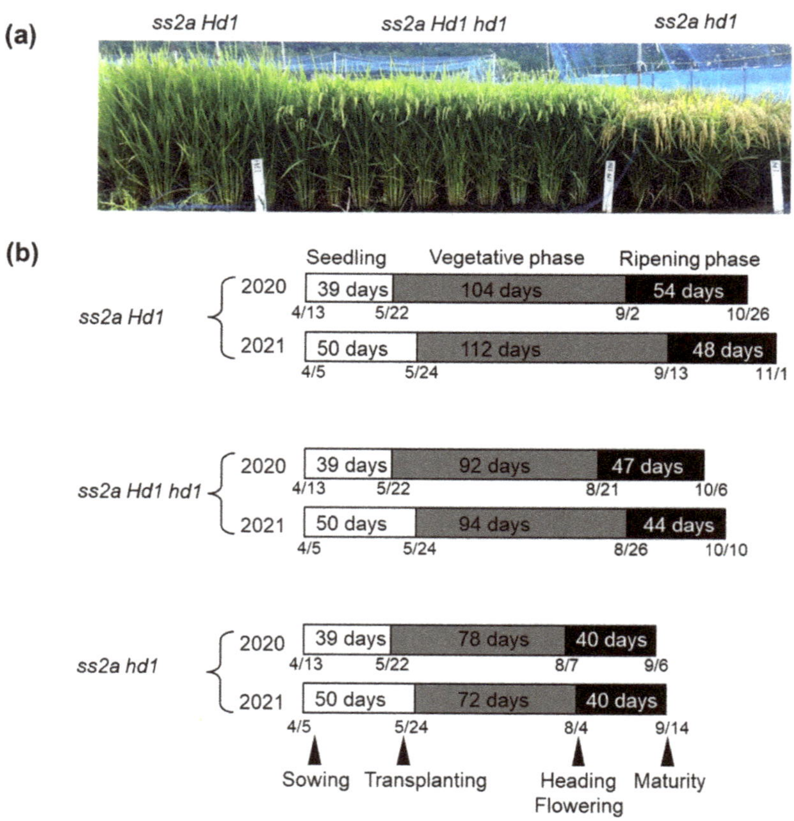

Figure 5. *Cont.*

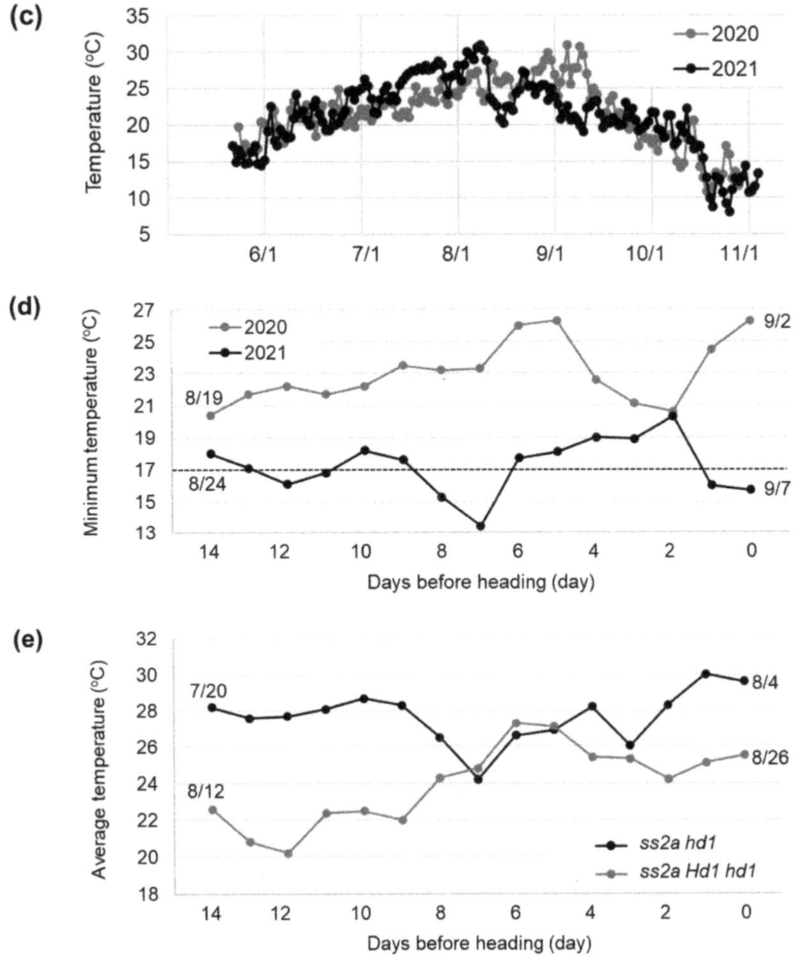

Figure 5. Comparison of the heading dates of *ss2a Hd1*, *ss2a Hd1 hd1*, and *ss2a hd1*. (**a**) Photo of NILs showing the differences in their heading dates. Note that *ss2a hd1* is mature, *ss2a Hd1 hd1* is at the mid-developmental stage, and *ss2a Hd1* is still flowering. (**b**) Differences among rice NILs in the number of days to heading and to maturity. Numbers (month/day) below the ribbon represent the actual dates of sowing, transplanting, heading, flowering, and maturity. (**c**) Average day temperature during the period from the end of May (transplanting) to the beginning of November (harvesting) in 2020 (gray) and 2021 (black). (**d**) Minimum temperature for 2 weeks before the heading date of *ss2a Hd1* in 2020 (gray) and 2021 (black). Dashed line indicates the threshold temperature (17 °C) that reduces the fertility rate. (**e**) Average temperature for 2 weeks before the heading date of *ss2a hd1* (black) and *ss2a Hd1 hd1* (gray) in 2021.

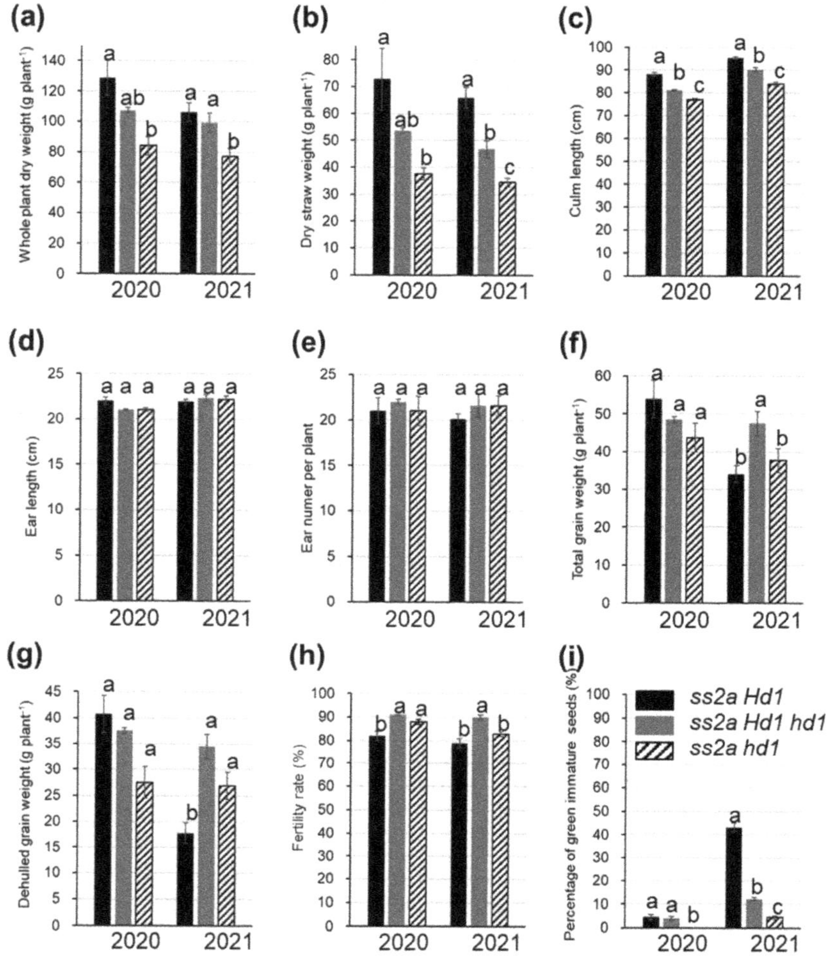

Figure 6. Agricultural traits of *ss2a Hd1* (black), *ss2a Hd1 hd1* (gray), and *ss2a hd1* (stripe) NILs. (**a**) Whole-plant dry weight, (**b**) dry straw weight per plant, (**c**) culm length, (**d**) ear length, (**e**) ear number per plant, (**f**) total grain weight per plant, (**g**) dehulled grain weight per plant, (**h**) fertility rate, (**i**) percentage of green immature seeds. Data represent mean ± standard error (SE). The three bars on the left represent data from 2020, and those on the right represent data from 2021. Data collected during the same harvest year were statistically analyzed via the Tukey-Kramer method ($p < 0.05$). The number of *ss2a Hd1*, *ss2a Hd1 hd1*, and *ss2a hd1* plants was 9, 20, and 8, respectively, in 2020, and 20 plants of each line were analyzed in 2021. Different lowercase letters above bars indicate significant differences.

The total grain weight and dehulled grain weight of *ss2a Hd1 hd1* were greater than those of *ss2a hd1* (Figure 6f,g). While those of *ss2a Hd1* differed between the 2 years, those values of *ss2a Hd1* were greater than those of *ss2a Hd1 hd1* in 2020 but lower than those of *ss2a hd1* in 2021. Some correlation was detected between plant biomass and grain yield; the higher the biomass, the better the yield, as long as the temperature during seed development remained optimal (Figures 5c and 6a,b,f,g). Reduction in the grain yield of *ss2a Hd1* in 2021 was likely caused by low temperature from mid-August to early September (Figure 5c,d). This delayed the heading date, which prolonged seed development and reduced starch synthesis, thus increasing the number of premature grains (Figure 6i). Low average day temperature is also known to prolong the seed maturation period [47]. In fact, the *ss2a Hd1* plants did not fully reach maturity in 2021 (Figure 6g, Tables S2 and S3). Although the effect of this phenomenon was minor, low temperature also led to reduced fertility rate (Figures 5d and 6h). Studies show that fertility rate declines when the minimum temperature remains under 17 °C for 2 weeks before the heading date [48] and when the temperature is too high [49]. These findings are consistent with the lower fertility rates of *ss2a Hd1* and *ss2a hd1* than that of *ss2a Hd1 hd1* (Figures 5e and 6h). Therefore, the functional *Hd1* allele is unsuitable for rice cultivars grown in high-latitude areas for the maintenance of stable grain quality and yield. However, if an increase in plant biomass is desired, especially for rice cultivars utilized as feed (straw) or for ethanol production, the functional *Hd1* allele is necessary for prolonging the vegetative phase.

The average weight of one dehulled seed of EM204 was only 16.5 mg [43], which is approximately 55% of that of *ss2a Hd1*. Therefore, the yield of *ss2a Hd1*, *ss2a Hd1 hd1*, and *ss2a hd*1 lines generated in the present study was greatly improved, owing to backcrossing with the high-yielding parental line Akita 63.

2.4. Effect of Hd1 Alleles on Apparent Amylose Content and GBSSI Expression Level

Apparent amylose content affects texture of cooked rice and rice products [50,51], and the abundance of GBSSI, which is responsible for amylose synthesis, is affected by the temperature during seed development [30]. Since the temperature during seed development varied considerably among NILs, depending on their heading dates (Figure 5c, Table S4), the apparent amylose content was measured via gel filtration chromatography using a series of single HW-55S and triple HW-50S Toyopearl columns (Table 2, Figure S2). Amylose was eluted in fraction I, and the long and short chains of amylopectin were eluted in fractions II and III, respectively (Figure S2).

Apparent amylose content of Akita 63 (17–18%) was relatively low (Table 2) compared with that of Kinmaze (22%) [43,52]. This is partly because Akita 63 flowered in early August when the temperature was high (average day temperature = 25–30 °C) during seed development, while Kinmaze flowered in early September under lower temperature (average day temperature = ~20 °C). Although the apparent amylose content of EM204 (24%) was higher than that of Kinmaze, both rice accessions flowered at a similar time (early September; average day temperature = ~20 °C). Therefore, the absence of SSIIa resulted in an increase of apparent amylose content in the Kinmaze background. Similarly, the amylose contents of all three NILs were significantly higher than that of Akita 63, as determined using a pairwise *t*-test (Table 2).

To determine whether the different heading dates of NILs affect apparent amylose content in the absence of SSIIa in the Akita 63 background, the apparent amylose contents of *ss2a Hd1*, *ss2a Hd1 hd1*, and *ss2a hd1* were compared (Table 2, Figure S2). The results showed that the apparent amylose contents of *ss2a Hd1*, *ss2a Hd1 hd1*, and *ss2a hd1* were 27.0%, 25.1%, and 22.1%, respectively, in 2020, and 28.0%, 26.6%, and 24.7%, respectively, in 2021 (Table 2, Figure S2). Thus, the amylose content of *ss2a Hd1* was the highest among the three lines and was significantly higher than that of *ss2a hd1*. We found that the earlier the heading date, the higher the seed development temperature and the lower the amylose content (Tables 2, S2 and S3, Figure S2). The apparent amylose content of *ss2a hd1* was 3–5% lower than that of *ss2a Hd1* and 4–8% higher than that of Akita 63 (Table 2). This suggests that loss of SSIIa mitigates the reduction in amylose content, even if the temperature during seed development is high.

To determine whether apparent amylose content is correlated with the GBSSI protein level, we performed western blotting of NILs (Figure 4). The amount of GBSSI protein showed a strong correlation with the apparent amylose content (Figure 4). Additionally, the GBSSI protein was the least abundant in Akita 63, and the level of GBSSI in *ss2a Hd1* was greater than that in *ss2a hd1* (Figure 4).

Table 2. Apparent amylose content and ratio of short to long chain of amylopectin in different rice accessions.

Rice Accession	Apparent Amylose Content (%) [1]		Ratio of Short to Long Chains of Amylopectin [1]	
	2020	2021	2020	2021
Akita 63	18.1 ± 0.8	17.1 ± 0.4	2.4 ± 0.1	2.7 ± 0.1
ss2a Hd1	27.0 ± 0.6 a*	28.0 ± 0.1 a*	3.1 ± 0.0 a*	3.2 ± 0.0 a*
ss2a Hd1 hd1	25.1 ± 0.6 ab*	26.6 ± 0.6 ab*	2.6 ± 0.0 b	2.9 ± 0.0 b
ss2a hd1	22.1 ± 0.4 b*	24.7 ± 0.6 b*	2.4 ± 0.0 b	3.0 ± 0.0 ab

[1] Apparent amylose content and the ratio of short to long chains of amylopectin were calculated from fraction I and fraction III/fraction II in Figure S2. Data represent the mean ± SE of three replicates. Different lowercase letters indicate significant differences among *ss2a Hd1*, *ss2a Hd1 hd1*, and *ss2a hd1* (Tukey-Kramer method; $p < 0.05$). Asterisk indicates significant differences relative to Akita 63 (t-test; $p < 0.05$).

2.5. Effect of Hd1 Alleles on Amylopectin Structure

The ratio of short amylopectin chains to long amylopectin chains (eluted in fraction III and fraction II, respectively, via gel filtration chromatography) was higher in *ss2a Hd1* than in *ss2a hd1* (Table 2). Therefore, the detailed amylopectin branch structure was analyzed via capillary electrophoresis using debranched starch purified from mature rice seeds (Figures 7 and S3). The differences in amylopectin structure were shown as a differential curve.

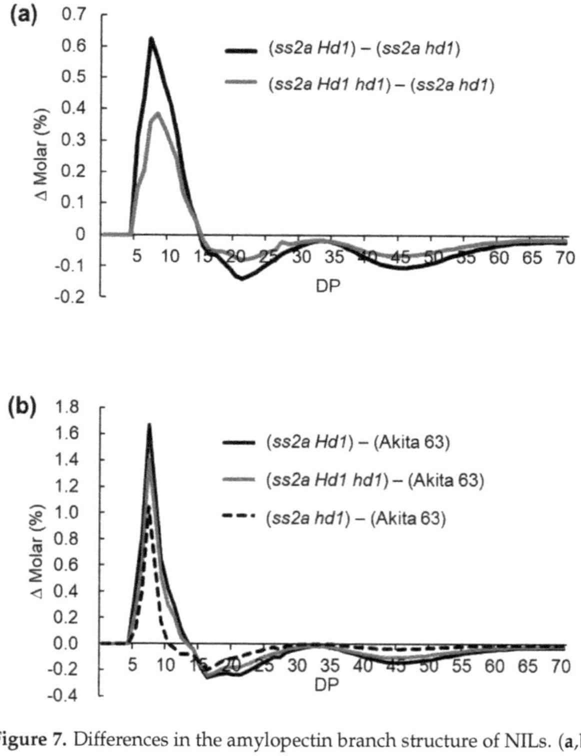

Figure 7. Differences in the amylopectin branch structure of NILs. (**a**,**b**) Subtraction curves showing the effects of *Hd1* alleles on amylopectin branch structure (**a**) and the effect of the loss of SSIIa on amylopectin structure (**b**). Each panel shows one typical representative data set of at least three replications. Data shown here were obtained from samples harvested in 2021, and the data of samples harvested in 2020 are shown in Figure S3.

To reveal the effect of heading date on amylopectin structure, values of chain length distribution of *ss2a hd1* were subtracted from those of *ss2a Hd1* or *ss2a Hd1 hd1* (Figures 7 and S3). The results showed that the number of short amylopectin chains (DP < 14) was larger in *ss2a Hd1* and *ss2a Hd1 hd1* seeds than in *ss2a hd1* seeds harvested in both years. While the number of long amylopectin chains (DP $\geq$ 15) was larger in *ss2a hd1* than in *ss2a Hd1* and *ss2a Hd1 hd1* (Figures 7a and S3a). The degree of difference was greater in *ss2a Hd1* than in *ss2a Hd1 hd1* (Figures 7a and S3a). The reason why *ss2a hd1* seeds contained fewer short amylopectin chains and more long amylopectin chains was probably because of a slight decrease in BEIIb levels in *ss2a hd1*, as shown via western blotting (Figure 4).

To reveal the effect of the loss of SSIIa on amylopectin structure, values of chain length distribution from Akita 63 were subtracted from those of *ss2a Hd1*, *ss2a Hd1 hd1*, or *ss2a hd1*. All three NILs, which lacked SSIIa, showed similar trends, i.e., a considerable increase in short amylopectin chains with DP 5–10 and a decrease in intermediate amylopectin chains with DP 12–24 (Figures 7b and S3b). These results are consistent with the role of SSIIa, which synthesizes intermediate chains [43].

2.6. Effect of Hd1 Alleles on the Thermal Properties of Starch

The gelatinization temperature of starch depends on the number of amylopectin branches with DP $\leq$ 24 [53,54]. An increase in short amylopectin branches lowers the gelatinization temperature [43], while an increase in long amylopectin branches raises the gelatinization temperature [52]. Therefore, we measured the gelatinization temperature of starch in *ss2a Hd1*, *ss2a Hd1 hd1*, or *ss2a hd1* using differential scanning calorimetry and compared the results with the gelatinization temperature of starch in Akita 63 (Table 3).

Table 3. Peak gelatinization temperature (T_p) of starch purified from rice grains harvested in 2020 and 2021, as analyzed via differential scanning calorimetry.

Rice Accession	**Tp (°C) [1]**	
	2020	**2021**
Akita 63	63.1 ± 0.0 a	62.0 ± 0.1 a
ss2a Hd1	50.3 ± 0.1 d	52.3 ± 0.2 d
ss2a Hd1 hd1	58.8 ± 0.1 c	55.5 ± 0.1 c
ss2a hd1	61.8 ± 0.1 b	57.2 ± 0.1 b

[1] Data represent the mean ± SE of three replicates. Different lowercase letters indicate significant differences (Tukey-Kramer method; $p < 0.05$).

The gelatinization temperature of lines lacking SSIIa (*ss2a Hd1*, *ss2a Hd1 hd1*, and *ss2a hd1*) was lower than that of Akita 63. To precisely evaluate the effect of the absence of SSIIa on starch gelatinization temperature, the gelatinization temperatures of starch in *ss2a hd1* and Akita 63 were compared since both rice accessions flowered in early August. The peak gelatinization temperature of *ss2a hd1* was 1.3–4.8 °C lower than that of Akita 63, although the heading dates of both these accessions were essentially the same. This is because *ss2a hd1* (owing to the loss of SSIIa) contained a higher number of short amylopectin chains with DP < 10 and lower number of chains with DP $\geq$ 10 than Akita 63 (Figure 7b). This suggests that the loss of SSIIa lowers the gelatinization temperature of starch, even under high temperature during seed development.

In addition, the peak gelatinization temperature of NILs followed the order *ss2a hd1* > *ss2a Hd1 hd1* > *ss2a Hd1*, although exact values of each line differed between the years (Table 3). This trend of the peak gelatinization temperature of NILs may be explained by differences in the chain length distribution of amylopectin among the NILs: the number of amylopectin chains with DP < 15 showed the order *ss2a Hd1* > *ss2a Hd1 hd1* > *ss2a hd1* and that of amylopectin chains DP > 15 followed the order *ss2a Hd1* < *ss2a Hd1 hd1* < *ss2a hd1* (Figure 7a). This suggests that gelatinization temperature is affected by the heading date: the higher the temperature during seed development, the higher the gelatinization temperature of starch, even in the absence of SSIIa (Figure 5, Tables 3 and S4).

3. Discussion

3.1. SNPs in Hd1

In this study, SNPs responsible for the differences in the heading dates of Kinmaze (the parental line of the *ss2a* null mutant EM204), Akita 63, and Akitakomachi were identified. Furthermore, the precise effects of different heading dates, determined by *Hd1*, *Hd1 hd1*, and *hd1*, on the agricultural traits and starch properties of rice were evaluated in NILs (lacking SSIIa) generated using Akita 63, an elite rice cultivar, as the recurrent parent. Sequencing analyses revealed that Akita 63 carries a loss-of-function *hd1* allele, while Kinmaze harbors a functional *Hd1* allele. The heading date of *ss2a hd1* was the earliest and 72–78 days after transplanting. Heading dates of *ss2a Hd1 hd1* and *ss2a Hd1* were 14–19 and 26–40 days later than those of *ss2a hd1*, respectively. These differences in the heading dates of *ss2a Hd1*, *ss2a Hd1 hd1*, and *ss2a hd1* were likely caused by the different *Hd1* alleles, although several other genes are also involved in determination of the heading date (Figure 1).

In addition, analyses of *Hd1* gene sequences using the basic local alignment search tool (BLAST; https://blast.ncbi.nlm.nih.gov/Blast.cgi, accessed on 20 June 2019) and the alignment of Hd1 amino acid sequences revealed that the *hd1* allele of Akita 63 and Akitakomachi is identical to that of the HS66 mutant (AB041841; [4]) and Sasanishiki (AB433218) (Figure S4a) but different from that of Kasalath (AB041839; [4], Figure 2d), Ginbouzu (AB041840; [4]), and Koshihikari (AB375859; [6]) (Figure S4b). The *Hd1* allele of Koshihikari is identical to that of Nipponbare, while Ginbouzu shares the same *Hd1* sequence as Kinmaze (MK449352; this study), Hoshinoyume (AB353276; [7]), and Hayamasari (AB353275; [7]) (Figure S4c). The PCR marker generated in this study (Table S1, Figure S1) as well as other PCR markers generated by Mo et al. [18] will serve as useful tools for determining the different types of *Hd1* alleles, which will accelerate the breeding of new rice cultivars with different heading dates. Different *Hd1* alleles have already been utilized to distribute the workload of the peak harvesting hours. For example, low-amylose rice lines harboring the Wx^{mq} gene, such as Milky Summer, Milky Queen, and Milky Autumn, are grown in the central to southern parts of Japan (https://www.naro.go.jp/publicity_report/press/laboratory/nics/079175.html, accessed on 28 June 2022). The choice of different *Hd1* alleles should be carefully considered, depending on the application (yield increase, starch property, or workload distribution).

3.2. Effect of Hd1 Alleles on Grain Yield

Differences in the heading date impacted the agricultural traits of NILs (Figure 6, Tables S2 and S3). The total grain yield of *ss2a Hd1 hd1* tended to be higher than that of *ss2a hd1*, although it was statistically insignificant due to a statistics outlier, while the total grain yield of *ss2a Hd1* varied depending on the year (Figure 6, Tables S2 and S3). The percentages of green immature grains were lowest in *ss2a hd1* and highest in *ss2a Hd1* (Figure 6, Tables S2 and S3). Presence of the *hd1* allele enabled efficient grain filling by promoting flowering at the appropriate temperature for starch biosynthesis during seed development, thus minimizing the time required for seed maturation and desiccation. However, because of the short vegetative period, the amount of stored photosynthetic products to be translocated from the culm might be decreased, which may lead to reduced yield. Thus, the heading date of *ss2a Hd hd1* seemed the most suitable for cultivation in Akita (Japan) as it showed stable high-level production of grains, judging from the limited data obtained under the extreme temperature conditions in 2020 and 2021, although the heterozygous allele (*Hd1 hd1*) would not be appropriate for commercial rice production as it would segregate in subsequent generations. The *ss2a hd1* NIL is also suitable for cultivation in northern Japan because the percentage of green immature grains of this genotype was the lowest, although its yield could be improved further. Increase in grain yield while maintaining seed quality should be possible by finetuning the combinations of other genes involved in the determination of the heading date, to ensure that the rice flowers in mid-August.

The *ss2a Hd1* NIL was not suitable for grain production because the heading date was too late and risked the early arrival of winter during seed development, which could lead to large yield differences between years. Moreover, if the heading date is delayed because of low temperature in August, there is a high chance that seed development may not be completed in time, resulting in drastic yield losses. However, *ss2a Hd1* showed the highest culm length and straw dry weight. Therefore, use of the *Hd1* allele would be beneficial for increasing the plant biomass, which could be used as feed for livestock or as a raw material for bioethanol production. Farmers generally prefer to grow rice varieties with relatively shorter culm length to avoid lodging. Short culm produces less waste, requires less fertilizer, and improves work efficiency. Therefore, cultivars with suitable *Hd1* alleles should be carefully considered, depending on whether the ultimate goal is to harvest grains or whole plants. The latitude and altitude of the planting area should also be taken into account when selecting rice cultivars with different *Hd1* alleles. Since *Hd1* functions by repressing heading under long days and promoting heading under short days [5,14,16], the effects of *Hd1* at different latitudes are expected to differ. The presence of the *hd1* allele likely prevents premature heading of rice plants grown near the equator and helps increase the tiller and ear numbers before transitioning to the reproductive phase. Growing *ss2a Hd1*, *ss2a Hd1 hd1*, and *ss2a hd1* genotypes at different latitudes and temperatures will provide additional useful information for achieving high yields in the respective regions.

3.3. Effect of Hd1 Alleles on Starch Structure

High temperature during seed development reduces apparent amylose content by reducing the abundance of GBSSI and increases the gelatinization temperature of starch by decreasing the abundance of BEIIb, thus affecting the quality of rice [28,30,40,42]. Compared with the effects of high temperature, the loss of SSIIa activity has opposite effects; the *ss2a* mutant shows higher apparent amylose content and lower gelatinization temperature compared with its parental line [43]. To reveal whether the loss of SSIIa can mitigate the above-described effect on starch under high temperature during seed development, the starch properties of *ss2a Hd1*, *ss2a Hd1 hd1*, and *ss2a hd1* NILs were evaluated since temperatures during the seed development of these lines were low, medium, and high, respectively, because of differences in their heading dates (Figure 5c and Table S4). The apparent amylose content of *ss2a hd1* was 22.1–24.7%, which was lower than that of *ss2a Hd1 hd1* (23.2–26.6%) and *ss2a Hd1* (27.0–28.0%) but higher than that of Akita 63 (17.1–18.1%) (Table 2). Nonetheless, both *ss2a hd1* and Akita 63 flowered at almost the same time and possessed an identical genetic background, except SSIIa. Therefore, the loss of SSIIa increased the apparent amylose content even if seed development occurred under high temperature. Increasing the apparent amylose content of rice grains can be used as one of the breeding strategies for increasing the health benefit of rice, since high apparent amylose content elevates the resistant starch content [52,55]. The *Hd1 hd1* and *Hd1* alleles are beneficial for increasing the apparent amylose content because these alleles delay flowering and facilitate seed development under cooler temperatures. However, to achieve high yield and avoid the risk of the early arrival of winter, heading dates should be no later than late August, especially if the rice is grown in the northern area of Japan.

3.4. Effect of Hd1 Alleles on Starch Gelatinization Temperature

The *ss2a hd1* NIL possessed a higher number of short amylopectin chains (DP < 15) than Akita 63, and its gelatinization temperature (57 °C) was lower than that of Akita 63 (62.0 °C) but higher than that of *ss2a Hd1 hd1* (55.5 °C) and *ss2a Hd1* (52.3 °C) (Figure 7, Table 3). One of the reasons why the gelatinization temperature of *ss2a hd1* was higher than that of *ss2a Hd1 hd1* and *ss2a hd1* might be the relatively lower abundance of the BEIIb protein under high temperature, which increased the number of long amylopectin branches (Figure 4). The balance between amylopectin branch generation and removal is important for controlling amylopectin structure, and loss of BEIIb can be mitigated by the additional loss of isoamylase 1 [56]. Therefore, reduction in the isoamylase 1 level may be one way to counterbalance the reduction in BEIIb level under high temperature during seed development. Alternatively, delaying the heading date of *ss2a hd1* offers a more practical way. Possible target genes for delaying the flowering time of *ss2a hd1* are *Ghd7* and *OsPRR37*, since combinations of the presence or absence of these genes and that of *Hd1* alleles allow the heading date of rice to be further finetuned [57,58]. Rice with a low gelatinization temperature is expected to retrograde slowly and be tasty. Thus, introduction of the *ss2a* allele into rice lines cultivated near the equator (with high temperature during seed development) may improve the quality of rice and rice products produced in tropical regions. Analysis of the retrogradation properties of *ss2a Hd1*, *ss2a Hd1 hd1*, and *ss2a hd1* lines will provide additional information for the use of these NILs in the food industry.

4. Materials and Methods

4.1. Plant Materials

Rice (*Oryza sativa* L.) *ss2a* mutant, EM204, was previously isolated from the NMU-mutagenized populations of the wild-type japonica cultivar, Kinmaze, which flowers late (early September) at high latitude [43]. EM204 harbors a mutation at the last nucleotide of intron 5, which inhibits splicing and results in the deficiency of 15 amino acids [43]. EM204 was backcrossed twice with the early-flowering, high-yielding elite japonica rice cultivar, Akita 63 [44]. The resulting F_1 seedlings were grown and self-pollinated to obtain the F_2 progeny. DNA was isolated from F_2 seedlings, and genotyping was performed as described previously [43]. The *ss2a Hd1 hd1* line was self-pollinated to obtain *ss2a Hd1*, *ss2a Hd1 hd1*, and *ss2a hd1* NILs. Theoretically, 87.5% of the genome in these three NILs was derived from Akita 63. Akitakomachi was obtained from Akita Prefectural Agricultural Experiment Station, Akita, Japan, and Kasalath and Nipponbare were obtained from the Genebank, National Agricultural and Food Research Organization, Tsukuba, Japan. All rice lines were grown in an experimental paddy field of Akita Prefectural University during the summer under natural light conditions.

4.2. Sequencing of the Hd1 Gene

Genomic DNA was isolated from leaves of Akita 63, Akitakomachi, and Kinmaze. Approximately 3 cm of young leaf was powdered with liquid nitrogen using a Multi-beads Shocker (Yasui Kikai, Osaka, Japan). The powder was extracted with 400 μL of 200 mM Tris-HCl, pH 7.5, 250 mM NaCl, 25 mM EDTA, 0.5% SDS. After centrifugation, 300 μL of the supernatant was mixed with an equal volume of isopropanol, let stand for 20 min or longer, and centrifuged. The DNA pellet was rinsed with 70% ethanol, dried, and resuspended in 25 μL of TE buffer containing 25 μL of 10 mM Tris-HCl, 1 mM EDTA. 1 μL of DNA was used for 10 μL PCR reaction. PCR amplification was carried out using the Quick Taq HS dye mix (TOYOBO, Osaka, Japan), dimethyl sulfoxide (DMSO; 5% final concentration), and sequence-specific primers (Table S1) under the following conditions: 94 °C for 2 min, and 38 cycles of 94 °C for 20 s, 50 °C for 20 s, and 68 °C for 20 s. The PCR products were sequenced at the Biotechnology Center in Akita Prefectural University, and the obtained sequences were aligned with that the *Hd1* gene of Nipponbare using Clustal Omega (https://www.ebi.ac.uk/Tools/msa/clustalo/, accessed on 20 June 2019) and analyzed using BLAST (https://blast.ncbi.nlm.nih.gov/Blast.cgi, accessed on 20 June 2019). The identified *Hd1* sequences were deposited in the NCBI GenBank database (https://www.ncbi.nlm.nih.gov/, accessed on 25 January 2019) under the following accession numbers: MK449350 (Akitakomachi), MK449351 (Akita 63), and MK449352 (Kinmaze).

4.3. Genotyping of Hd1 and SS2a Alleles

The *SSIIa* gene was genotyped as described [43]. To genotype the *Hd1* gene, PCR was performed using the Quick Taq HS dye mix (TOYOBO, Osaka, Japan), 5% DMSO, and sequence-specific primers (5′-GGCATGTATTTTGGTGAAGTCG-3′ and 5′-GTTGT CGTAGTACGAATTGTACCCGAC-3′) under the following conditions: 94 °C for 2 min, and 30 cycles of 94 °C for 20 s, 60 °C for 20 s, and 68 °C for 20 s. This enabled successful amplification, since the region was enriched in guanine and cytosine. PCR products were separated via electrophoresis on 15% acrylamide gel in 1× TBE buffer. The expected sizes of the PCR products were 170 bp for *Hd1* and 130 bp for *hd1*.

4.4. Field Experiments and Agricultural Traits

All rice lines were sown and transplanted on the same day, with a spacing of 20 cm between plants and 25 cm between rows. A total of 35 plants each of *ss2a Hd1* and *ss2a Hd1* genotypes, 100 plants of the *ss2a Hd1 hd1* genotype, and 20 plants each of the parental lines were grown according to the local agricultural practices. Heading date was recorded when 50% of plants of a given genotype initiated heading. Maturation date was recorded when 90% of the panicles turned yellow. Plant height and ear length were measured prior to harvesting. After 2 weeks of desiccation, whole-plant dry weight and total grain weight were measured, and dry straw weight was calculated by subtracting the total grain weight from the whole-plant dry weight. Total grain weight was measured including empty seeds. Grains were dehulled and sieved through a mesh with 1.9 mm pore size using a sieving machine, TEST Grain Selector (TWSB, Satake, Tokyo, Japan), and the weight of grains above 1.9 mm thickness and width was measured as total dehulled grain weight. Fertility rate was calculated by counting and subtracting the number of empty seeds from the total number of seeds. Quality of brown rice was analyzed using the VIRGO Rice Grain Selector (ES-V; Shizuoka Seiki, Shizuoka, Japan) by detecting the green premature seeds. Data were obtained in 2020 and 2021.

4.5. Meteorological Data

Meteorological data for 2021 were obtained from the Japan Meteorological Agency. Daily temperature data were extracted, and average temperature during seed development was calculated.

4.6. Western Blot Analysis

Three mature seeds of each rice genotype were ground to a fine powder, and total protein was extracted using 20 volumes (*w*/*v*) of buffer containing 125 mM Tris-HCl (pH 6.8), 8 M urea, 4% (*w*/*v*) SDS, 5% (*v*/*v*) β-mercaptoethanol, and 0.05% (*w*/*v*) bromophenol blue. After centrifugation, proteins in the supernatants were subjected to sodium dodecyl sulfate-polyacrylamide gel electrophoresis (SDS-PAGE) on 7.5% acrylamide gel and blotted onto a membrane. Membranes were incubated with the following primary antibodies: anti-SSI (1:3000 dilution [59]), anti-SSIIa (1:1000 dilution [60], anti-GBSSI (1:5000 [59]), and anti-BEIIb (1:5000 [35]). Subsequently, secondary antibody incubation and protein detection were performed as described previously [60].

4.7. Measurement of Apparent Amylose Content and Short to Long Chain Amylopectin Ratio

Starch was purified using the cold-alkaline method as described previously [61,62]. Purified starch was debranched using *Pseudomonas* isoamylase (Hayashibara, Okayama, Japan) and analyzed via gel filtration chromatography (Toyopearl HW-55S and HW-50S×3; Tosoh, Tokyo, Japan) [63–65]. Amylose (fraction I), long amylopectin chains (fraction II), short amylopectin chains (fraction III), and apparent amylose content were quantified as described previously [63–65].

4.8. Analysis of Amylopectin Structure

Debranched purified starch was fluorescently labeled and analyzed via capillary electrophoresis (P/ACE MDQ Plus Carbohydrate System; AB Sciex, Framingham, MA, USA), as described [66].

4.9. Measurement of Gelatinization Temperature

The thermal properties of purified starch were analyzed via differential scanning calorimetry (Seiko Instrument 6100; Seiko, Chiba, Japan) as described previously [59,67].

5. Conclusions

This study precisely evaluated the agricultural traits and starch properties of rice NILs (*ss2a hd1*, *ss2a Hd1 hd1*, and *ss2a Hd1*) lacking SSIIa and showing different heading dates, although the data were limited to two harvest years. These NILs were generated by crossing the elite rice cultivar Akita 63 (as the recurrent parent) with the *ss2a* null mutant EM204. Sequencing analyses revealed that Akita 63 carries a loss-of-function *hd1* allele, while Kinmaze (the parental line of EM204) possesses a functional *Hdl* allele. The *ss2a hd1* NIL was the first to initiate heading (early August), while the heading dates of *ss2a Hd1 hd1* and *ss2a Hd1* were approximately 2 and 4 weeks later, respectively, than that of *ss2a hd1*. The time required to reach maturity was the shortest in *ss2a hd1*, which reached maturation in mid-September, while the harvesting dates of *ss2a Hd1 hd1* and *ss2a Hd1* were approximately 4 and 6 weeks later, respectively, than that of *ss2a hd1*. Although *ss2a hd1* showed the lowest whole-plant dry weight, it also produced the lowest number of green immature seeds. Analyses of starch properties showed that the amylose content of *ss2a hd1* was lower than that of *ss2a Hd1*, but its gelatinization temperature was higher. Overall, this study provides useful information about the different heading dates, agricultural traits, and starch properties of rice accessions.

Supplementary Materials: The following supporting information can be downloaded at: https://www.mdpi.com/article/10.3390/ijms231810783/s1.

Author Contributions: Conceptualization, N.C. and N.F.; methodology, N.C. and K.H.; resources, N.C., S.M., N.F.O. and Y.H.; writing—original draft preparation, N.C.; writing—review and editing, N.C. and N.F.; supervision, N.F.; project administration, N.F.; funding acquisition, N.C. and N.F. All authors have read and agreed to the published version of the manuscript.

Funding: This research was funded by the Science and Technology Research Promotion Program for Agriculture, Forestry and Fisheries and Food Industry (25033AB and 28029C awarded to N.F.), the President's Funds of Akita Prefectural University (N.F. and N.C.), Grant-in-Aid for JSPS fellows from Japan Society for the Promotion of Science (#15J40176 and JP18J40020 awarded to N.C.), and Japan Society for the Promotion of Science (#16K18571, JP18K14438, and 20K05961 awarded to N.C.; and 19H01608 awarded to N.F.).

Institutional Review Board Statement: Not applicable.

Informed Consent Statement: Not applicable.

Data Availability Statement: The sequence data generated in this study are freely available from the NCBI GenBank database (accession numbers: MK449352.1, MK449351.1, MK449350.1).

Acknowledgments: The authors thank Yuko Nakaizumi (Akita Prefectural University) for growing the rice plants. The authors also thank Toshihiro Kumamaru for providing EM204. *Pseudomonas* isoamylase, used for debranching amylopectin, was a kind gift from Hayashibara Co., Ltd.

Conflicts of Interest: The authors declare no conflict of interest.

References

1. Wu, C.C.; Wei, F.J.; Chiou, W.Y.; Tsai, Y.C.; Wu, H.P.; Gotarkar, D.; Wei, Z.H.; Lai, M.H.; Hsing, Y.C. Studies of rice *Hd1* haplotypes worldwide reveal adaptation of flowering time to different environments. *PLoS ONE* **2020**, *15*, e0239028. [CrossRef] [PubMed]
2. Izawa, T. Adaptation of flowering-time by natural and artificial selection in Arabidopsis and rice. *J. Exp. Bot.* **2007**, *58*, 3091–3097. [CrossRef] [PubMed]
3. Kim, S.R.; Torollo, G.; Yoon, M.R.; Kwak, J.; Lee, C.K.; Prahalada, G.D.; Choi, I.R.; Yeo, U.S.; Jeong, O.Y.; Jena, K.K.; et al. Loss-of-Function Alleles of *Heading date 1* (*Hd1*) are associated with adaptation of temperate *japonica* rice plants to the tropical region. *Front. Plant Sci.* **2018**, *9*, 1827. [CrossRef] [PubMed]
4. Yano, M.; Katayose, Y.; Ashikari, M.; Yamanouchi, U.; Monna, L.; Fuse, T.; Baba, T.; Yamamoto, K.; Umehara, Y.; Nagamura, Y.; et al. *Hd1*, a major photoperiod sensitivity quantitative trait locus in rice, is closely related to the Arabidopsis flowering time gene *CONSTANS*. *Plant Cell* **2000**, *12*, 2473–2484. [CrossRef] [PubMed]
5. Tamaki, S.; Matsuo, S.; Wong, H.L.; Yokoi, S.; Shimamoto, K. Hd3a protein is a mobile flowering signal in rice. *Science* **2007**, *316*, 1033–1036. [CrossRef]

6. Matsubara, K.; Kono, I.; Hori, K.; Nonoue, Y.; Ono, N.; Shomura, A.; Mizubayashi, T.; Yamamoto, S.; Yamanouchi, U.; Shirasawa, K.; et al. Novel QTLs for photoperiodic flowering revealed by using reciprocal backcross inbred lines from crosses between japonica rice cultivars. *Theor. Appl. Genet.* **2008**, *117*, 935–945. [CrossRef]
7. Nonoue, Y.; Fujino, K.; Hirayama, Y.; Yamanouchi, U.; Lin, S.Y.; Yano, M. Detection of quantitative trait loci controlling extremely early heading in rice. *Theor. Appl. Genet.* **2008**, *116*, 715–722. [CrossRef]
8. Takahashi, Y.; Teshima, K.M.; Yokoi, S.; Innan, H.; Shimamoto, K. Variations in Hd1 proteins, Hd3a promoters, and Ehd1 expression levels contribute to diversity of flowering time in cultivated rice. *Proc. Natl. Acad. Sci. USA* **2009**, *106*, 4555–4560. [CrossRef]
9. Matsubara, K.; Yamanouchi, U.; Nonoue, Y.; Sugimoto, K.; Wang, Z.X.; Minobe, Y.; Yano, M. *Ehd3*, encoding a plant homeodomain finger-containing protein, is a critical promoter of rice flowering. *Plant J.* **2011**, *66*, 603–612. [CrossRef]
10. Taoka, K.; Ohki, I.; Tsuji, H.; Furuita, K.; Hayashi, K.; Yanase, T.; Yamaguchi, M.; Nakashima, C.; Purwestri, Y.A.; Tamaki, S.; et al. 14-3-3 proteins act as intracellular receptors for rice Hd3a florigen. *Nature* **2011**, *476*, 332–335. [CrossRef]
11. Nemoto, Y.; Nonoue, Y.; Yano, M.; Izawa, T. *Hd1*, a CONSTANS ortholog in rice, functions as an Ehd1 repressor through interaction with monocot-specific CCT-domain protein Ghd7. *Plant J.* **2016**, *86*, 221–233. [CrossRef] [PubMed]
12. Hori, K.; Matsubara, K.; Yano, M. Genetic control of flowering time in rice: Integration of Mendelian genetics and genomics. *Theor. Appl. Genet.* **2016**, *129*, 2241–2252. [CrossRef] [PubMed]
13. Song, S.; Chen, Y.; Liu, L.; Wang, Y.; Bao, S.; Zhou, X.; Teo, Z.W.; Mao, C.; Gan, Y.; Yu, H. OsFTIP1-mediated regulation of florigen transport in rice is negatively regulated by the ubiquitin-like domain kinase OsUbDKγ4. *Plant Cell* **2017**, *29*, 491–507. [CrossRef] [PubMed]
14. Du, A.; Tian, W.; Wei, M.; Yan, W.; He, H.; Zhou, D.; Huang, X.; Li, S.; Ouyang, X. The DTH8-Hd1 Module Mediates Day-Length-Dependent Regulation of Rice Flowering. *Mol. Plant.* **2017**, *10*, 948–961. [CrossRef] [PubMed]
15. Zhu, C.; Peng, Q.; Fu, D.; Zhuang, D.; Yu, Y.; Duan, M.; Xie, W.; Cai, Y.; Ouyang, Y.; Lian, X.; et al. The E3 ubiquitin ligase HAF1 modulates circadian accumulation of EARLY FLOWERING3 to control heading date in rice under long-day conditions. *Plant Cell* **2018**, *30*, 2352–2367. [CrossRef]
16. Kojima, S.; Takahashi, Y.; Kobayashi, Y.; Monna, L.; Sasaki, T.; Araki, T.; Yano, M. *Hd3a*, a rice ortholog of the Arabidopsis FT gene, promotes transition to flowering downstream of *Hd1* under short-day conditions. *Plant Cell Physiol.* **2002**, *43*, 1096–1105. [CrossRef] [PubMed]
17. Takahashi, Y.; Shimamoto, K. *Heading date 1* (*Hd1*), an ortholog of Arabidopsis CONSTANS, is a possible target of human selection during domestication to diversify flowering times of cultivated rice. *Genes Genet. Syst.* **2011**, *86*, 175–182. [CrossRef]
18. Mo, Y.; Lee, C.M.; Park, H.M.; Ha, S.K.; Kim, M.J.; Kwak, J.; Lee, H.S.; Lee, J.H.; Jeung, J.U. *Hd1* allele types and their associations with major agronomic traits in Korean rice cultivars. *Plants* **2021**, *10*, 2408. [CrossRef]
19. Hizukuri, S. Polymodal distribution of the chain lengths of amylopectins, and its significance. *Carbohydr. Res.* **1986**, *147*, 342–347. [CrossRef]
20. Takeda, Y.; Hizukuri, S.; Takeda, C.; Suzuki, A. Structures of branched molecules of amyloses of various origins, and molar fractions of branched and unbranched molecules. *Carbohydr. Res.* **1987**, *165*, 139–145. [CrossRef]
21. Jane, J.; Chen, Y.Y.; Lee, L.F.; Mcpherson, A.E.; Wong, K.S.; Radosavljevic, M.; Kasemsuwan, T. Effects of amylopectin branch chain length and amylose content on the gelatinization and pasting properties of starch. *Cereal. Chem.* **1999**, *76*, 629–637. [CrossRef]
22. Tanaka, N.; Fujita, N.; Nishi, A.; Satoh, H.; Hosaka, Y.; Ugaki, M.; Kawasaki, S.; Nakamura, Y. The structure of starch can be manipulated by changing the expression levels of starch branching enzyme IIb in rice endosperm. *Plant Biotechnol. J.* **2004**, *2*, 507–516. [CrossRef] [PubMed]
23. Xu, Y.; Lin, Q.; Li, X.; Wang, F.; Chen, Z.; Wang, J.; Li, W.; Fan, F.; Tao, Y.; Jiang, Y.; et al. Fine-tuning the amylose content of rice by precise base editing of the *Wx* gene. *Plant Biotechnol. J.* **2021**, *19*, 11–13. [CrossRef] [PubMed]
24. Sano, Y. Differential regulation of *waxy* gene expression in rice endosperm. *Theor. Appl. Genet.* **1984**, *68*, 467–473. [CrossRef]
25. Wang, Z.Y.; Zheng, F.Q.; Shen, G.Z.; Gao, J.P.; Snustad, D.P.; Li, M.G.; Zhang, J.L.; Hong, M.M. The amylose content in rice endosperm is related to the post-transcriptional regulation of the waxy gene. *Plant J.* **1995**, *7*, 613–622. [CrossRef] [PubMed]
26. Cai, X.L.; Wang, Z.Y.; Xing, Y.Y.; Zhang, J.L.; Hong, M.M. Aberrant splicing of intron 1 leads to the heterogeneous 5′ UTR and decreased expression of waxy gene in rice cultivars of intermediate amylose content. *Plant J.* **1998**, *14*, 459–465. [CrossRef]
27. Isshiki, M.; Morino, K.; Nakajima, M.; Okagaki, R.J.; Wessler, S.R.; Izawa, T.; Shimamoto, K.A. Naturally occurring functional allele of the rice waxy locus has a GT to TT mutation at the 5′ splice site of the first intron. *Plant J.* **1998**, *15*, 133–138. [CrossRef]
28. Hirano, H.Y.; Sano, Y. Enhancement of *Wx* gene expression and the accumulation of amylose in response to cool temperatures during seed development in rice. *Plant Cell Physiol.* **1998**, *39*, 807–812. [CrossRef]
29. Larkin, P.D.; Park, W.D. Transcript accumulation and utilization of alternate and non-consensus splice sites in rice granule-bound starch synthase are temperature-sensitive and controlled by a single-nucleotide polymorphism. *Plant Mol. Biol.* **1999**, *40*, 719–727. [CrossRef]
30. Zhang, H.; Duan, L.; Dai, J.S.; Zhang, C.Q.; Li, J.; Gu, M.H.; Liu, Q.Q.; Zhu, Y. Major QTLs reduce the deleterious effects of high temperature on rice amylose content by increasing splicing efficiency of *Wx* pre-mRNA. *Theor. Appl. Genet.* **2014**, *27*, 273–282. [CrossRef]
31. Fujita, N. Starch biosynthesis in rice endosperm. *AGri-Biosci. Monogr.* **2014**, *4*, 1–18. [CrossRef]

32. Crofts, N.; Nakamura, Y.; Fujita, N. Critical and speculative review of the roles of multi-protein complexes in starch biosynthesis in cereals. *Plant Sci.* **2017**, *262*, 1–8. [CrossRef] [PubMed]
33. Nakamura, Y.; Francisco, P.B., Jr.; Hosaka, Y.; Sato, A.; Sawada, T.; Kubo, A.; Fujita, N. Essential amino acids of starch synthase IIa differentiate amylopectin structure and starch quality between japonica and indica rice varieties. *Plant Mol. Biol.* **2005**, *58*, 213–227. [CrossRef] [PubMed]
34. Nakamura, Y. Towards a better understanding of the metabolic system for amylopectin biosynthesis in plants: Rice endosperm as a model tissue. *Plant Cell Physiol.* **2002**, *43*, 718–725. [CrossRef] [PubMed]
35. Nishi, A.; Nakamura, Y.; Tanaka, N.; Satoh, H. Biochemical and genetic analysis of the effects of *amylose-extender* mutation in rice endosperm. *Plant Physiol.* **2001**, *127*, 459–472. [CrossRef]
36. Butardo, V.M.; Fitzgerald, M.A.; Bird, A.R.; Gidley, M.J.; Flanagan, B.M.; Larroque, O.; Resurreccion, A.P.; Laidlaw, H.K.; Jobling, S.A.; Morell, M.K.; et al. Impact of down-regulation of starch branching enzyme IIb in rice by artificial microRNA- and hairpin RNA-mediated RNA silencing. *J. Exp. Bot.* **2011**, *.62*, 4927–4941. [CrossRef] [PubMed]
37. Nakata, M.; Miyashita, T.; Kimura, R.; Nakata, Y.; Takagi, H.; Kuroda, M.; Yamaguchi, T.; Umemoto, T.; Yamakawa, H. MutMapPlus identified novel mutant alleles of a rice starch branching enzyme IIb gene for fine-tuning of cooked rice texture. *Plant Biotechnol. J.* **2018**, *16*, 111–123. [CrossRef]
38. Sun, Y.; Jiao, G.; Liu, Z.; Zhang, X.; Li, J.; Guo, X.; Du, W.; Du, J.; Francis, F.; Zhao, Y.; et al. Generation of high-amylose rice through CRISPR/Cas9-mediated targeted mutagenesis of starch branching enzymes. *Front. Plant Sci.* **2017**, *8*, 298. [CrossRef]
39. Jiang, H.; Dian, W.; Wu, P. Effect of high temperature on fine structure of amylopectin in rice endosperm by reducing the activity of the starch branching enzyme. *Phytochemistry* **2003**, *63*, 53–59. [CrossRef]
40. Yamakawa, H.; Hirose, T.; Kuroda, M.; Yamaguchi, T. Comprehensive expression profiling of rice grain filling-related genes under high temperature using DNA microarray. *Plant Physiol.* **2007**, *144*, 258–277. [CrossRef]
41. Ohdan, T.; Sawada, T.; Nakamura, Y. Effects of temperature on starch branching enzyme properties of rice. *J. Appl. Glycosci.* **2011**, *58*, 19–26. [CrossRef]
42. Kato, K.; Suzuki, Y.; Hosaka, Y.; Takahashi, R.; Kodama, I.; Sato, K.; Kawamoto, T.; Kumamaru, T.; Fujita, N. Effect of high temperature on starch biosynthetic enzymes and starch structure in japonica rice cultivar 'Akitakomachi' (*Oryza sativa* L.) endosperm and palatability of cooked rice. *J. Cereal Sci.* **2019**, *87*, 209–214. [CrossRef]
43. Miura, S.; Crofts, N.; Saito, Y.; Hosaka, Y.; Oitome, N.F.; Watanabe, T.; Kumamaru, T.; Fujita, N. Starch synthase IIa-deficient mutant rice line produces endosperm starch with lower gelatinization temperature than Japonica rice cultivars. *Front. Plant Sci.* **2018**, *9*, 645. [CrossRef]
44. Makino, A.; Kaneta, Y.; Obara, M.; Ishiyama, K.; Kanno, K.; Kondo, E.; Suzuki, Y.; Mae, T. High yielding ability of a large-grain rice cultivar, Akita 63. *Sci. Rep.* **2020**, *10*, 12231. [CrossRef] [PubMed]
45. Takeuchi, Y. Developing Isogenic Lines of Japanese Rice Cultivar 'Koshihikari' with Early and Late Heading. *JARQ* **2011**, *45*, 15–22. [CrossRef]
46. Takeuchi, Y.; Ando, I.; Nemoto, H.; Kato, H.; Hirabayashi, H.; Ohta, H.; Ishii, T.; Maeda, H.; Takemoto-kuno, Y.; Imbe, T.; et al. Breeding of "Milky Summer", an isogenic line of the rice cultivar "Milky Queen" with modified heading. *Bull. NARO Inst. Crop. Sci.* **2013**, *14*, 77–95.
47. Shibata, M. Present conditions and subjects of rice breeding for cold-tolerance in Japan. *JARQ* **1970**, *5*, 4.
48. Satake, T. Determination of the most sensitive stage to sterile-type cool injury in rice plants. *Res. Bull. Hokkaido Natl. Agric. Exp. Stn.* **1976**, *113*, 1–43.
49. Endo, M.; Tsuchiya, T.; Hamada, K.; Kawamura, S.; Yano, K.; Ohshima, M.; Higashitani, A.; Watanabe, M.; Kawagishi-Kobayashi, M. High temperatures cause male sterility in rice plants with transcriptional alterations during pollen development. *Plant Cell Physiol.* **2009**, *50*, 1911–1922. [CrossRef]
50. Zhou, L.J.; Sheng, W.T.; Wu, J.; Zhang, C.Q.; Liu, Q.q.; Deng, Q.Y. Differential expressions among five *Waxy* alleles and their effects on the eating and cooking qualities in specialty rice cultivars. *J. Integr. Agric.* **2015**, *14*, 1153–1162. [CrossRef]
51. Shi, S.; Wang, E.; Li, C.; Cai, M.; Cheng, B.; Cao, C.; Jiang, Y. Use of protein content, amylose content, and RVA parameters to evaluate the taste quality of rice. *Front. Nutr.* **2022**, *13*, 8. [CrossRef]
52. Miura, S.; Koyama, N.; Crofts, N.; Hosaka, Y.; Abe, M.; Fujita, N. Generation and starch characterization of non-transgenic BEI and BEIIb double mutant rice (*Oryza sativa*) with ultra-high level of resistant starch. *Rice* **2021**, *14*, 3. [CrossRef] [PubMed]
53. Fujita, N. Manipulation of rice starch properties for application. In *Starch*; Nakamura, Y., Ed.; Springer: Tokyo, Japan, 2015; pp. 335–369.
54. Hayashi, M.; Kodama, M.; Nakamura, Y.; Fujita, N. Thermal and pasting properties, morphology of starch granules, and crystallinity of endosperm starch in the rice SSI and SSIIIa double-mutant. *J. Appl. Glycosci.* **2015**, *62*, 81–86. [CrossRef]
55. Tsuiki, K.; Fujisawa, H.; Itoh, A.; Sato, M.; Fujita, N. Alterations of starch structure lead to increased resistant starch of steamed rice: Identification of high resistant starch rice lines. *J. Cereal Sci.* **2016**, *68*, 88–92. [CrossRef]
56. Nagamatsu, S.; Wada, T.; Matsushima, R.; Fujita, N.; Miura, S.; Crofts, N.; Hosaka, Y.; Yamaguchi, O.; Kumamaru, T. Mutation in BEIIb mitigates the negative effect of the mutation in ISA1 on grain filling and amyloplast formation in rice. *Plant Mol. Biol.* **2022**, *108*, 497–512. [CrossRef] [PubMed]

57. Gao, H.; Jin, M.; Zheng, X.M.; Chen, J.; Yuan, D.; Xin, Y.; Wang, M.; Huang, D.; Zhang, Z.; Zhou, K.; et al. Days to heading 7, a major quantitative locus determining photoperiod sensitivity and regional adaptation in rice. *Proc. Natl. Acad. Sci. USA* **2014**, *111*, 16337–16342. [CrossRef]
58. Lee, S.J.; Kang, K.; Lim, J.H.; Paek, N.C. Natural alleles of *CIRCADIAN CLOCK ASSOCIATED1* contribute to rice cultivation by fine-tuning flowering time. *Plant Physiol.* **2022**, *190*, 640–656. [CrossRef]
59. Fujita, N.; Yoshida, M.; Asakura, N.; Ohdan, T.; Miyao, A.; Hirochika, H.; Nakamura, Y. Function and characterization of starch synthase I using mutants in rice. *Plant Physiol.* **2006**, *140*, 1070–1084. [CrossRef]
60. Crofts, N.; Abe, N.; Oitome, N.F.; Matsushima, R.; Hayashi, M.; Tetlow, I.J.; Emes, M.J.; Nakamura, Y.; Fujita, N. Amylopectin biosynthetic enzymes from developing rice seed form enzymatically active protein complexes. *J. Exp. Bot.* **2015**, *66*, 4469–4482. [CrossRef]
61. Yamamoto, K.; Sawada, S.; Onogaki, T. Properties of rice prepared by alkali method with various conditions. *J. Jpn. Soc. Starch. Sci.* **1973**, *20*, 99–104. [CrossRef]
62. Yamamoto, K.; Sawada, S.; Onogaki, T. Effects of quality and quantity of alkaline solution on the properties of the rice starch. *J. Jap. Soc. Starch Sci.* **1981**, *28*, 241–244. [CrossRef]
63. Horibata, T.; Nakamoto, M.; Fuwa, H.; Inouchi, N. Structural and physicochemical characteristics of endosperm starches of rice cultivars recently bred in Japan. *J. Appl. Glycosci.* **2004**, *51*, 303–313. [CrossRef]
64. Fujita, N.; Yoshida, M.; Kondo, T.; Saito, K.; Utsumi, Y.; Tokunaga, Y.; Nishi, A.; Satoh, H.; Park, J.H.; Jane, J.L.; et al. Characterization of SSIIIa-deficient mutants of rice: The function of SSIIIa and pleiotropic effects by SSIIIa deficiency in the rice endosperm. *Plant Physiol.* **2007**, *144*, 2009–2023. [CrossRef] [PubMed]
65. Toyosawa, Y.; Kawagoe, Y.; Matsushima, R.; Crofts, N.; Ogawa, M.; Fukuda, M.; Kumamaru, T.; Okazaki, Y.; Kusano, M.; Saito, K.; et al. Deficiency of starch synthase IIIa and IVb alters starch granule morphology from polyhedral to spherical in rice endosperm. *Plant Physiol.* **2016**, *170*, 1255–1270. [CrossRef] [PubMed]
66. Fujita, N.; Hasegawa, H.; Taira, T. The isolation and characterization of *waxy* mutant of diploid wheat (*Triticum monococcum* L.). *Plant Sci.* **2001**, *160*, 595–602. [CrossRef]
67. Fujita, N.; Kubo, A.; Suh, D.; Wong, K.; Jane, J.; Ozawa, K.; Takaiwa, F.; Inaba, Y.; Nakamura, Y. Antisense inhibition of isoamylase alters the structure of amylopectin and the physicochemical properties of starch in rice endosperm. *Plant Cell Physiol.* **2003**, *44*, 607–618. [CrossRef] [PubMed]

International Journal of
Molecular Sciences

Article

The Function of DNA Demethylase Gene ROS1a Null Mutant on Seed Development in Rice (*Oryza Sativa*) Using the CRISPR/CAS9 System

Faiza Irshad [1,†], Chao Li [1,2,†], Hao-Yu Wu [1,†], Yan Yan [1] and Jian-Hong Xu [1,2,3,*]

1 Zhejiang Key Laboratory of Crop Germplasm, Institute of Modern Seed Industry, Zhejiang University, Hangzhou 310058, China; faizairshad2012@yahoo.com (F.I.); 0621121@zju.edu.cn (C.L.); why1625@zju.edu.cn (H.-Y.W.); yanyan9569@126.com (Y.Y.)
2 Shandong (Linyi) Institute of Modern Agriculture, Zhejiang University, Linyi 276000, China
3 Hainan Institute, Zhejiang University, Sanya 572025, China
* Correspondence: jhxu@zju.edu.cn
† These authors contributed equally to this work.

Abstract: The endosperm is the main nutrient source in cereals for humans, as it is a highly specialized storage organ for starch, lipids, and proteins, and plays an essential role in seed growth and development. Active DNA demethylation regulates plant developmental processes and is ensured by cytosine methylation (5-meC) DNA glycosylase enzymes. To find out the role of *OsROS1a* in seed development, the null mutant of *OsROS1a* was generated using the CRISPR/Cas9 system. The null mutant of *OsROS1a* was stable and heritable, which affects the major agronomic traits, particularly in rice seeds. The null mutant of *OsROS1a* showed longer and narrower grains, and seeds were deformed containing an underdeveloped and less-starch-producing endosperm with slightly irregularly shaped embryos. In contrast to the transparent grains of the wild type, the grains of the null mutant of *OsROS1a* were slightly opaque and rounded starch granules, with uneven shapes, sizes, and surfaces. A total of 723 differential expression genes (DEGs) were detected in the null mutant of *OsROS1a* by RNA-Seq, of which 290 were downregulated and 433 were upregulated. The gene ontology (GO) terms with the top 20 enrichment factors were visualized for cellular components, biological processes, and molecular functions. The key genes that are enriched for these GO terms include starch synthesis genes (*OsSSIIa* and *OsSSIIIa*) and cellulose synthesis genes (*CESA2*, *CESA3*, *CESA6*, and *CESA8*). Genes encoding polysaccharides and glutelin were found to be downregulated in the mutant endosperm. The glutelins were further verified by SDS-PAGE, suggesting that glutelin genes could be involved in the null mutant of *OsROS1a* seed phenotype and *OsROS1a* could have the key role in the regulation of glutelins. Furthermore, 378 differentially alternative splicing (AS) genes were identified in the null mutant of *OsROS1a*, suggesting that the *OsROS1a* gene has an impact on AS events. Our findings indicated that the function on rice endosperm development in the null mutant of *OsROS1a* could be influenced through regulating gene expression and AS, which could provide the base to properly understand the molecular mechanism related to the *OsROS1a* gene in the regulation of rice seed development.

Keywords: rice (*Oryza sativa*); *OsROS1a*; CRISPR/Cas9; seed storage protein; starch; RNA-Seq; alternative splicing

check for updates

Citation: Irshad, F.; Li, C.; Wu, H.-Y.; Yan, Y.; Xu, J.-H. The Function of DNA Demethylase Gene ROS1a Null Mutant on Seed Development in Rice (*Oryza Sativa*) Using the CRISPR/CAS9 System. *Int. J. Mol. Sci.* **2022**, *23*, 6357. https://doi.org/10.3390/ijms23126357

Academic Editor: Yong-Gu Cho

Received: 18 May 2022
Accepted: 4 June 2022
Published: 7 June 2022

Publisher's Note: MDPI stays neutral with regard to jurisdictional claims in published maps and institutional affiliations.

Copyright: © 2022 by the authors. Licensee MDPI, Basel, Switzerland. This article is an open access article distributed under the terms and conditions of the Creative Commons Attribution (CC BY) license (https://creativecommons.org/licenses/by/4.0/).

1. Introduction

The development of seeds is an essential process in the angiosperm life cycle. It involves embryo and endosperm development. The rice endosperm provides energy and materials for seed germination and development, which contains seed storage proteins (SSPs), starch, lipids, and additional trace substances, and occupies most of the space. The production and quality of seed is directly determined by endosperm development at the filling stage. Amino acids, sugars, and other key metabolites' storage are important for the development of rice endosperm and affect the quality and milling yield [1]. These metabolites are distributed to numerous biosynthetic pathways, mainly the metabolism of starch and the biosynthesis and storage of proteins. In addition, it is also accountable for proteins and starch synthesis in defined amounts and ratios [2].

Rice (*Oryza sativa*) is an excellent material for studying the biosynthesis of SSPs, as it is one of the few plants that synthesize and accumulate both major classes of SSPs, i.e., prolamins and glutelins. The glutelins account for more than 60% of total SSPs in rice, which are encoded by 15 genes, while prolamins are 20% to 30% of total SSPs and are encoded by 34 genes [3,4]. Based on the sequence similarity of the amino acid, glutelins can be divided into four groups (GluA, GluB, GluC, and GluD) [3], and prolamins into three types, 10-, 13-, and 16-kDa [4]. Glutelins are a form of precursor protein (proglutelin), and they synthesize in the endoplasmic reticulum (ER) and transfer the protein body II (PBII) into the protein storage vacuoles (PSV) through the Golgi apparatus [5,6]. Eventually, they are prepared into mature 20-kDa basic and 37-kDa acidic subunits, which are linked by disulfide bonds [7].

The DNA methylation profile study revealed that preferentially expressed genes in the endosperm are mainly coding for major SSPs and starch synthesizing enzymes, and the most important mechanisms for gene activation in rice endosperm are CG and CHG hypomethylation [8]. DNA methylation, an evolutionarily conserved epigenetic mechanism, controls many biological processes such as the imprinting of genes, expression of the tissue-specific genes, stress responses, and transposable elements' inactivation. The cytosine methylation (5-meC) occurs in three contexts (CG, CHG, and CHH, in which H represents A, T, or C) in plants, and is dynamically regulated by balanced methylation and demethylation [9]. In plants, the active form of DNA demethylation is initialized by the REPRESSOR OF SILENCING (ROS1) transglucosylase gene family, including ROS1, DEMETER (DME), DEMETER-like 2 (DML2), and DML3 [10–12]. They all are capable to excise 5-mC, regardless of whether the methylation is in CG, CHG, or CHH form [13–15]. Previous studies showed that the loss function of *OsROS1a* produced sterile rice through defects in both female and male gametogenesis [16,17]. *OsROS1a* demethylates both the vegetative cell genome and central cell genome that is vital for viable seed production [18]. The mutation of *OsROS1* generated a new transcript of 21-nt insertion, which increased the number of aleurone cell layers of rice seed by initializing hypermethylation of *rice seed beta-zipper1* (*RISBZ1*) and *rice prolamin-box binding factor* (*RPBF*) [19,20]. Furthermore, *ROS1* genes are also involved in the seed development of rice, wheat, and barley by epigenetic influence on the accumulation of SSPs [21,22].

Alternative splicing (AS) is a key regulatory mechanism that directly contributes to the structural and functional diversity of mRNA and proteins [23]. Advancements in high-throughput technology enabled a global analysis of AS, which has been widely discovered in plants, including rice [24], maize (*Zea mays*) [25,26], *Arabidopsis* (*Arabidopsis thaliana*) [27], cotton (*Gossypium raimondii*) [28], pineapple (*Ananas comosus*) [29], and soybean (*Glycine max*) [30]. Furthermore, stage-dependent AS events are probably significantly important, and considerably affect the grain yield significantly. One of the FLOWERING LOCUST (FT) homologues' gene *FT2* undergoes AS and results in two isoforms in brachypodium (*Brachypodium distachyon*), which work on different functions in the same flowering regulating pathway [31]. In rice, *OsbZIP74* mRNA can be alternatively spliced under treatment with ER stress-inducing agents, which is induced by heat stress, and involved in plant resistance against pathogens or parasites [32]. Based on RNA-Seq, a total of 16,995 AS events of lncRNAs were identified in tomato root, leaf, and flower tissues [33]. More than 1000 genes that experienced AS events were identified in the nitrogen-treated maize roots, and one of the transcription factor *ZmNLP6* isoforms were found to have the strong ability to activate downstream genes [34], suggesting that AS plays a vital role in plant growth and development. There has been emerging evidence showing that DNA methylation can regulate AS. Knockdown of DNA methyltransferase 3 (*Dnmt3*) in honey bees reduced global genomic methylation levels and induced global and diverse changes in AS in fat tissue [35]. The deficiency of DNA methylation in mouse embryonic stem cells has been proven to influence the splicing of more than 20% of alternative exons [36]. However, the effect of the DNA methylation pattern on AS has not been studied in rice.

To better define the role of *OsROS1a* in seed development, the null mutant of *OsROS1a* was generated using the CRISPR/Cas9 system, with 75-nt deletion and 1-nt substitution in the coding region, which destroy the permuted version of a methylated CpG-discriminating CXXC (Per-CXXC) domain, causing alteration in grain size. RNA-Seq analysis revealed that the polysaccharides and glutelin coding genes are downregulated in the 15 days after pollination (DAP) endosperm of the mutant, and 378 genes that experienced AS events were identified, indicating that the function of the rice endosperm development in the null mutant of *OsROS1a* could be influenced through gene expression and AS.

2. Results

2.1. The Phenotypes of the Null Mutant of OsROS1a

In order to well describe the role of *OsROS1a* in active DNA demethylation and seed development, we created null mutants of *OsROS1a*, S1 and S2 using the CRISPR/Cas9 system. The homozygous mutant, S1contains an in-frame deletion of 25 amino acid residues from the 1798th amino acid and an amino acid A to T substitution in the Per-CXXC domain, which could lead to the complete loss function of this domain, while the RNA recognition motif fold (RRMF) domain, started from the 1831st amino acid, was maintained (Figure 1). Another biallele gene-editing event was also obtained that has one allele with the same editing as S1 and the other with 86-nt deletion. As the frameshift mutation of *OsROS1a* produces sterile rice, the S2 mutant of T_1 generation has the same genotype as S1. Therefore, the S1 mutant was used for further analysis.

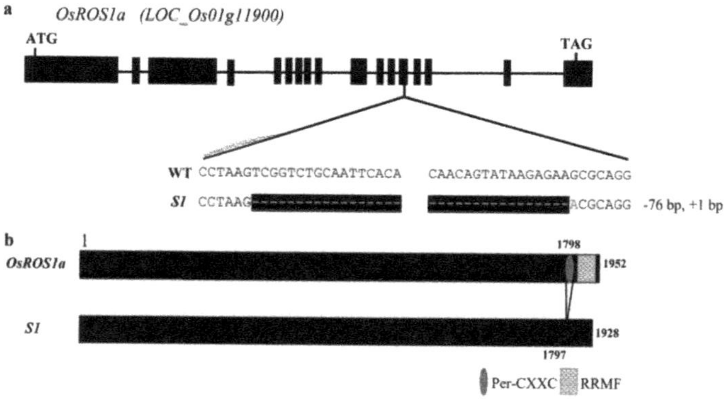

Figure 1. CRISPR/Cas9-induced *OsROS1a* gene editing. (**a**) Schematic of *OsROS1a* gene structure. Exons and introns are denoted as black blocks and lines, respectively. The translation initiation codon (ATG) and the termination codon (TAG) are shown. The recovered mutated allele is shown below the WT reference sequences. The target sites' nucleotides are indicated in black capital letters. The white dashes indicate the deleted nucleotides, and the substitution nucleotide A is shown in red. (**b**) The protein structure of WT and the null mutant S1. The predicted structure of WT contained the Per-CXXC domain and RRMF domain, while the Per-CXXC domain is not predicted inthe null mutant S1.

Microscopic analysis of rice anther showed that T_0 generation of the null mutant of *OsROS1a* displays partial male sterility with smaller anthers than WT (Figure 2a,g). The S1 mutant anthers had fewer pollen grains as compared to WT, and only few pollen grains were stained by a I_2-KI solution (Figure 2b,h). To further examine the cellular defects in the S1 mutant, transverse section analysis was performed on the WT and S1 mutant anthers, which were observed at four different developmental stages. At stage 6 (the microspore-mother-cells stage), epidermis, endothecium, tapetum, and the microspore mother cells were clearly visible, and no obvious difference was observed between WT and S1 mutant (Figure 2c,i). However, at stage 9 (the young microspore stage), the microspores in WT were globular, deeply stained, and densely distributed, while the microspores in the S1 mutant were lighter, scarcer, and more irregularly shaped (Figure 2d,j). At stage 10 (the vacuolated pollen stage), as compared to WT microspores that were round and vacuolated, half of the S1 mutant microspores had a normal round shape, comparable to WT, while the remaining microspores were degraded and irregularly shaped (Figure 2e,k). At stage 13 (the mature pollen stage), the WT anther locule was full of mature pollen grains with entirely formed pollen walls and accumulated starch. However, there were fewer microspores in the S1 mutant and some of them appeared to be degenerated and partially sterile (Figure 2f,l). These results are consistent with the pollen staining, indicating that the S1 mutant has partial pollen sterility.

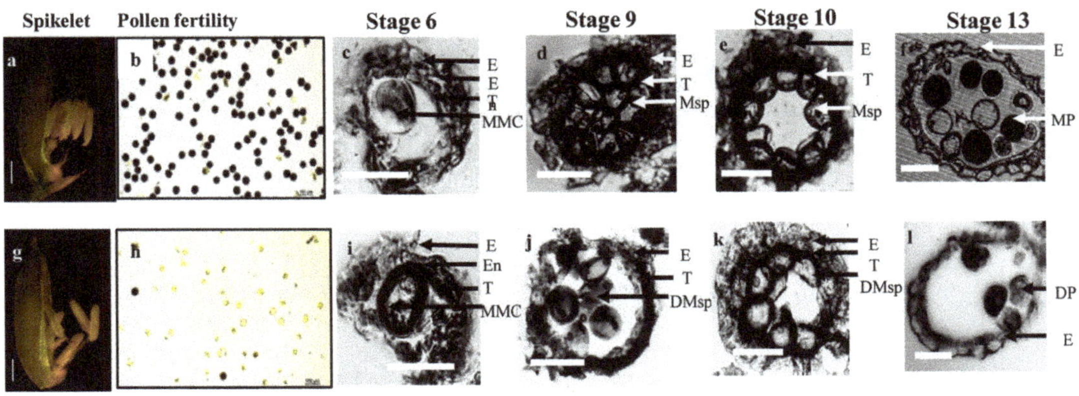

Figure 2. The phenotype of S1 mutant. The spikelet of WT (**a**) and S1 mutant (**g**). The pollen fertility of WT (**b**) and the S1 mutant (**h**) by iodine staining magnified 20×. Transverse section analysis of the anther development in WT (**c–f**) and the S1 mutant (**i–l**). Locules from the anther section of WT and at stage 6 (the microspore mother cells stage), stage 9 (the young microspore stage), Stage 10 (the vacuolated pollen stage), and stage 13 (the mature pollen stage). E, epidermis; En, endothecium; T, tapetum; MMC, microspore mother cells; Msp, microspores; MP, mature pollen; DMsp, degenerated microspores; DP, degenerated pollen. Scale bar: 50 μm.

T_1 generation of the S1 mutant plants were used to examine various agronomic traits. The plant height, panicle length, and the primary branch number were not significantly changed (Figure 3). The plant height was only reduced by 2.4% in the S1 mutant as compared to WT (Figure 3a,c). Similarly, in the case of panicle length, only a 12.8% reduction was observed in the S1 mutant as compared to WT (Figure 3b,e). Likewise, the number of primary branches was reduced by 12.5% (Figure 3f). However, the tiller number was significantly increased (43.9%) (Figure 3d) and the seed fertility percentage was significantly reduced (23.0%) in the null mutant of *OsROS1a* when compared to WT (Figure 3g).

The null mutant of *OsROS1a* significantly changed the grain size. The seed length was significantly increased by 5.8% and the seed width was significantly decreased by 4.8% in the S1 mutant as compared to WT (Figure 4). To know whether the grain shape and size can affect the seed storage materials, the transverse section analysis was performed for the dehusked mature grains, which involved cutting the seed transversely into two halves. No detectable difference was observed in the aleurone layer between seeds of the S1 mutant and WT. However, the null mutant S1 consisted of deformed seeds containing underdeveloped and less-starch-producing endosperms (Figure 5a), and the shape of embryos was slightly irregular when compared to WT. In contrast to the semitransparent grains of the WT, the S1 mutantgrains were slightly opaque (Figure 5b). The starch granules were further observed by SEM, and the starch granules of WT grains had sharp edges, flat surfaces, and compound and similarly sized polygonal granules. In contrast, the S1 mutant starch granules were variable in size and shape, with rounded and irregular surfaces (Figure 5c). Furthermore, the total starch content was decreased from 72.73% in WT to 69.58% in the S1 mutant, while the amylose content was increased from 11.24% in WT to 12.79% in the null mutant of *OsROS1a* (Figure 6).

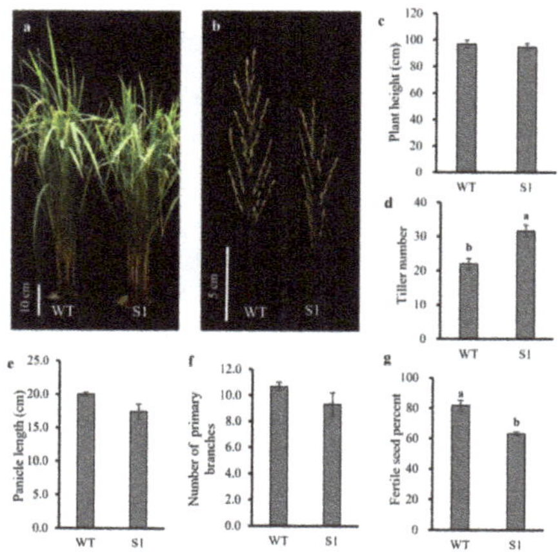

Figure 3. Phenotypes of WT and the S1 mutant. (**a**) Phenotype comparison of the WT and the S1 mutant(**b**) The panicle of WT and the S1 mutant. The statistic calculation of agronomic traits of (**c**) plant height, (**d**) tiller number, (**e**) panicle length, (**f**) number of primary branches, and (**g**) fertile seed percentage. Lettering indicates statistical significance at $p \leq 0.01$. Data are means ± standard deviation (SD) (n = 3).

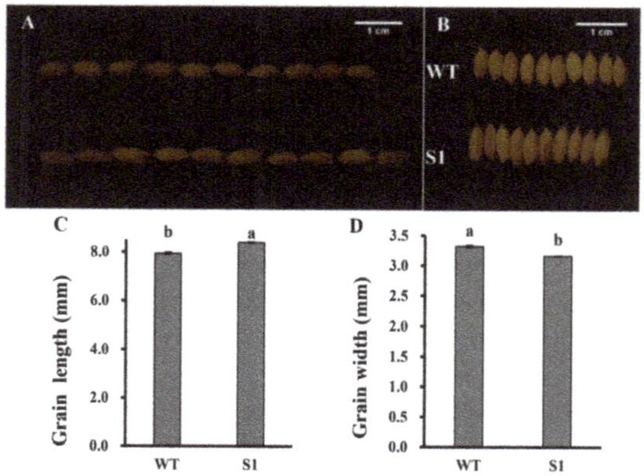

Figure 4. Comparison of seed length (**A**) and seed width (**B**), and the statistical analysis of seed length (**C**) and seed width (**D**) between WT and the S1 mutant. Lettering indicates statistical significance at $p \leq 0.01$. Data are means ± SD (n = 3).

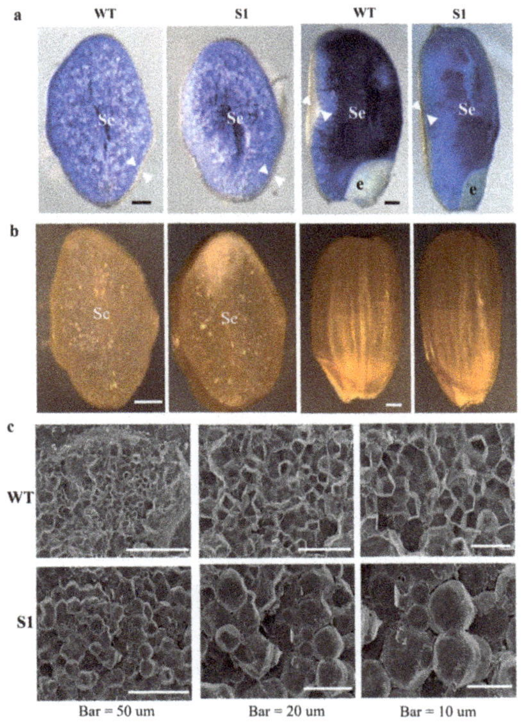

Figure 5. Genetic screening of *OsROS1a* gene editing mutant by half-seed assay. (**a**) Transversally and longitudinally sectioned of WT and the S1-mutant dehusked, mature grains, stained with the Evans blue dye. Arrowheads indicate the aleurone. se, starchy endosperm; e, embryo. (**b**) Transversely sectioned and nonsectioned mature grains showed the opaque endosperm in the S1 mutant, as compared to the semitransparent endosperm in WT. Scale bar, 0.5 mm. (**c**) Scanning electron micrographs of the endosperm in transverse sections are shown with increasing magnification from left to right. Scanning electron microscopy (SEM) of WT grain revealed similarly sized polygonal starch granules with sharp edges; smooth, flat surfaces; and compound starch granules while the S1 granules are rounded, variable in size and shape, and have uneven surfaces.

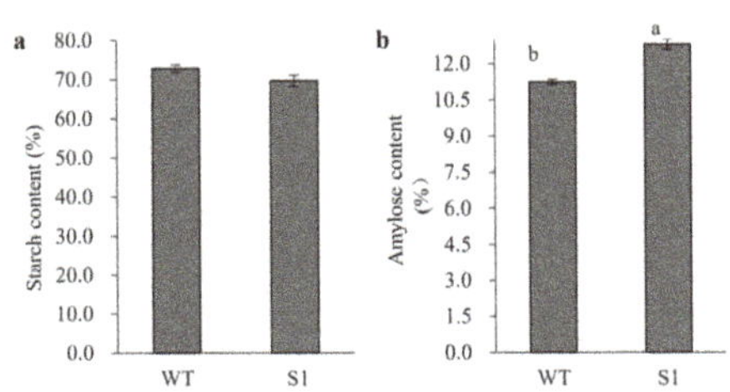

Figure 6. Total starch and amylose content in the mature seed of WT and the S1 mutant. (**a**) Total starch content in the seeds of WT and S1 mutant. (**b**) Total amylose content in seeds of WT and S1 mutant. Lettering indicates the statistical difference at $p \leq 0.01$. Data are means $\pm$ SD with three replicates.

2.2. RNA-Seq Identifies Responsive Genes in the Null Mutant of OsROS1a

Because the null mutant of *OsROS1a* can affect the rice grain, 15 DAP immature endosperms of the S1 mutant and WT were harvested for RNA-Seq. In total, more than 45,000,000 clean reads were generated for each sample, and the QC30 ratios were all above

94.2%, indicating the transcriptome was of high quality. The abundance of the expressed genes was then quantified using FPKM.

DEGs between WT and the S1 mutant were screened out based on the threshold of the $\log_2$ fold change being either ≥ 1 or ≤ -1 and a p-value < 0.05. Under such criteria, 723 DEGs were identified in total, and among them 290 (40.1%) genes were downregulated and 433 (59.9%) genes were upregulated in the S1 mutant (Figure 7). Furthermore, a GO term enrichment analysis was performed using agriGO [37]. The GO terms with the top 20 enrichment factors were observed, which consisted of eleven biological processes (BP), six molecular functions (MF), and three cellular components (CC). A total of ten out of eleven enriched BP terms are connected to polysaccharide synthetic. In CC, the GO term "macromolecular complex" (GO:0032991) was enriched, including ten glutelin genes (Figure 8).

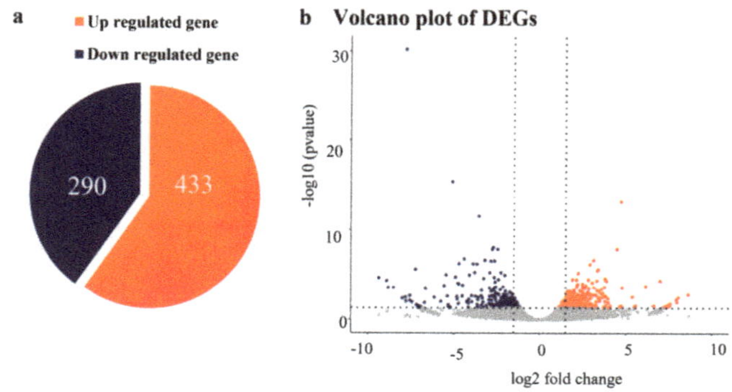

Figure 7. Differential expression of 15 DAP endosperm from WT and the S1 mutant. (**a**) The number of upregulated genes and downregulated genes. (**b**) Volcano plot of differentially expressed genes with the threshold of $|\log_2 \text{foldchange}| > 1$. Orange indicates upregulated genes, and blue indicates downregulated genes.

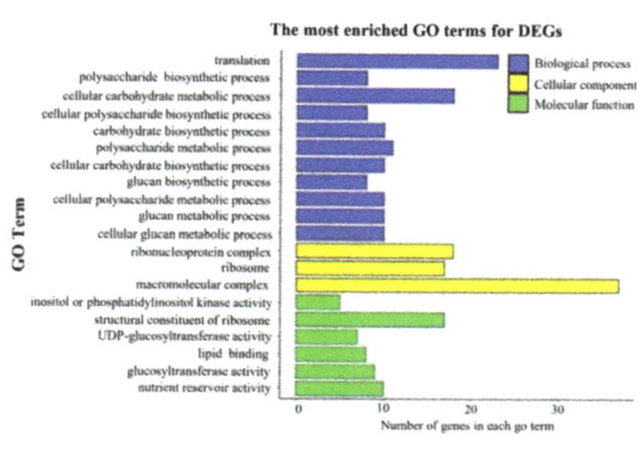

Figure 8. GO enrichment of DEGs. GO-term enrichment analysis was performed using agriGO [37], and the top 20 GO terms were shown that belonged to biological processes, cellular components, and molecular functions.

We found that the expressions of seven starch synthesis- and cellulose-synthesis-related genes were significantly reduced, and most of the starch-synthesis-related genes were downregulated in the S1 mutant (Figure 9, Table 1). Nine of the ten glutelin genes were significantly downregulated, which were then confirmed by SDS-PAGE in dry seeds. The 37-40-kDa acidic and 20-kDa basic two subunits of rice glutelin storage proteins were significantly decreased in the S1 mutant as compared to WT (Figure 10).

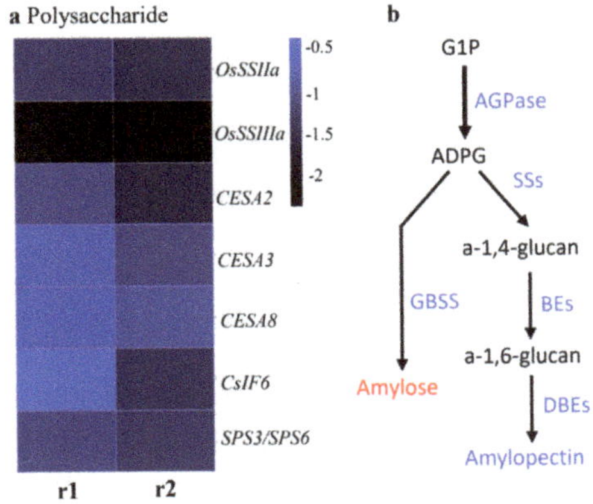

Figure 9. The expression of polysaccharide-related genes and starch synthesis pathway. (**a**) The heat plot of differential expressed polysaccharide-related genes. (**b**) The starch synthesis pathway. The downregulated genes were shown in blue colors, amylose in red showed increased content, and amylopectin in blue showed decreased content.

Table 1. Expression values of starch and cellulose-synthesis-related genes.

Gene Name	S1-r1	S1-r2
OsSSIIa	−1.358472011	−1.538954326
OsSSIIIa	−2.395417526	−2.467690768
CESA2	−1.163670981	−1.655416381
CESA3	−0.825392927	−1.18339011
CESA8	−0.792395169	−1.006805408
Cs1F6	−0.450447312	−1.428153108
sps3/sps6	−1.168312567	−1.35323233

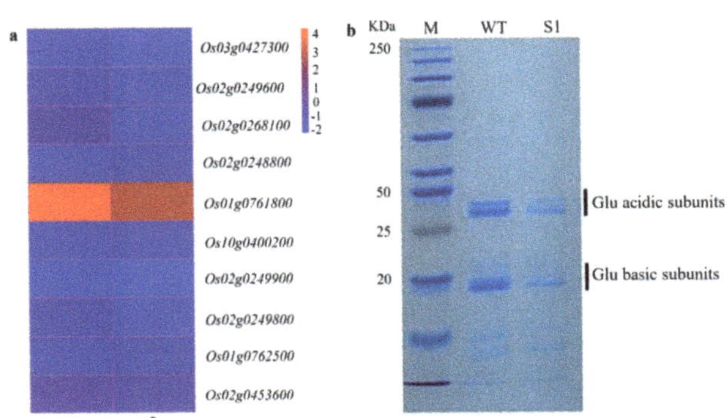

Figure 10. The gene expression and protein accumulation of glutelins. (**a**) The heat plot of significantly differentially expressed glutelin genes based on the RNA-Seq data. (**b**) SDS-PAGE analysis of glutelins. Glutelins extracted from mature seeds of WT and the S1 mutant were separated on a 12% SDS-PAGE gel and stained with Coomassie Brilliant Blue (CBB).

2.3. RNA-Seq Identifies Differentially AS Events in the Null Mutant of OsROS1a

As DNA methylation can regulate AS, to further investigate whether the 25 amino acid deletion of complete Per-CXXC domain deletion in the demethylase gene *OsROS1a* can regulate AS events occurring in rice 15 DAP endosperm or not, rMATs were used to identify both annotated and novel AS events in the S1 mutant and the expression level of

AS events was defined as "exon inclusion level". In total, 378 differentially (234 up- and 144 downregulated) AS genes were identified, which belonged to SE (Skipped exon), A5SS (Alternative 5′ splice site), A3SS (Alternative 3′ splice site), MXE (Mutually exclusive exon), and IR (Intron retained), which are five major AS types (Figure 11a). Among these, the majority of AS genes occurred via SE, which contains 271 genes (71.69%), and the second most common occurred via IR (58/378, 15.34%) (Figure 11b).

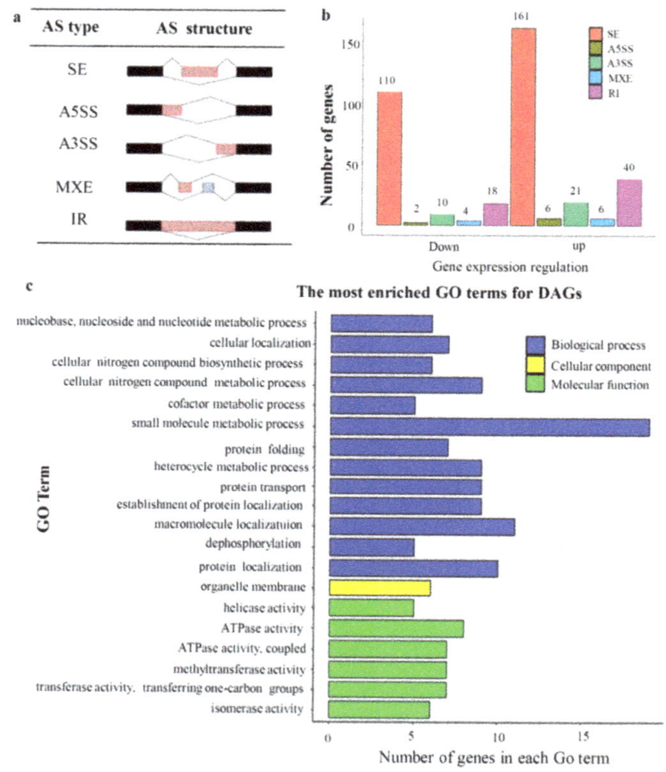

Figure 11. Alternative splicing affected by the mutation of *OsROS1a*. (**a**) Five types of alternative splicing events. (**b**) Statistics of differential alternative spliced genes between WT and the S1 null mutant. (**c**) GO enrichment of differential alternative spliced genes.

To study the trends in functions of the AS genes, GO enrichment analyses were carried out for genes with differential AS events in rice 15 DAP endosperm. The top 20 enriched GO terms included the largest number of enriched GO terms of 13 in BP, only 1 GO term in CC, and 6 in MF (Figure 11c). In BP, cellular nitrogen compound metabolic process (GO:0034641), nucleobase, nucleoside, nucleotide metabolic process (GO:0055086), and small molecule metabolic process (GO:0044281) are all relevant to small molecule metabolism; cellular localization (GO:0051641), protein localization (GO:0008104), and macromolecule localization (GO:0033036) are all relevant to cellular compound localization. In addition, we found that genes enriched to protein localization (GO:0008104), macromolecule localization (GO:0033036), and establishment of protein localization (GO:0045184) share quite a few common genes, indicating that such a function is likely to be influenced by *OsROS1a* through AS regulation in rice endosperm.

3. Discussion

In cereals, the endosperm is a highly specialized storage organ for starch, proteins, and lipids, and plays a crucial role in seed growth and development. Active DNA demethylation in plants is initialized by the ROS1 transglucosylase gene family, including DME, ROS1, DML2, and DML3, which regulates plant developmental processes [10–12]. The loss function of *OsROS1a* produced sterile rice through defects in both female and male gametogenesis [16,17]. Furthermore, only two heterozygous lines with one allele of 9-bp and 6-bp produced few seeds, and all frameshift mutants, including six having only truncated RRMF domain, failed to produce seeds [17], suggesting that the RRMF domain is necessary for fertility. We generated the null mutant of *OsROS1a* with 75-bp deletion and 1-bp substitution, which resulted in the complete loss of the Per-CXXC domain but the complete retention of the RRMF domain (Figure 1), which showed semisterile pollen and reduced seed fertility, indicating that the Per-CXXC domain also plays a role in rice fertility. Besides the fertility, the null mutant S1 also altered the rice grain morphology, including long and narrow grain, deformed seeds containing underdeveloped endosperm and a lower amount of starch, slightly irregular shape of the embryo, and slightly opaque grain (Figures 4 and 5). The mutation of *OsROS1a* with a 21-nt insertion generated a new transcript *mOsROS1a* with the insertion of seven amino acid residues (CSNVMRQ) in the RRMF domain to make the thick aleurone and improve rice grain nutrition, which resulted from the hypermethylation and reduced expression of *RISBZ1* and *RPBF* [19,20]. While the expression of *RISBZ1* and *RPBF* was similar between WT and the S1 mutant, this suggested that the Per-CXXC domain cannot alter the expression of these two important TFs. Overexpression of *BiP* suppresses SSPs and starch content and displays the opaque phenotype with shrunken and floury features in rice seeds [38,39].

A total of 723 DEGs were identified in the S1mutant by RNA-Seq analysis. most of these genes are related to starch and SSPs synthesis, suggests that the synthesis of starch, cellulose, and other types of polysaccharide in the endosperm are regulated by *OsROS1a*. The top 20 enrichment factors in GO terms were observed, in which 11 terms consisted of BP, 6 were in MF, and 3 terms were in CC. The BP occupies the largest proportion of GO terms. Out of 11 enriched BP terms, 10 are connected to polysaccharide synthesis. We identified the key genes in starch synthesis, such as *starch synthase IIa* (*OsSSIIa*) and *OsSSIIIa*, and in cellulose synthesis, such as *cellulose synthase A2* (*CESA2*), *CESA3*, *CESA6*, and *CESA8*, that are enriched for these terms, providing more evidence for the hypothesis. *OsSSI* and *OsSSIIIa* contribute to a huge portion of the overall SS enzyme activity in rice during endosperm development [40]. SSIIIa protein linked with other proteins in rice endosperm [41]. Similarly, the *SSIIIa* proportion is also present in a large complex containing *Pyruvate orthophosphate dikinase* (*PPDK*), *ADP-glucose pyrophosphorylase* (*AGPase*), *SSIIa*, and *the starch branching enzyme gene IIa* (*SBEIIa*) and *SBEIIb* in maize [42]. Likewise, *OsSSIIa* enzymes have abundant gene expression in the starch filling stage [43,44], and modify the quality of rice starch [45]. The GO term "macromolecular complex" (GO:0032991) was enriched in the cellular compound and nine glutelin genes related to this term were significantly downregulated in the S1 mutant, which was further confirmed by proteins with SDS-PAGE analysis. The decreased expression of these starch and protein-associated genes in our study suggested that it might be involved in the null mutant of *OsROS1a* seed phenotype and *OsROS1a* might have a main role in the regulation of these genes. The study of the DNA methylation profile during the development of endosperm revealed that genes expressed preferentially in endosperm were related to key storage proteins and starch synthesizing enzymes, are normally hypomethylated [8], and suggested that the demethylation of CG and CHG was the main mechanism for gene activation.

AS is an essential post-transcriptional process that generates several mRNA variants from a single pre-mRNA molecule and improves the genome coding and regulatory potential. Several studies have been carried out to find the AS events in different plant species, tissues, and environmental conditions [32–34]. However, the effect of the DNA methylation pattern on AS has not been reported in rice. Advancements in high-throughput technology enabled a global analysis of AS to study its functional characteristics in response to stress [46]. In this study, we identified 378 differentially AS genes in the 15 DAP endosperm of the S1 mutant, which belonged to SE, A5SS, A3SS, MXE, and IR, the five major AS types (Figure 11a), and the SE (71.69%) and IR (15.34%) were the top two types (Figure 11b). Whereas in *Arabidopsis*, AS had the low level of SE (<5%) and the high level of IR (65%) [47]. The SE and IR were predominant AS events in maize endosperm containing 28.33% and 29.85%, respectively [48]. Earlier studies have revealed that AS changes are affected by developmental and environmental factors [49,50]. The SE type is comparatively low (5%) in most maize tissues, which can be increased by up to more than 27% under abiotic stress. Furthermore, during seed development of maize, AS isoforms alter considerably in seed, embryo, and endosperm [25]. Our results revealed that AS genes of the A5SS and the MXE types were relatively less common and less influenced by the mutation of *OsROS1a* during endosperm development and that the function modification of *OsROS1a* did not lead to AS type-specific. The mutation of the *OsROS1a* gene showed a major impact on the alternation of AS events of downregulated genes and the impacts are not AS type-specific in the 15 DAP rice endosperm. Several GO terms indicated identical functions were enriched. Moreover, we identified that *OsROS1a* might influence the genes which are enriched to "protein localization" (GO:0008104), "macromolecule localization" (GO:0033036), and "establishment of protein localization" (GO:0045184) through AS regulation in the endosperm. These results suggested that rice endosperm development mediated by *OsROS1a* gene could be influenced through AS regulating.

4. Materials and Methods

4.1. Generation of Null Mutants of ROS1a in Rice Using CRISPR/Cas9 System

The CRISPR/Cas9 binary vector, pYLCRISPR/Cas9 Pubi-H, was used for targeted genome editing [51]. Two single guide RNA (sgRNA) were designed to precisely target the 13th exon of *OsROS1a*. Two sgRNA intermediate vectors, pYLsgRNA-OsU6a and pYLsgRNA-OsU6b, were used for gene targeting on rice OsU6a and OsU6b small nuclear RNA promoters, respectively. The sgRNA sequences were cloned into a binary vector that contained sgRNA and Cas9 expression cassettes, and the resulting construct was transformed into rice cultivar Nipponbare by *Agrobacterium* infection of callus explants.

4.2. Pollen Fertility Examination and Histochemical Assay

Floral organs were photographed with a dissecting microscope. For the analysis of pollen fertility, the anther from the WT (wild type, Nipponbare) and *osros1a* mutant plants were sampled from the spikelets just before flowering and the 1% potassium iodide (I_2-KI) solution was used to stain the pollen grains. A Nikon eclipse Ni fluorescence microscope was used to be visualized, and the stained pollen grains was photographed. For transverse section analysis, spikelets of various anther development stages from WT and the S1 mutant, according to the previous study [52], were collected and frozen in an OCT (Tissue-Tek; Sakura Finetek, Torrance, CA, USA) compound. Samples were sectioned with the help of the freezing microtome (Thermo Shandon Cryotome FE, Shandon, China). Then, 0.05% toluidine blue dye was used to stain the crossed section and the microscopic images were captured by a Nikon eclipse Ni fluorescence microscope.

4.3. Agronomic Traits Analyses

The agronomic traits of plant height, panicle length, tiller number, fertile seed percentage, and the primary and secondary branches' number in the main panicle of the *osros1a* mutant and WT were measured for plant grown in the greenhouse.

4.4. Half-Seed Assay and Scanning Electron Microscopy (SEM)

For Evans Blue staining-based half-seed assay, mature grains from WT and the null mutant S1 were dehusked and transversally sectioned into two halves using a razor blade, then dipped in 0.1% (w/v) Evans Blue solution for 10 min, and by using distilled water, were washed three times. All samples were observed and photographed under a dissecting microscope (Nikon). Further, to study the mature seeds' morphological changes and their starch granules, the WT and the S1 mutant seeds were transversely sectioned. Samples were then examined and photographed by SEM.

4.5. RNA Extraction, Library Preparation, and RNA-Seq

The S1 mutant was cultivated in the greenhouse, and the extraction of total RNA from the mutant and WT endosperm was conducted at 15 DAP. For RNA-Seq analysis, 1 μg RNA per sample was used. The mRNA was enriched using magnetic beads with oligo-dT, then divalent cations in NEBNext First Strand Synthesis Reaction Buffer (5X) were used to break it into short fragments and then reversed transcription into cDNA. DNA fragments 3′-ends were adenylated, and a hairpin loop structure NEBNext Adaptor was ligated to prepare for hybridization. After purification, terminal modification, fragments selection, and the amplification of PCR with Phusion High-Fidelity DNA polymerase and another round of purification, two libraries (one for WT and one for the mutant) were constructed for Next-Generation Sequencing (NGS). The Agilent Bioanalyzer 2100 system was used to assess the library quality.

A cBot Cluster Generation System was utilized for performing clustering of the index-coded samples, using TruSeq PE Cluster Kit v3-cBot-HS (Illumia) as per the manufacturer's instructions. After cluster generation, the preparations of the library were sequenced on an Illumina Novaseq platform and 150 bp paired-end reads were generated.

4.6. Quality Control and RNA-Seq Analysis

The clean reads were obtained by removing reads containing ploy-N, adapter, and low-quality from raw data. Hisat2 (v2.0.5, Daehwan Kim, Dallas, TX, USA) was adopted to build the reference genome index and align the pair-end clean read to the genome [53]. We chose IRGSP1.0 (https://rapdb.dna.affrc.go.jp/download/irgsp1.html, accessed on 20 December 2020), as the reference genome for its high quality and comprehensive annotation. Samtools (v1.4.1, Heng Li, London, UK) was then used to convert the SAM (Sequence Align Mapping) file to its binary coding form BAM file [54], which was smaller and faster to perform downstream analysis on.

To count the reads numbers mapped for each gene, Feature Counts (v1.5.0-p3, Yang Liao, Melbourne, Australia) was used [55]. The FPKM (Fragments Per Kilobase of transcript sequence per Millions base pairs sequenced) of each gene was then calculated based on the gene length and reads count mapped to this gene. DESeq2 R package v1.26.0 was then used to determine DEGs (Differential Expression Genes) [56]. The threshold for significant differential expression was set as fold change $\geq$2 and p-value $\leq$ 0.05. GO (Gene Ontology) enrichment analysis of DEGs was implemented using agriGO (v2.0, Tian Tian, Beijing, China) [37].

4.7. SDS-PAGE Analysis

Storage protein analysis was conducted by SDS-PAGE with the Laemmli method. Mature rice seeds of the S1 mutant and WT were ground with the help of mortar and pestle to make it fine powder. Rice glutelin was extracted from powdered seeds (25 mg) by 0.2% NaOH after stepwise removal of albumins (with deionized water), globulins (with 0.5 M NaCl and 50 mM Tris-HCl; PH 6.8), and prolamins (with 70% (v/v) alcohol). SDS-PAGE of 12% gel was conducted by using Standard Twin Mini Gel Unit (CA, USA) for 2.5 h at 120 V. The glutelin proteins were detected by staining the gel with a staining solution having Coomassie Brilliant Blue R-250 (Sigma) for 30 min and then 20% methanol, and 5% acetic acid in DW solution was used for destaining. The protein bands of the sample were measured by comparing it with the protein ladder (Thermo Scientific Page Ruler Prestained Protein Ladder 10-250 KDa) in the electropherogram.

4.8. Identification of Differential AS Events

To investigate the difference in AS pattern between WT and mutant, rMATs (v4.1.1, Shihao Shen, CA, USA) was used to identify both annotated and novel AS events in the mutant [57], and the expression level of AS events was defined as "exon inclusion level". The threshold was set to 0.0001 and AS events with a p-value above 0.05 were filtered out for further analysis. As a result, five major types of AS events, including SE, A5SS, A3SS, MXE, and IR, were identified. GO enrichment analysis of DAGs (Differential Alternative splicing Genes) was implemented using DAVID [58].

5. Conclusions

In summary, we identified that the loss function of the *OsROS1a* gene altered the rice grain size, having longer grain and a reduced width as compared to WT. The null mutant of *OsROS1a* seeds were deformed containing underdeveloped, less-starch-producing endosperm and a slightly irregularly shaped embryo. Furthermore, RNA-Seq analysis showed that many genes encoding polysaccharides and glutelins were found to be downregulated in the endosperm of the S1 mutant, suggesting that it might be involved in the S1 mutant seed phenotype and *OsROS1a* might have a main role in the regulation of these genes. Moreover, AS analysis revealed that the *OsROS1a* gene has a major impact on the alternation of AS events of these downregulated genes. These findings have provided a base to properly understand the molecular mechanism related to the *OsROS1a* gene in the regulation of rice seed development.

Author Contributions: Conceptualization, J.-H.X.; methodology, F.I., C.L. and H.-Y.W.; validation, C.L. and Y.Y.; formal analysis, F.I., C.L., H.-Y.W. and J.-H.X.; investigation, J.-H.X.; resources, J.-H.X.; data curation, F.I., C.L., H.-Y.W., Y.Y. and J.-H.X.; writing—original draft preparation, F.I. and H.-Y.W.; writing—review and editing, J.-H.X. and C.L.; supervision, J.-H.X.; funding acquisition, J.-H.X. All authors have read and agreed to the published version of the manuscript.

Funding: This research was funded by Shandong (Linyi) Institute of Modern Agriculture, Zhejiang University.

Institutional Review Board Statement: Not applicable.

Informed Consent Statement: Not applicable.

Data Availability Statement: The data presented in this study are available within the article.

Acknowledgments: We are grateful to Li Xinxin for her obtaining *osros1a* mutants, and Yao-Guang Liu for his kindly providing CRISPR/Cas9 binary vector.

Conflicts of Interest: The authors declare no conflict of interest.

References

1. Xu, S.B.; Li, T.; Deng, Z.Y.; Chong, K.; Xue, Y.; Wang, T. Dynamic proteomic analysis reveals a switch between central carbon metabolism and alcoholic fermentation in rice filling grains. *Plant Physiol.* **2008**, *148*, 908–925. [CrossRef]
2. Venu, R.; Sreerekha, M.; Nobuta, K.; Belo, A.; Ning, Y.; An, G.; Meyers, B.C.; Wang, G.L. Deep sequencing reveals the complex and coordinated transcriptional regulation of genes related to grain quality in rice cultivars. *BMC Genom.* **2011**, *12*, 190. [CrossRef]
3. Kawakatsu, T.; Yamamoto, M.P.; Hirose, S.; Yano, M.; Takaiwa, F. Characterization of a new rice glutelin gene GluD-1 expressed in the starchy endosperm. *J. Exp. Bot.* **2008**, *59*, 4233–4245. [CrossRef]
4. Xu, J.H.; Messing, J. Amplification of prolamin storage protein genes in different subfamilies of the Poaceae. *Theor. Appl. Genet.* **2009**, *119*, 1397–1412. [CrossRef]
5. Tian, L.; Okita, T.W. mRNA-based protein targeting to the endoplasmic reticulum and chloroplasts in plant cells. *Curr. Opin. Plant Biol.* **2014**, *22*, 77–85. [CrossRef] [PubMed]
6. Tian, L.; Chou, H.L.; Zhang, L.; Hwang, S.K.; Starkenburg, S.R.; Doroshenk, K.A.; Kumamaru, T.; Okita, T.W. RNA-Binding Protein RBP-P Is Required for Glutelin and Prolamine mRNA Localization in Rice Endosperm Cells. *Plant Cell* **2018**, *30*, 2529–2552. [CrossRef]
7. Yamagata, H.; Tanaka, K. The Site of Synthesis and Accumulation of Rice Storage Proteins. *Plant Cell Physiol.* **1986**, *27*, 135–145.
8. Zemach, A.; Kim, M.Y.; Silva, P.; Rodrigues, J.A.; Dotson, B.; Brooks, M.D.; Zilberman, D. Local DNA hypomethylation activates genes in rice endosperm. *Proc. Natl. Acad. Sci. USA* **2010**, *107*, 18729–18734. [CrossRef] [PubMed]

9. Zhang, H.; Lang, Z.; Zhu, J.K. Dynamics and function of DNA methylation in plants. *Nat. Rev. Mol. Cell Biol.* **2018**, *19*, 489–506. [CrossRef]
10. Gong, Z.; Morales-Ruiz, T.; Ariza, R.R.; Roldan-Arjona, T.; David, L.; Zhu, J.K. ROS1, a repressor of transcriptional gene silencing in Arabidopsis, encodes a DNA glycosylase/lyase. *Cell* **2002**, *111*, 803–814. [CrossRef]
11. Choi, Y.; Gehring, M.; Johnson, L.; Hannon, M.; Harada, J.J.; Goldberg, R.B.; Jacobsen, S.E.; Fischer, R.L. DEMETER, a DNA glycosylase domain protein, is required for endosperm gene imprinting and seed viability in arabidopsis. *Cell* **2002**, *110*, 33–42. [CrossRef]
12. Ortega-Galisteo, A.P.; Morales-Ruiz, T.; Ariza, R.R.; Roldan-Arjona, T. Arabidopsis DEMETER-LIKE proteins DML2 and DML3 are required for appropriate distribution of DNA methylation marks. *Plant Mol. Biol.* **2008**, *67*, 671–681. [CrossRef]
13. Penterman, J.; Zilberman, D.; Huh, J.H.; Ballinger, T.; Henikoff, S.; Fischer, R.L. DNA demethylation in the Arabidopsis genome. *Proc. Natl. Acad. Sci. USA* **2007**, *104*, 6752–6757. [CrossRef]
14. Brooks, S.C.; Fischer, R.L.; Huh, J.H.; Eichman, B.F. 5-methylcytosine recognition by Arabidopsis thaliana DNA glycosylases DEMETER and DML3. *Biochemistry* **2014**, *53*, 2525–2532. [CrossRef]
15. Zhu, J.K. Active DNA demethylation mediated by DNA glycosylases. *Annu. Rev. Genet.* **2009**, *43*, 143–166. [CrossRef]
16. Ono, A.; Yamaguchi, K.; Fukada-Tanaka, S.; Terada, R.; Mitsui, T.; Iida, S. A null mutation of ROS1a for DNA demethylation in rice is not transmittable to progeny. *Plant J.* **2012**, *71*, 564–574. [CrossRef]
17. Xu, Y.; Wang, F.; Chen, Z.; Wang, J.; Li, W.; Fan, F.; Tao, Y.; Jiang, Y.; Zhu, Q.H.; Yang, J. CRISPR/Cas9-targeted mutagenesis of the OsROS1 gene induces pollen and embryo sac defects in rice. *Plant. Biotechnol. J.* **2020**, *18*, 1999–2001. [CrossRef]
18. Kim, M.Y.; Ono, A.; Scholten, S.; Kinoshita, T.; Zilberman, D.; Okamoto, T.; Fischer, R.L. DNA demethylation by ROS1a in rice vegetative cells promotes methylation in sperm. *Proc. Natl. Acad. Sci. USA* **2019**, *116*, 9652–9657. [CrossRef]
19. Liu, J.; Wu, X.; Yao, X.; Yu, R.; Larkin, P.J.; Liu, C.M. Mutations in the DNA demethylase OsROS1 result in a thickened aleurone and improved nutritional value in rice grains. *Proc. Natl. Acad. Sci. USA* **2018**, *115*, 11327–11332. [CrossRef]
20. Lang, Z.; Gong, Z. A role of OsROS1 in aleurone development and nutrient improvement in rice. *Proc. Natl. Acad. Sci. USA* **2018**, *115*, 11659–11660. [CrossRef] [PubMed]
21. La, H.; Ding, B.; Mishra, G.P.; Zhou, B.; Yang, H.; Bellizzi Mdel, R.; Chen, S.; Meyers, B.C.; Peng, Z.; Zhu, J.K.; et al. A 5-methylcytosine DNA glycosylase/lyase demethylates the retrotransposon Tos17 and promotes its transposition in rice. *Proc. Natl. Acad. Sci. USA* **2011**, *108*, 15498–15503. [CrossRef]
22. Wen, S.; Wen, N.; Pang, J.; Langen, G.; Brew-Appiah, R.A.; Mejias, J.H.; Osorio, C.; Yang, M.; Gemini, R.; Moehs, C.P.; et al. Structural genes of wheat and barley 5-methylcytosine DNA glycosylases and their potential applications for human health. *Proc. Natl. Acad. Sci. USA* **2012**, *109*, 20543–20548. [CrossRef] [PubMed]
23. Ast, G. How did alternative splicing evolve? *Nat. Rev. Genet.* **2004**, *5*, 773–782. [CrossRef]
24. Shankar, R.; Bhattacharjee, A.; Jain, M. Transcriptome analysis in different rice cultivars provides novel insights into desiccation and salinity stress responses. *Sci. Rep.* **2016**, *6*, 23719. [CrossRef]
25. Thatcher, S.R.; Danilevskaya, O.N.; Meng, X.; Beatty, M.; Zastrow-Hayes, G.; Harris, C.; Van Allen, B.; Habben, J.; Li, B. Genome-Wide Analysis of Alternative Splicing during Development and Drought Stress in Maize. *Plant Physiol.* **2016**, *170*, 586–599. [CrossRef]
26. Thatcher, S.R.; Zhou, W.; Leonard, A.; Wang, B.B.; Beatty, M.; Zastrow-Hayes, G.; Zhao, X.; Baumgarten, A.; Li, B. Genome-wide analysis of alternative splicing in Zea mays: Landscape and genetic regulation. *Plant Cell* **2014**, *26*, 3472–3487. [CrossRef]
27. Filichkin, S.A.; Priest, H.D.; Givan, S.A.; Shen, R.; Bryant, D.W.; Fox, S.E.; Wong, W.K.; Mockler, T.C. Genome-wide mapping of alternative splicing in Arabidopsis thaliana. *Genome Res.* **2010**, *20*, 45–58. [CrossRef]
28. Li, Q.; Xiao, G.; Zhu, Y.X. Single-nucleotide resolution mapping of the Gossypium raimondii transcriptome reveals a new mechanism for alternative splicing of introns. *Mol. Plant* **2014**, *7*, 829–840. [CrossRef]
29. Wai, C.M.; Powell, B.; Ming, R.; Min, X.J. Analysis of Alternative Splicing Landscape in Pineapple (*Ananas comosus*). *Trop. Plant Biol.* **2016**, *9*, 150–160. [CrossRef]
30. Shen, Y.; Zhou, Z.; Wang, Z.; Li, W.; Fang, C.; Wu, M.; Ma, Y.; Liu, T.; Kong, L.A.; Peng, D.L.; et al. Global dissection of alternative splicing in paleopolyploid soybean. *Plant Cell* **2014**, *26*, 996–1008. [CrossRef]
31. Qin, Z.; Wu, J.; Geng, S.; Feng, N.; Chen, F.; Kong, X.; Song, G.; Chen, K.; Li, A.; Mao, L.; et al. Regulation of FT splicing by an endogenous cue in temperate grasses. *Nat. Commun.* **2017**, *8*, 14320. [CrossRef]
32. Lu, S.J.; Yang, Z.T.; Sun, L.; Sun, L.; Song, Z.T.; Liu, J.X. Conservation of IRE1-regulated bZIP74 mRNA unconventional splicing in rice (*Oryza sativa* L.) involved in ER stress responses. *Mol. Plant* **2012**, *5*, 504–514. [CrossRef]
33. Yang, Z.; Yang, Z.; Yang, C.; Wang, Z.; Chen, D.; Xie, Y.; Wu, Y. Identification and genetic analysis of alternative splicing of long non-coding RNAs in tomato initial flowering stage. *Genomics* **2020**, *112*, 897–907. [CrossRef]
34. Wang, Y.; Xu, J.; Ge, M.; Ning, L.; Hu, M.; Zhao, H. High-resolution profile of transcriptomes reveals a role of alternative splicing for modulating response to nitrogen in maize. *BMC Genom.* **2020**, *21*, 353. [CrossRef] [PubMed]
35. Li-Byarlay, H.; Li, Y.; Stroud, H.; Feng, S.; Newman, T.C.; Kaneda, M.; Hou, K.K.; Worley, K.C.; Elsik, C.G.; Wickline, S.A.; et al. RNA interference knockdown of DNA methyl-transferase 3 affects gene alternative splicing in the honey bee. *Proc. Natl. Acad. Sci. USA* **2013**, *110*, 12750–12755. [CrossRef]

36. Yearim, A.; Gelfman, S.; Shayevitch, R.; Melcer, S.; Glaich, O.; Mallm, J.P.; Nissim-Rafinia, M.; Cohen, A.H.; Rippe, K.; Meshorer, E.; et al. HP1 is involved in regulating the global impact of DNA methylation on alternative splicing. *Cell Rep.* **2015**, *10*, 1122–1134. [CrossRef]
37. Tian, T.; Liu, Y.; Yan, H.; You, Q.; Yi, X.; Du, Z.; Xu, W.; Su, Z. agriGO v2.0: A GO analysis toolkit for the agricultural community, 2017 update. *Nucleic Acids Res.* **2017**, *45*, W122–W129. [CrossRef]
38. Yasuda, H.; Hirose, S.; Kawakatsu, T.; Wakasa, Y.; Takaiwa, F. Overexpression of BiP has inhibitory effects on the accumulation of seed storage proteins in endosperm cells of rice. *Plant Cell Physiol.* **2009**, *50*, 1532–1543. [CrossRef] [PubMed]
39. Wakasa, Y.; Yasuda, H.; Oono, Y.; Kawakatsu, T.; Hirose, S.; Takahashi, H.; Hayashi, S.; Yang, L.; Takaiwa, F. Expression of ER quality control-related genes in response to changes in BiP1 levels in developing rice endosperm. *Plant J.* **2011**, *65*, 675–689. [CrossRef]
40. Fujita, N.; Yoshida, M.; Asakura, N.; Ohdan, T.; Miyao, A.; Hirochika, H.; Nakamura, Y. Function and characterization of starch synthase I using mutants in rice. *Plant Physiol.* **2006**, *140*, 1070–1084. [CrossRef]
41. Crofts, N.; Abe, N.; Oitome, N.F.; Matsushima, R.; Hayashi, M.; Tetlow, I.J.; Emes, M.J.; Nakamura, Y.; Fujita, N. Amylopectin biosynthetic enzymes from developing rice seed form enzymatically active protein complexes. *J. Exp. Bot.* **2015**, *66*, 4469–4482. [CrossRef]
42. Hennen-Bierwagen, T.A.; Lin, Q.; Grimaud, F.; Planchot, V.; Keeling, P.L.; James, M.G.; Myers, A.M. Proteins from multiple metabolic pathways associate with starch biosynthetic enzymes in high molecular weight complexes: A model for regulation of carbon allocation in maize amyloplasts. *Plant Physiol.* **2009**, *149*, 1541–1559. [CrossRef]
43. Hirose, T.; Terao, T. A comprehensive expression analysis of the starch synthase gene family in rice (*Oryza sativa* L.). *Planta* **2004**, *220*, 9–16. [CrossRef]
44. Ohdan, T.; Francisco, P.B., Jr.; Sawada, T.; Hirose, T.; Terao, T.; Satoh, H.; Nakamura, Y. Expression profiling of genes involved in starch synthesis in sink and source organs of rice. *J. Exp. Bot.* **2005**, *56*, 3229–3244. [CrossRef]
45. Nakamura, Y.; Francisco, P.B., Jr.; Hosaka, Y.; Sato, A.; Sawada, T.; Kubo, A.; Fujita, N. Essential amino acids of starch synthase IIa differentiate amylopectin structure and starch quality between japonica and indica rice varieties. *Plant Mol. Biol.* **2005**, *58*, 213–227. [CrossRef]
46. Mastrangelo, A.M.; Marone, D.; Laido, G.; De Leonardis, A.M.; De Vita, P. Alternative splicing: Enhancing ability to cope with stress via transcriptome plasticity. *Plant Sci.* **2012**, *185–186*, 40–49. [CrossRef]
47. Ner-Gaon, H.; Halachmi, R.; Savaldi-Goldstein, S.; Rubin, E.; Ophir, R.; Fluhr, R. Intron retention is a major phenomenon in alternative splicing in Arabidopsis. *Plant J.* **2004**, *39*, 877–885. [CrossRef]
48. Qu, J.; Ma, C.; Feng, J.; Xu, S.; Wang, L.; Li, F.; Li, Y.; Zhang, R.; Zhang, X.; Xue, J.; et al. Transcriptome Dynamics during Maize Endosperm Development. *PLoS ONE* **2016**, *11*, e0163814. [CrossRef]
49. Szakonyi, D.; Duque, P. Alternative Splicing as a Regulator of Early Plant Development. *Front. Plant Sci.* **2018**, *9*, 1174. [CrossRef]
50. Calixto, C.P.G.; Tzioutziou, N.A.; James, A.B.; Hornyik, C.; Guo, W.; Zhang, R.; Nimmo, H.G.; Brown, J.W.S. Cold-Dependent Expression and Alternative Splicing of Arabidopsis Long Non-coding RNAs. *Front. Plant Sci.* **2019**, *10*, 235. [CrossRef]
51. Ma, X.; Zhang, Q.; Zhu, Q.; Liu, W.; Chen, Y.; Qiu, R.; Wang, B.; Yang, Z.; Li, H.; Lin, Y.; et al. A Robust CRISPR/Cas9 System for Convenient, High-Efficiency Multiplex Genome Editing in Monocot and Dicot Plants. *Mol. Plant* **2015**, *8*, 1274–1284. [CrossRef] [PubMed]
52. Zhang, D.; Luo, X.; Zhu, L. Cytological analysis and genetic control of rice anther development. *J. Genet. Genom.* **2011**, *38*, 379–390. [CrossRef]
53. Kim, D.; Paggi, J.M.; Park, C.; Bennett, C.; Salzberg, S.L. Graph-based genome alignment and genotyping with HISAT2 and HISAT-genotype. *Nat. Biotechnol.* **2019**, *37*, 907–915. [CrossRef] [PubMed]
54. Li, H.; Handsaker, B.; Wysoker, A.; Fennell, T.; Ruan, J.; Homer, N.; Marth, G.; Abecasis, G.; Durbin, R. Genome Project Data Processing, S. The Sequence Alignment/Map format and SAMtools. *Bioinformatics* **2009**, *25*, 2078–2079. [CrossRef]
55. Liao, Y.; Smyth, G.K.; Shi, W. Feature Counts: An efficient general purpose program for assigning sequence reads to genomic features. *Bioinformatics* **2014**, *30*, 923–930. [CrossRef]
56. Love, M.I.; Huber, W.; Anders, S. Moderated estimation of fold change and dispersion for RNA-seq data with DESeq2. *Genome Biol.* **2014**, *15*, 550. [CrossRef] [PubMed]
57. Shen, S.; Park, J.W.; Lu, Z.X.; Lin, L.; Henry, M.D.; Wu, Y.N.; Zhou, Q.; Xing, Y. rMATS: Robust and flexible detection of differential alternative splicing from replicate RNA-Seq data. *Proc. Natl. Acad. Sci. USA* **2014**, *111*, E5593–E5601. [CrossRef] [PubMed]
58. Huang da, W.; Sherman, B.T.; Lempicki, R.A. Systematic and integrative analysis of large gene lists using DAVID bioinformatics resources. *Nat. Protoc.* **2009**, *4*, 44–57. [CrossRef]

International Journal of *Molecular Sciences*

MDPI

Article

Loss of Function of the RRMF Domain in OsROS1a Causes Sterility in Rice (*Oryza sativa* L.)

Jian-Hong Xu [1,2,3,*,†], Faiza Irshad [2,†], Yan Yan [2] and Chao Li [2,3,*]

1 Hainan Institute, Zhejiang University, Sanya 572025, China
2 Zhejiang Key Laboratory of Crop Germplasm, College of Agriculture & Biotechnology, Zhejiang University, Hangzhou 310058, China
3 Shandong (Linyi) Institute of Modern Agriculture, Zhejiang University, Linyi 276000, China
* Correspondence: jhxu@zju.edu.cn (J.-H.X.); 0621121@zju.edu.cn (C.L.)
† These authors contributed equally to this work.

Abstract: For crop seed production, the development of anthers and male fertility are the main agronomic traits and key biological processes for flowering plants. Active DNA demethylation regulates many plant developmental processes and is ensured by 5-meC DNA glycosylase enzymes. To find out the role of *OsROS1a*, *OsROS1a* gene editing mutants were generated using the CRISPR/Cas9 system. The *osros1a* mutants had shrink spikelets, smaller anthers and pollen grains, and were not stained by iodine staining showing a significant reduction in total soluble sugar and starch contents as compared to wildtype (WT), which caused complete male sterility. Similarly, the expression of genes involved in pollen and anther development was decreased in *osros1a* mutants as compared to WT. Furthermore, bisulfite sequencing showed that the CG and CHG methylation of the *OsPKS2* gene promoter was significantly increased in the *osros1a* mutant, which caused a reduced expression of *OsPKS2* in *osros1a* mutants. DNA methylation of the *TDR* gene promoter was similar between WT and *osros1a* mutants, indicating that the DNA methylation effect by *OsROS1a* was gene specific. The expression of *OsROS1a* in the mutants was not changed, but it produced a frame-shift mutation to truncate the Pem-CXXC and RRMF domains. Combined with previous studies, our findings suggested that the RRMF domain in OsROS1a is the functional domain and loss of RRMF for *OsROS1a* causes sterility in rice.

Keywords: *ROS1*; DNA demethylation; CRISPR/Cas9; pollen fertility; bisulfite sequencing; rice (*Oryza satia* L.)

check for updates

Citation: Xu, J.-H.; Irshad, F.; Yan, Y.; Li, C. Loss of Function of the RRMF Domain in OsROS1a Causes Sterility in Rice (*Oryza sativa* L.). *Int. J. Mol. Sci.* **2022**, 23, 11349. https://doi.org/10.3390/ijms231911349

Received: 24 August 2022
Accepted: 19 September 2022
Published: 26 September 2022

Publisher's Note: MDPI stays neutral with regard to jurisdictional claims in published maps and institutional affiliations.

Copyright: © 2022 by the authors. Licensee MDPI, Basel, Switzerland. This article is an open access article distributed under the terms and conditions of the Creative Commons Attribution (CC BY) license (https://creativecommons.org/licenses/by/4.0/).

1. Introduction

DNA methylation plays a crucial role in plant growth, development, and stress responses [1]. In plants, DNA methylation usually occurs at the cytosine (C) base in CG, CHG and CHH configurations (H = A, C, or T), and is dynamically regulated by balanced methylation and demethylation [1]. Plant 5-meC DNA glycosylases are placed in the DEMETER-like (DML) family, which is related to the HhH-GPD superfamily and is the main functionally-varied group of DNA glycosylases [2]. DML proteins ranging from 1100 to over 2000 residues are unusually large DNA glycosylases, which are bifunctional enzymes having both DNA glycosylase and apurinic/apyrimidinic (AP) lyase activities [3]. Specific DNA glycosylases, including Repressor of silencing 1 (ROS1) [4], DEMETER (DME) [5], DML2 and DML3 [6] catalyze the cytosine demethylation in *Arabidopsis*. DNA demethylation mediated by DME is important for the reproduction of plants, and the inheritance of loss-of-function paternal or maternal mutant *dme* alleles resulting in a reduced sperm transmission or seed abortion, respectively [5,7]. present only in dicots, but not in monocots. ROS1 encodes a DNA glycosylase/lyase that represses DNA methylation of promoter DNA [8]. The loss-of-function of *ROS1* causes DNA hypermethylation of CG and CHH and decreases the related gene expression in *Arabidopsis* [3,4]. The rice genome contains four *ROS1* paralog genes (*OsROS1a*, *OsROS1b*, *OsROS1c*, and *OsROS1d*) that mediate DNA demethylation [9]. *OsROS1a* is highly expressed in rice anthers and pistils [10]. The disrupted *OsROS1a* mutants exhibited severe defects in both male and female gametogenesis to produce a sterile phenotype [10,11]. The rice vegetative cell genome is actively demethylated by *OsROS1a* and is vital for viable seed production, and sperm non-CG methylation is indirectly promoted by DNA methylation in the vegetative cell, which suggests that dynamic DNA methylation reprogramming occurs during plant embryogenesis [12]. The aleurone layer is the most nutritious part of cereal grains which stores lipids, vitamins, proteins, and minerals. The point mutation in the fourteenth intron of *OsROS1* generates a new transcript *mOsROS1* with a 21-nt insertion, which can lead to DNA hypermethylation and suppresses the expression of two transcription factors, *RISBZ1* and *RPBF*, increasing the number of aleurone cell layers [13,14]. Knockout or knockdown of *OsROS1b* causes DNA hypermethylation of *Tos17* retrotransposons in rice, and overexpression of *OsROS1b* extensively reduces DNA methylation of the rice genome [15]. Furthermore, *ROS1* is involved in the seed development of rice, wheat and barley by its epigenetic influence on the accumulation of seed storage proteins (SSPs) [15,16].

Pollen fertility is important for successful seed production in flowering plants. However, defective development of the anthers can lead to either an absence or formation of non-functional pollen grains [17]. Successful development of male reproductive organs includes events such as specification of the meristem, cell differentiation, cell-to-cell communication, meiosis and mitosis [18–21]. The development of pollen is controlled specifically by four sporophytic cell layers of the anther (tapetum, middle layer, endothecium and epidermis), which surround the gametophytic pollen grains [22]. Many genes involved in the development of pollen and anthers have been identified in rice. A plant-specific type III polyketide synthase gene *OsPKS2* is involved in the normal development of pollen wall formation and mutation in the *OsPKS2* caused male sterility in rice [23]. The rice tapetum degeneration retardation gene (*TDR*) has an essential role in regulating the transcriptional network for the development and degradation of tapetum, and mutation in this gene caused male sterility [24,25]. MEIOSIS ARRESTED AT LEPTOTENE1 (*MEL1*) gene has a crucial role in microsporogenesis, abnormal accumulation of *MEL1* can make a semi-sterile phenotype in rice [26]. Cytochrome P450 family member *CYP704B2* is specifically expressed in the tapetum and microspores, which is required for the synthesis of cutin monomers, and its mutant *cyp704B2* shows a male sterile phenotype with a swollen sporophytic tapetal layer and aborted pollen grains [27]. Pollen development, being extremely susceptible to the cellular environment, is regulated by the successive expression of genes specifically expressed in reproductive tissues. Male gametogenesis also shows abnormalities and becomes inactive if any imbalances occur in the expression of anther and microspores development-related genes.

This study aimed to reveal the underlying mechanism of how *OsROS1a* regulates pollen fertility. The CRISPR/Cas9 system was used to generate *OsROS1a* gene editing mutants, and the *osros1a* S6, S7, and S16 mutants exhibited complete male sterile, which resulted from the disrupted Pem-CXXC and RNA recognition motif fold (RRMF) domains. The qRT-PCR results showed that the expression of genes involved in pollen and anther development, and starch biosynthesis decreased in these mutants as compared to WT. Furthermore, bisulfite sequencing showed that the CG and CHG methylation of the *OsPKS2* gene promoter was significantly increased in the *osros1a* mutants. These findings could provide new insights to reveal the DNA demethylation of *OsROS1a* on sterility in rice.

2. Results

2.1. Phylogenetic Analysis of OsROS1a Glycosylase Domain

OsROS1a contains 17 exons that encode a protein having 1952 amino acids (aa), and 3′-and 5′-UTRs of 607 and 73 bp, respectively (Figure 1A). The *OsROS1a* contains three key domains of DNA glycosylase, Per-CXXC and RRMF (Figure 1B), having lengths of 141, 31 and 102 aa, respectively. The location of these domains in the OsROS1a CDS protein sequences was 1472–1612 for DNA glycosylase, 1797–1827 for Per-CXXC, and 1831–1932 for RRMF (Figure 1C).

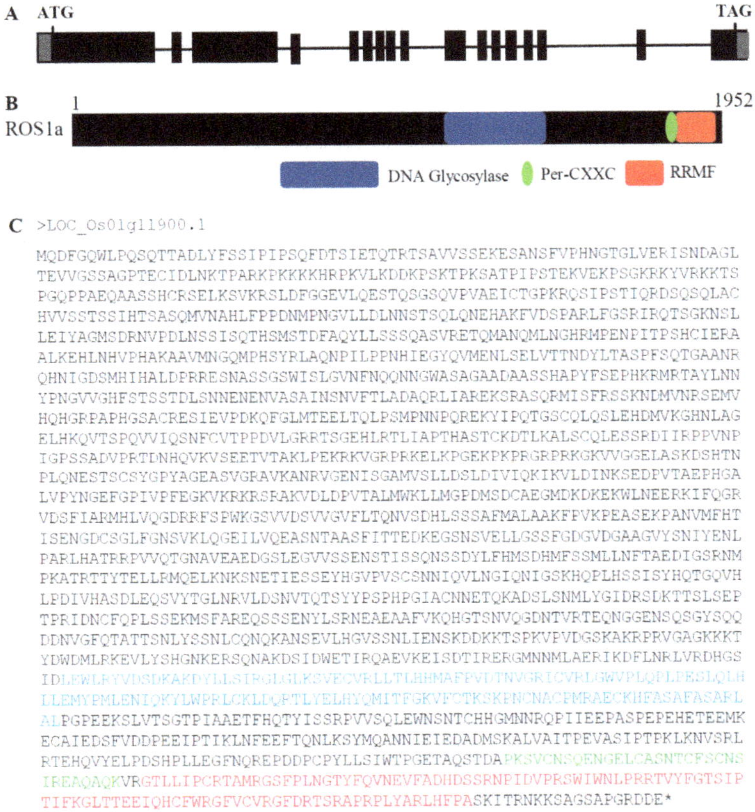

Figure 1. Schematic of *OsROS1a* gene structure and protein features. (**A**) Exons and introns of the *OsROS1a* gene are denoted as black blocks and lines, respectively. The translation initiation codon (ATG) and termination codon (TAG) are shown. (**B**) The protein structure of the *OsROS1a* gene. The open reading frame (ORF) is 1952 amino acid residues. (**C**) The protein sequences of the *OsROS1a* gene. The DNA glycosylase, Per-CXXC and RRMF domains are shown in blue, green and red colors, and all others showed in black.

To determine the evolutionary relationship of *OsROS1a* genes in different species, its protein sequences were used, and 159 homologous gene copies were identified in 52 plant species. The conserved DNA glycosylase domain (151 amino acids) of 159 genes was then used for constructing the phylogenetic tree with the neighbor-joining (NJ) method (Figure 2). The phylogenetic tree can be clustered into 4 clades, monocots *ROS1* (mROS1), dicots *DME* (dDME), dicots *ROS1* (dROS1) and dicots *DML2* (dDML2) (Figure 2). Alignment of multiple sequences of deduced amino acid sequences revealed that all homologous *ROS1* genes contain the DNA glycosylase domain. The evolutionary analysis discovered that DNA demethylase gene copy number varied greatly among different species or different families of the same species. The homologous copy number varies from 1 to 6 in different plants.

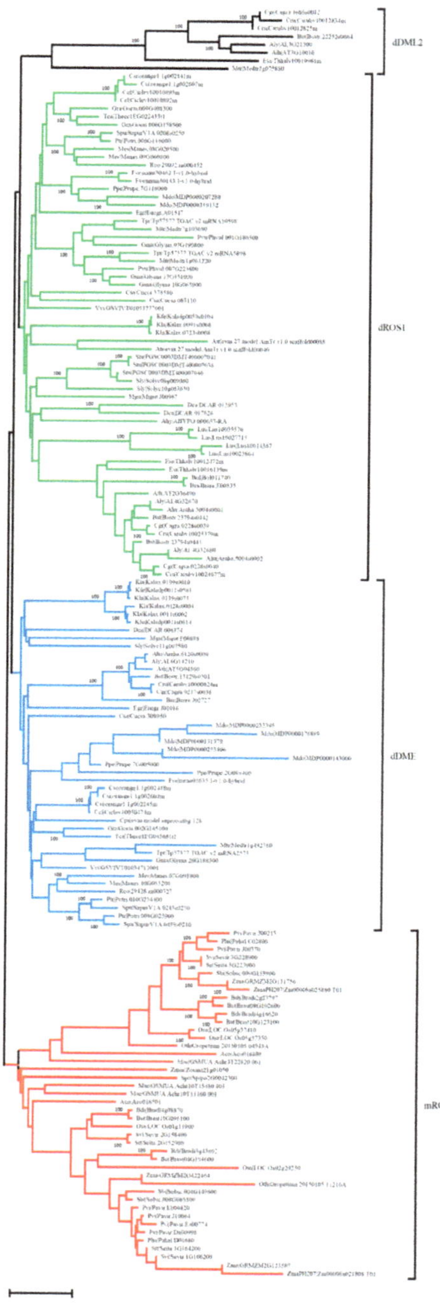

Figure 2. Phylogenetic analysis of the *ROS1* and *ROS1*-like DNA glycosylase family. A phylogenetic tree based on conserved DNA glycosylase domain was constructed. Monocot ROS1 (mROS1) proteins are shown in red color, while dicot DME (dDME), dicot ROS1 (dROS1) and dicot DML2 (dDML2) proteins are shown in blue, green and black, respectively. Protein sequences were downloaded from Phytozome (https://phytozome-next.jgi.doe.gov) (accessed on 19 October 2018).

The sequence logos of the identified DNA glycosylase domain in all 52 species were generated using the WebLogo program to further verify the conservation of aa residues (Figure 3A). A total of three conserved motifs were determined, including the Helix-hairpin-Helix (H-h-H), GPD and [4Fe-4S] motifs. The motifs length ranged from 21 to 26 aa (Figure 3B). The [4Fe-4S] cluster motif contained four cysteine residues that function to keep a [4Fe-4S] cluster and was essential for 5mC excision.

A

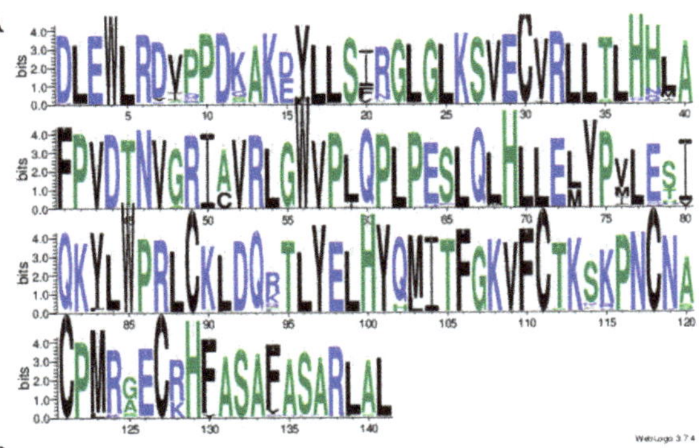

B

Motif	Length (aa)	Consensus sequence
H-h-H motif	25	KAKDYLLSIRGLGLKSVECVRLLTL
GPD motif	21	PVDTNVGRICVRLGWVPLQPL
[4Fe-4S] motif	26	GKVFCTKSKPNCNACPMRAECKHFAS

Figure 3. Conserved domain analysis of DNA glycosylase. (**A**) The conserved domain analysis of DNA glycosylase by the WebLogo program. The letter height shows the amino acid residue at each position designating the conservation degree. On the *x*-axis, numbers represent the sequence position in the corresponding conserved domains. The *y*-axis indicates the information content calculated in bits. (**B**) The consensus sequences of three motifs in the glycosylase domain.

2.2. Generation of OsROS1a Gene Editing Mutants by CRISPR/Cas9

Previous studies showed that the disrupted *OsROS1a* exhibited severe defects in both male and female gametogenesis to produce a sterile phenotype [10,11]. To well describe the function of *OsROS1a* in active DNA demethylation on pollen and seed development, the stable *OsROS1a* gene editing mutants were created using the CRISPR/Cas9 system, and three stable mutants (S6, S7, S16) were obtained. The S6 and S16 mutants are homozygous mutations on both target sides, while S16 had a homozygous mutation in the first target and a biallelic mutation in the second target (Figure 4A and Figure S1). All mutants caused a frame shift and generated premature stop codons at the thirteenth exon of *OsROS1a* (Figure 4B), which made a frame shift from the 1798th amino acid. The S6 and S16 mutants created a stop codon at the 1807th amino acid, while S7 terminated translation by a stop codon at the 1820th amino acid by a one base deletion in the first target (Figure 4B). Both frame shifts occurred before Pem-CXXC and RRMF domains that altered the function of cytosine DNA demethylation.

2.3. The osros1a Mutations Cause Complete Male Sterility

All three obtained *osros1a* mutants were completely male sterile, exhibiting shrunk spikelet morphology and smaller anthers compared to WT controls (Figure 5A–C). To understand the male sterile phenotype of *osros1a* mutants, we then examined the starch-staining capacity of pollen grains and found that the pollen grains of the mutants were not stained by I_2-KI solution, indicating that they contained little or no starch. While the pollen grains were stained dark within WT controls (Figure 5D–F), and the fertile seed percentage in all mutants were 0% compared to WT controls (77%) (Figure 5G). Furthermore, at the harvest stage, agronomic traits were examined and no significant difference was found between the WT and *osros1a* mutants (Figure S2). The plant height only increased by 2.3% and 0.4% in S6 and S16 mutants, respectively, when compared to WT controls. Similarly, the primary branch number per panicle increased in S6 and S16 by 2.7% and 5.4% compared to WT controls. While in the case of panicle length, a 2.9% decrease was observed only in S16 compared to WT controls. Moreover, the tiller number decreased by 1.3% in S16 mutants only compared to WT (Figure S2).

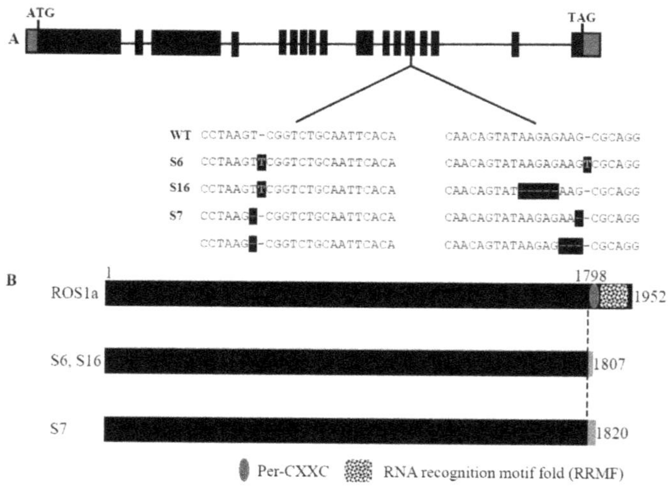

Figure 4. CRISPR/Cas9-induced *OsROS1a* gene modification in rice. (**A**) Schematic of *OsROS1a* gene structure. Exons and introns are denoted as black blocks and lines, respectively. The translation initiation codon (ATG) and termination codon (TAG) are shown. The recovered mutated alleles are shown below the wild-type reference sequence. The target sites nucleotides are indicated in black capital letters. The white dashes with black backgrounds indicate the deleted nucleotides. The white capital letters with black backgrounds indicate the inserted nucleotides. (**B**) Alignment of the protein sequence of *ROS1a* in mutant and wild type. The open reading frame (ORF) is 1952 amino acid residues in length in wild type. The dotted line indicates the frame shift mutation position because of the insertion or deletion of nucleotides due to the gene editing in mutants. The gray color indicates the positions where the premature stop codon was generated in the mutant alleles.

To further investigate the cellular defects in the *osros1a* mutants, we performed transverse section analysis on the anthers of the WT and *osros1a* mutant. Previous studies have divided the development of rice anthers into 14 stages based on the morphological landmarks of cellular events [28,29]. The anther sections were observed at developmental stages 6, 9, 10 and 13 under light microscopy. At stage 6 normal epidermis, endothecium, middle layer, tapetum, and microspore mother cells (MMC) were found in both WT and the *osros1a* S16 mutant (Figure 6A,E). At stage 9, the microspores were released from tetrads, tapetal cells had deeply stained cytoplasm and the middle layer was hardly visible, and the middle layer appeared degenerated and almost invisible in both mutant and WT. However, the microspores of the WT were globular, whereas those of the *osros1a* mutant were irregularly shaped (Figure 6B,F). At stage 10, in contrast to the WT microspores that were vacuolated and round, the *osros1a* mutant microspores appeared degraded and irregularly shaped (Figure 6C,G). At stage 13, the WT anther locule was full of mature pollen grains with completely formed pollen walls and starch accumulation, which can be deeply stained with toluidine blue. By contrast, the pollen grains of the *osros1a* mutant were aborted (Figure 6D,H), indicating that pollen viability and function are mainly affected by the accumulation of starch and lipids [28,30].

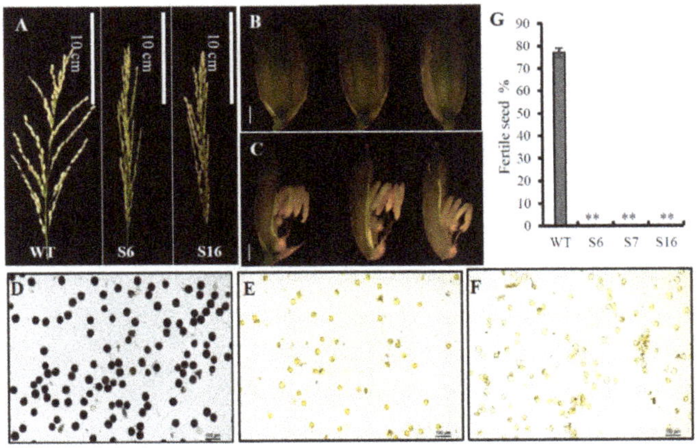

Figure 5. The male fertility phenotype of WT and the *osros1a* mutants. (**A**) Comparison of the mature panicles between WT and *osros1a* mutants showed that *osros1a* mutants failed to produce seeds. Scale bars 10 cm. Comparison of (**B**) the spikelets and (**C**) the anthers phenotype between WT and *osros1a* mutants before anthesis. Scale bars 1 mm. (**D–F**) The pollen grains of WT and *osros1a* mutants staining with I_2-KI staining, respectively. Compared to (**D**) WT, (**E**) S6, and (**F**) S16 failed to form viable pollen. Magnified 20X. (**G**) The fertile seed percentage in WT and S6, S7, S16 mutants. ** denote significant differences from WT at $p < 0.01$ level.

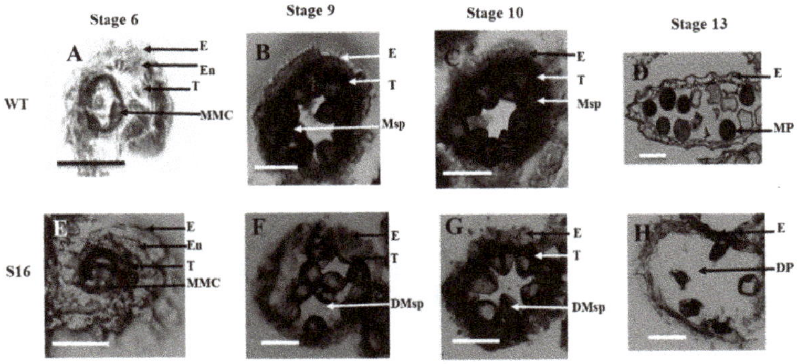

Figure 6. Transverse section analysis of the anther development in WT and *osros1a* S16 mutant. Locules from the anther section of WT and S16 at (**A**,**E**) stage 6 (the microspore mother cells stage), (**B**,**F**) stage 9 (the young microspore stage), (**C**,**G**) stage 10 (the vacuolated pollen stage), and (**D**,**H**) stage 13 (the mature pollen stage). WT sections are shown in (**A–D**); *osros1a* S16 mutant in (**E–H**). E, epidermis; T, tapetum; En, endothecium; MMC, microspore mother cells; Msp, microspores; MP, mature pollen; DMsp, degenerated microspores; DP, the degenerated pollen. Scale bar: 50 µm.

Sugar supply and starch accumulation are essential for pollen grain development that will further affect seed maturation [31–33]. Therefore, the total soluble sugar and starch contents were analyzed in the anthers to investigate whether the *OsROS1a* gene can affect sugar and starch synthesis in pollen grains. The reduction of 36.2% and 56.5% soluble sugar was observed in S6 and S16 mutants, respectively, compared to WT controls (Figure 7A), and similarly 37.1% and 57.9% starch content was significantly ($p < 0.01$) decreased in both mutants, respectively, compared to WT controls (Figure 7B) This was consistent with the pollen staining and transverse section analysis (Figures 5 and 6), confirming that the *osros1a* mutation caused complete male sterile in rice.

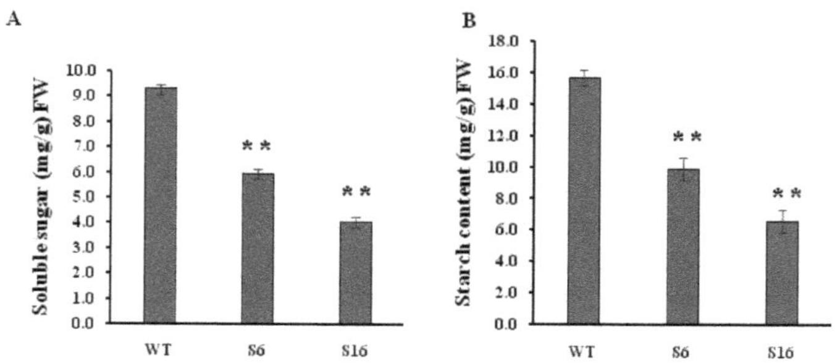

Figure 7. Soluble sugar and starch contents in mature anthers of WT and *osros1a* mutants. (**A**) Total soluble sugar content in mature anthers of WT and *osros1a* mutants. (**B**) Starch content of mature anthers in WT and *osros1a* mutants. Anthers were sampled at stage 13 (the mature pollen stage). Data are means ± SD ($n = 3$). ** denote significant differences from WT at $p < 0.01$ level.

2.4. The Reduction Expression of Genes Involved in Anther and Pollen Development in Osros1a Mutants

To identify the potential targets responsible for the *osros1a* mutant phenotype, qRT-PCR was performed to validate the expression of anther and pollen development-related genes in young panicles. The results showed that the expression of *OsPKS2* and *CYP704B2* genes was reduced in S6 and S16 mutants compared to WT controls (Figure 8B,D). It might be suggested that *OsROS1a* may involve in the activation of these genes by the process of DNA demethylation. Interestingly, the *OsROS1a* gene in mutants had half the expression level of WT controls (Figure 8A), suggesting that the frame-shift mutations that altered the Pem-CXXC and RRMF domains could cause the sterility phenotype in rice. The expression of genes related to soluble sugar and starch synthesis were further investigated to validate the previously obtained soluble sugar and starch content analysis results and found that the expression of *CSA* was decreased in mutants compared to WT controls (Figure 8G), and the expression of other genes was not significantly changed (Figure 8C,E,F,H) These results are consistent with the pollen staining and transverse section analysis (Figures 5 and 6), showing that the *osros1a* mutant promotes the pollen sterility phenotype.

2.5. DNA Hypermethylation of OsPKS2 Gene Promoter in Osros1a Mutants

To reveal the function of *OsROS1a* in DNA demethylation in pollen, the DNA methylation level of the 221 bp *OsPKS2* gene promoter was investigated (Figure 9A). The methylation levels at both CG and CHG sites of the *OsPKS2* promoter were increased in *osros1a* mutants compared to WT controls (Figure 9B,C). In CG residues, 96.15% and 100% methylation levels were observed in S6 and S16 mutants, respectively, while only 51.79% in WT controls. In CHG residues, methylation levels increased from 54.29% in WT controls to 66.15% and 67.69% in S6 and S16 mutants, respectively. However, no obvious difference was observed in CHH residues between WT and S16, whilst a decrease was observed in S6 mutants (Figure 9B,C). These results suggest that DNA hypermethylation in the promoter region of *OsPKS2* may repress the gene expression. While DNA methylation of the *TDR* promoter was not significantly changed between WT and *osros1a* mutants in all three contexts (Figure S3).

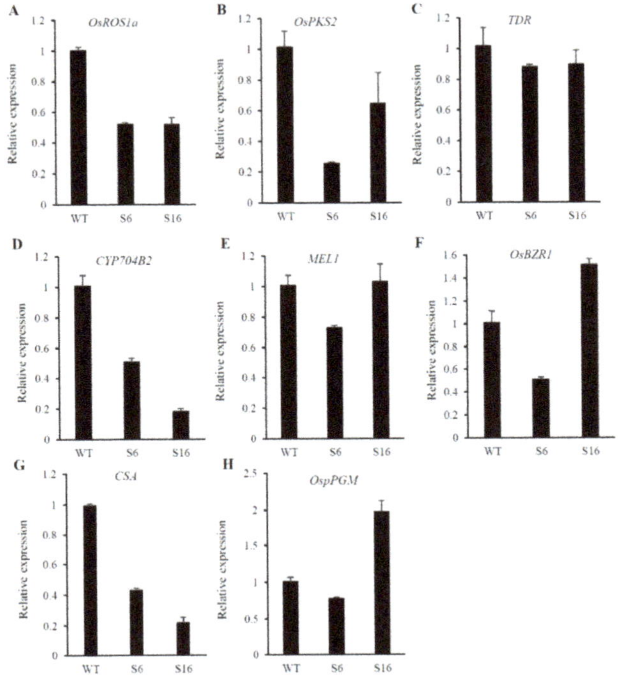

Figure 8. Expression patterns of anther and pollen development-related genes (**A–E**) and starch synthesis-related genes were analyzed by qRT-PCR. The young panicles (at stage 10, the vacuolated pollen stage) of WT, *osros1a* S6 and S16 mutants were sampled for RNA extraction. Data are shown as means ± standard deviations, and *OsACTIN* was used as an internal control.

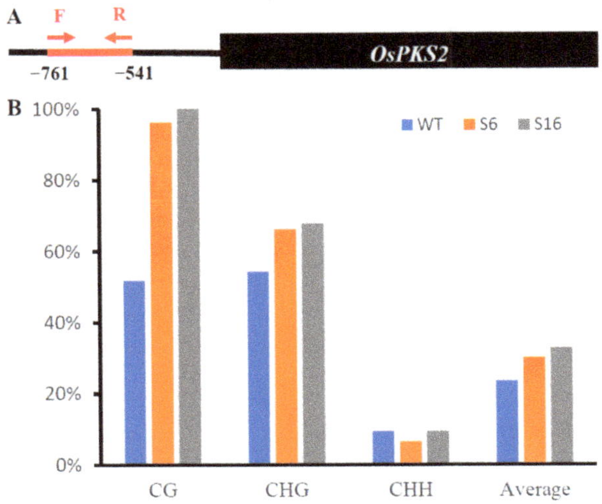

Figure 9. *Cont.*

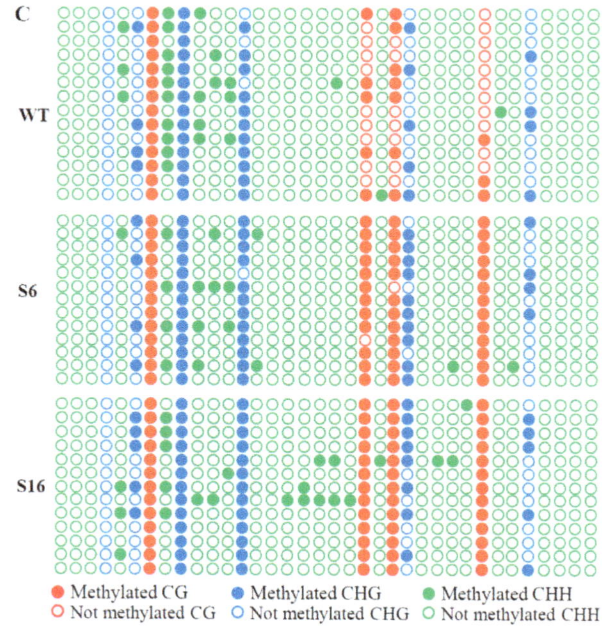

Figure 9. The DNA methylation of *OsPKS2* gene promoter in young panicles. (**A**) The diagram of bisulfite sequencing of 221 bp *OsPKS2* promoter. (**B**) The statistics of bisulfite sequencing of 221 bp *OsPKS2* promoter in WT and *osros1a* mutants. (**C**) The dot plot comparison of DNA methylation between WT and *osros1a* mutants.

3. Discussion

The loss function of the *OsROS1a* gene caused complete male sterility with degenerated pollen (Figures 5 and 6), which was consistent with previous studies [10–12]. Starch is the key storage reservoir in matured pollen grains in cereal crops, which is important for pollen germination, pollen tube growth, supplying energy and a carbon skeleton. In pollen grains, the starch granule accumulation starts at stage 11 and prolongs throughout stage 12 until maturation [29]. *OsBZR1* gene was expressed in anther vascular tissue, tapetum and the developing seeds, and *OsBZR1* directly promotes *CSA* expression thus activating downstream gene expression [30]. Similar to sugar partitioning in rice, *CSA* is a key transcriptional regulator during male reproductive development, and mutation in the *CSA* results in carbohydrate level reduction in the later anthers and male sterility [33]. The deficiency of starch synthesis in pollen grains could cause male sterility [34–37]. Our study showed that the soluble sugar and starch contents were significantly reduced in pollen from *osros1a* mutants compared to WT controls (Figure 7), which could be the reason for male sterility affected by disrupted *OsROS1a* gene function.

Previous research has revealed that *OsPKS2* [23], *TDR* [25], *MEL1* [26], and *CYP704B2* [27] are involved in anther and pollen development and mutation in these genes caused rice male sterility. The expression of *OsPKS2* and *CYP704B2* genes were reduced in *osros1a* mutant anther compared to WT controls (Figure 8), suggesting that *OsROS1a* may be involved in the activation of these genes, loss-of-function of *OsROS1a* could degenerate the tapetum and form abnormal pollen walls causing male sterility (Figures 5 and 6). The DNA demethylase genes contain a conserved DNA glycosylase domain, which can be classified into monofunctional and bifunctional types. DNA glycosylase is bifunctional in plants removing the 5-meC base and then cleaving the DNA backbone at the abasic site [38]. ROS1 is a bifunctional DNA glycosylase that can execute DNA demethylation, especially gene promoter DNA methylation by a base excision repair pathway [8,38]. The reduced expression of pollen and anther development, and starch- and soluble sugar-related genes in *osros1a* mutants demonstrated that *OsROS1a* may regulate the activation of these genes by controlling their DNA methylation. Bisulfite sequencing results showed that the CG and CHG methylations in the *OsPKS2* gene promoter were significantly increased in the *osros1a* mutants when compared to WT controls, but not CHH methylation (Figure 6), suggesting that *OsROS1a* facilitated CG and CHG demethylation. However, DNA methylation of the *TDR* promoter was not significantly changed between WT and *osros1a* mutants (Figure S2), indicating that the demethylation action of the *OsROS1* gene on pollen and anther development-related genes was not global, but gene-specific, which was also found in SSP genes in maize [39].

The expression of the *OsROS1a* gene in the mutants was half that of WT controls (Figure 8), and 1807, 1807 and 1820 amino acids were predicted for S6, S16 and S7, respectively, which resulted in the frame-shift mutation to truncate the Pem-CXXC and RRMF domains (Figures 1 and 3). The whole DNA glycosylase domain, followed by an EndIII_4Fe-4S domain and two putative nuclear localization signals are complete [15]. The C-terminal of DME and ROS1 including divergent, circularly Per-CXXC and RRMF domains are conserved [40], which has a function in excising 5-meC in vitro [5,41,42]. The zinc finger CxxC (ZF-CxxC) in DNA methyltransferase 1 (*DNMT1*) can block the catalytic activity of *DNMT1* specifically on non-methylated DNA [43]. Furthermore, the *OsROS1a* knock-in mutant disrupted the whole gene affecting both male and female gametophytes [10], while truncation the RRMF domain in *OsROS1a* knock-out mutants using CRISPR/Cas9 induced pollen and embryo sac defects in rice [11]. Recently, a 75 bp deletion caused the complete loss of the Per-CXXC domain, but retention of the whole RRMF domain, which showed partially sterile pollen in rice [44]. All these results suggest that the RRMF domain in OsROS1a could be the main factor in the sterility in rice. The RNA-Seq and whole-genome bisulfite sequencing will be the future work, which will clearly clarify the molecular mechanism of how the *OsROS1a* gene affects the sterility phenotype in rice through its DNA demethylation function.

4. Materials and Methods

4.1. Identification of OsROS1a Homologous Genes and Phylogenetic Analysis

The gene structure (Exon-intron distribution) and protein (CDS) sequence of the DNA demethylase gene *OsROS1a* was obtained from Phytozome (https://phytozome-next.jgi.doe.gov/) (accessed on 19 October 2018). The conserved domain sequence position of the OsROS1a protein was searched using the Inter Pro software. The protein sequences of OsROS1a were used to identify the homologous gene copies in 52 plant species by BLAST search. The protein sequences of all identified *OsROS1a* homologous genes were aligned via Muscle by the default settings with manual alterations. The phylogenetic analysis was done by the neighbor-joining (NJ) method using MAGA-X with the default parameters, and the bootstrap test with 1000 replicates was made to determine the confidence of the evolutionary tree [45]. The conserved DNA glycosylase domains (151 amino acid sequences) were analyzed by the WebLogo program (http://weblogo.berkeley.edu/) (accessed on 15 January 2019).

4.2. OsROS1a Gene Editing Mutants Generated by CRISPR/Cas9

The CRISPR/Cas9 system was used for generating *osros1a* mutants [46]. Two single guide RNAs (sgRNAs) were designed to precisely target the thirteenth exon of the *OsROS1a* gene by using the web tool (http://cbi.hzau.edu.cn/crispr/) (accessed on 19 December 2016), which were ligated with rice OsU6a and OsU6b small nuclear RNA promoters, respectively. The sgRNAs were cloned into a binary vector that contained the sgRNA and Cas9 expression cassettes. Then the binary vector was transferred into rice (Nipponbare) calluses by following the method of *Agrobacterium*-mediated transformation of *Japonica* Rice.

Genomic DNA was isolated from the leaves of the rice transgenic plants to detect the mutation types. The Cas9-specific primers were used to identify the successful transgenic plants by PCR and Figure S4A showed the gel electrophoresis image. And the target gene-specific primers were used to amplify and sequence the DNA fragments containing the target sequences (Figure S4B). Then the DSDecode (http://skl.scau.edu.cn/dsdecode/) (accessed on 12 September 2017), a web-based tool, was used to further analyze the genotypes of the targeted mutations.

4.3. Pollen Fertility Examination and Histochemical Assay

Floral organs were photographed with a dissecting microscope. The anthers from the wild type (WT) and *osros1a* mutants were sampled from the spikelets just before flowering and a 1% potassium iodide (I_2-KI) solution was used to stain the pollen grains. A Nikon eclipse Ni fluorescence microscope was used to visualize and photograph the stained pollen grains. For transverse section analysis, spikelets of various anther development stages according to the previous study [29] were collected and frozen in OCT compound (Tissue-Tek) (Sakura Finetek, Torrance, CA USA). Samples were sectioned with the help of a freezing microtome (Shandon Cryotome FE) (ThermoFisher Scientific, Waltham, MA USA). Then, 0.05% toluidine blue dye was used to stain the crossed sections and microscopic images were captured by a Nikon eclipse Ni fluorescence microscope.

4.4. Total Soluble Sugar and Starch Contents Analysis

The 0.05 g anthers from WT and *osros1a* mutants were weighed and ground in 5 mL of 80% ethanol according to the Anthrone method [47]. The mixed samples were centrifuged at 8000 rpm and 1 mL supernatant was obtained. After centrifugation, the pellets were used for starch analysis. The 5 mL anthrone was mixed with 1 mL supernatant and heated at 100 °C in a water bath for 10 min. The mixture was cool down, and the sugar content was then determined by a spectrophotometer at a wavelength of 620 nm. Starch analysis was done by perchloric acid digestion and starch content was determined spectrophotometrically at 630 nm wavelength.

4.5. Gene Expression by RT-PCR Analysis

Total RNA was extracted from young panicles of WT and *osros1a* mutants using TRIzol® Reagent (ThermoFisher, Scientific, Waltham, MA USA). RT reagent Kit PrimeScript™ having gDNA Eraser (Takara, Japan) was used to eliminate genomic DNA residues, then cDNA was synthesized. For RT-PCR reactions, the first strand of cDNA was used with gene-specific primers and the *OsACTIN* gene was used as an internal control (Table S1).

4.6. DNA Methylation Analysis

For bisulfite sequencing of *OsPKS2* and *TDR* gene promoters, genomic DNA from young panicles was extracted by the CTAB method. The EpiTect Bisulfite Kit from Qiagen was used for bisulfite conversion of genomic DNA following the manufacturer's instructions. The primers were designed using Methyl Primer Express v1.0 (Applied biosystem) (Table S1). DNA samples were amplified by PCR and separated by 1.5% agarose gel. Then, PCR products were purified with the gel purification kit, and then cloned into the pMD19-T vector and sequenced. Kismeth tool (http://katahdin.mssm.edu/kismeth/revpage.pl) (accessed on 21 October 2019) was used for DNA methylation analysis.

Supplementary Materials: The following supporting information can be downloaded at: https://www.mdpi.com/article/10.3390/ijms231911349/s1, Figure S1: DNA sequencing chromatograms of *osros1a* mutants; Figure S2: Phenotype comparison of the WT with *osros1a* mutants; Figure S3: The DNA methylation of the *TDR* gene promoter in young panicle; Figure S4: PCR of *OsROS1a* gene editing plants (A) with Cas9 primers and (B) with knockout target-specific primers; Table S1: Primers used in this study.

Author Contributions: Conceptualization, J.-H.X.; methodology, J.-H.X., F.I., and C.L.; validation, C.L. and Y.Y.; formal analysis, J.-H.X., F.I., and C.L.; investigation, J.-H.X.; resources, J.-H.X.; data curation, J.-H.X., F.I., Y.Y. and C.L.; writing—original draft preparation, F.I. and C.L.; writing—review and editing, J.-H.X. and C.L.; supervision, J.-H.X.; funding acquisition, J.-H.X. All authors have read and agreed to the published version of the manuscript.

Funding: This work was supported by the projects from Zhejiang Zhengjingyuan Pharmacy Chain Co., Ltd. (H20151699 and H20151788) and Shandong (Linyi) Institute of Modern Agriculture, Zhejiang University.

Institutional Review Board Statement: Not applicable.

Informed Consent Statement: Not applicable.

Data Availability Statement: Not applicable.

Acknowledgments: We are grateful to Xinxin Li for her obtaining *osros1a* mutants, and Yao-Guang Liu for kindly providing the CRISPR/Cas9 binary vector.

Conflicts of Interest: The authors declare no conflict of interest.

References

1. Zhang, H.; Lang, Z.; Zhu, J.K. Dynamics and function of DNA methylation in plants. *Nat. Rev. Mol. Cell Biol.* **2018**, *19*, 489–506. [CrossRef] [PubMed]
2. Nash, H.M.; Bruner, S.D.; Scharer, O.D.; Kawate, T.; Addona, T.A.; Spooner, E.; Lane, W.S.; Verdine, G.L. Cloning of a yeast 8-oxoguanine DNA glycosylase reveals the existence of a base-excision DNA-repair protein superfamily. *Curr. Biol.* **1996**, *6*, 968–980. [CrossRef]
3. Zhu, J.; Kapoor, A.; Sridhar, V.V.; Agius, F.; Zhu, J.K. The DNA glycosylase/lyase ROS1 functions in pruning DNA methylation patterns in *Arabidopsis*. *Curr. Biol.* **2007**, *17*, 54–59. [CrossRef]
4. Agius, F.; Kapoor, A.; Zhu, J.K. Role of the Arabidopsis DNA glycosylase/lyase ROS1 in active DNA demethylation. *Proc. Natl. Acad. Sci. USA* **2006**, *103*, 11796–11801. [CrossRef]
5. Choi, Y.; Gehring, M.; Johnson, L.; Hannon, M.; Harada, J.J.; Goldberg, R.B.; Jacobsen, S.E.; Fischer, R.L. DEMETER, a DNA glycosylase domain protein, is required for endosperm gene imprinting and seed viability in *Arabidopsis*. *Cell* **2002**, *110*, 33–42. [CrossRef]
6. Penterman, J.; Zilberman, D.; Huh, J.H.; Ballinger, T.; Henikoff, S.; Fischer, R.L. DNA demethylation in the *Arabidopsis* genome. *Proc. Natl. Acad. Sci. USA* **2007**, *104*, 6752–6757. [CrossRef]
7. Schoft, V.K.; Chumak, N.; Choi, Y.; Hannon, M.; Garcia-Aguilar, M.; Machlicova, A.; Slusarz, L.; Mosiolek, M.; Park, J.S.; Park, G.T.; et al. Function of the DEMETER DNA glycosylase in the *Arabidopsis thaliana* male gametophyte. *Proc. Natl. Acad. Sci. USA* **2011**, *108*, 8042–8047. [CrossRef]
8. Gong, Z.; Morales-Ruiz, T.; Ariza, R.R.; Roldan-Arjona, T.; David, L.; Zhu, J.K. ROS1, a repressor of transcriptional gene silencing in *Arabidopsis*, encodes a DNA glycosylase/lyase. *Cell* **2002**, *111*, 803–814. [CrossRef]
9. Zemach, A.; Kim, M.Y.; Silva, P.; Rodrigues, J.A.; Dotson, B.; Brooks, M.D.; Zilberman, D. Local DNA hypomethylation activates genes in rice endosperm. *Proc. Natl. Acad. Sci. USA* **2010**, *107*, 18729–18734. [CrossRef]
10. Ono, A.; Yamaguchi, K.; Fukada-Tanaka, S.; Terada, R.; Mitsui, T.; Iida, S. A null mutation of ROS1a for DNA demethylation in rice is not transmittable to progeny. *Plant J.* **2012**, *71*, 564–574. [CrossRef]
11. Xu, Y.; Wang, F.; Chen, Z.; Wang, J.; Li, W.; Fan, F.; Tao, Y.; Jiang, Y.; Zhu, Q.H.; Yang, J. CRISPR/Cas9-targeted mutagenesis of the *OsROS1* gene induces pollen and embryo sac defects in rice. *Plant Biotechnol. J.* **2020**, *18*, 1999–2001. [CrossRef] [PubMed]
12. Kim, M.Y.; Ono, A.; Scholten, S.; Kinoshita, T.; Zilberman, D.; Okamoto, T.; Fischer, R.L. DNA demethylation by ROS1a in rice vegetative cells promotes methylation in sperm. *Proc. Natl. Acad. Sci. USA* **2019**, *116*, 9652–9657. [CrossRef] [PubMed]
13. Liu, J.; Wu, X.; Yao, X.; Yu, R.; Larkin, P.J.; Liu, C.M. Mutations in the DNA demethylase OsROS1 result in a thickened aleurone and improved nutritional value in rice grains. *Proc. Natl. Acad. Sci. USA* **2018**, *115*, 11327–11332. [CrossRef] [PubMed]
14. Lang, Z.; Gong, Z. A role of OsROS1 in aleurone development and nutrient improvement in rice. *Proc. Natl. Acad. Sci. USA* **2018**, *115*, 11659–11660. [CrossRef]
15. La, H.; Ding, B.; Mishra, G.P.; Zhou, B.; Yang, H.; Bellizzi Mdel, R.; Chen, S.; Meyers, B.C.; Peng, Z.; Zhu, J.K.; et al. A 5-methylcytosine DNA glycosylase/lyase demethylates the retrotransposon Tos17 and promotes its transposition in rice. *Proc. Natl. Acad. Sci. USA* **2011**, *108*, 15498–15503. [CrossRef]
16. Wen, S.; Wen, N.; Pang, J.; Langen, G.; Brew-Appiah, R.A.; Mejias, J.H.; Osorio, C.; Yang, M.; Gemini, R.; Moehs, C.P.; et al. Structural genes of wheat and barley 5-methylcytosine DNA glycosylases and their potential applications for human health. *Proc. Natl. Acad. Sci. USA* **2012**, *109*, 20543–20548. [CrossRef]
17. Guo, J.X.; Liu, Y.G. Molecular control of male reproductive development and pollen fertility in rice. *J. Integr. Plant Biol.* **2012**, *54*, 967–978. [CrossRef]
18. Scott, R.J.; Spielman, M.; Dickinson, H.G. Stamen structure and function. *Plant Cell* **2004**, *16* (Suppl. 1), S46–S60. [CrossRef]

19. McCormick, S. Control of male gametophyte development. *Plant Cell* **2004**, *16* (Suppl. 1), S142–S153. [CrossRef]
20. Ma, H. Molecular genetic analyses of microsporogenesis and microgametogenesis in flowering plants. *Annu. Rev. Plant Biol.* **2005**, *56*, 393–434. [CrossRef]
21. Wilson, Z.A.; Zhang, D.B. From Arabidopsis to rice: Pathways in pollen development. *J. Exp. Bot.* **2009**, *60*, 1479–1492. [CrossRef] [PubMed]
22. Suzuki, G. Recent progress in plant reproduction research: The story of the male gametophyte through to successful fertilization. *Plant Cell Physiol.* **2009**, *50*, 1857–1864. [CrossRef] [PubMed]
23. Zou, T.; Liu, M.; Xiao, Q.; Wang, T.; Chen, D.; Luo, T.; Yuan, G.; Li, Q.; Zhu, J.; Liang, Y.; et al. OsPKS2 is required for rice male fertility by participating in pollen wall formation. *Plant Cell Rep.* **2018**, *37*, 759–773. [CrossRef] [PubMed]
24. Zhang, D.S.; Liang, W.Q.; Yuan, Z.; Li, N.; Shi, J.; Wang, J.; Liu, Y.M.; Yu, W.J.; Zhang, D.B. Tapetum degeneration retardation is critical for aliphatic metabolism and gene regulation during rice pollen development. *Mol. Plant* **2008**, *1*, 599–610. [CrossRef]
25. Li, N.; Zhang, D.S.; Liu, H.S.; Yin, C.S.; Li, X.X.; Liang, W.Q.; Yuan, Z.; Xu, B.; Chu, H.W.; Wang, J.; et al. The rice tapetum degeneration retardation gene is required for tapetum degradation and anther development. *Plant Cell* **2006**, *18*, 2999–3014. [CrossRef] [PubMed]
26. Lian, J.P.; Yang, Y.W.; He, R.R.; Yang, L.; Zhou, Y.F.; Lei, M.Q.; Zhang, Z.; Huang, J.H.; Cheng, Y.; Liu, Y.W.; et al. Ubiquitin-dependent Argonauteprotein MEL1 degradation is essential for rice sporogenesis and phasiRNA target regulation. *Plant Cell* **2021**, *33*, 2685–2700. [CrossRef]
27. Li, H.; Pinot, F.; Sauveplane, V.; Werck-Reichhart, D.; Diehl, P.; Schreiber, L.; Franke, R.; Zhang, P.; Chen, L.; Gao, Y.; et al. Cytochrome P450 family member CYP704B2 catalyzes the {omega}-hydroxylation of fatty acids and is required for anther cutin biosynthesis and pollen exine formation in rice. *Plant Cell* **2010**, *22*, 173–190. [CrossRef]
28. Zhang, D.B.; Wilson, Z.A. Stamen specification and anther development in rice. *Chin. Sci. Bull.* **2009**, *54*, 2342–2353. [CrossRef]
29. Zhang, D.; Luo, X.; Zhu, L. Cytological analysis and genetic control of rice anther development. *J. Genet. Genomics* **2011**, *38*, 379–390. [CrossRef]
30. Zhu, X.; Liang, W.; Cui, X.; Chen, M.; Yin, C.; Luo, Z.; Zhu, J.; Lucas, W.J.; Wang, Z.; Zhang, D. Brassinosteroids promote development of rice pollen grains and seeds by triggering expression of Carbon Starved Anther, a MYB domain protein. *Plant J.* **2015**, *82*, 570–581. [CrossRef]
31. Zhang, H.; Xu, C.; He, Y.; Zong, J.; Yang, X.; Si, H.; Sun, Z.; Hu, J.; Liang, W.; Zhang, D. Mutation in CSA creates a new photoperiod-sensitive genic male sterile line applicable for hybrid rice seed production. *Proc. Natl. Acad. Sci. USA* **2013**, *110*, 76–81. [CrossRef] [PubMed]
32. Zhang, Y.; Liang, W.; Shi, J.; Xu, J.; Zhang, D. MYB56 encoding a R2R3 MYB transcription factor regulates seed size in *Arabidopsis thaliana*. *J. Integr. Plant Biol.* **2013**, *55*, 1166–1178. [CrossRef] [PubMed]
33. Zhang, H.; Liang, W.; Yang, X.; Luo, X.; Jiang, N.; Ma, H.; Zhang, D. Carbon starved anther encodes a MYB domain protein that regulates sugar partitioning required for rice pollen development. *Plant Cell* **2010**, *22*, 672–689. [CrossRef] [PubMed]
34. Dickinson, D.B. Rapid starch synthesis associated with increased respiration in germinating lily pollen. *Plant Physiol.* **1968**, *43*, 1–8. [CrossRef] [PubMed]
35. Wen, L.Y.; Chase, C.D. Mitochondrial gene expression in developing male gametophytes of male-fertile and S male-sterile maize. *Sex. Plant Rep.* **1999**, *11*, 323–330. [CrossRef]
36. Datta, R.; Chamusco, K.C.; Chourey, P.S. Starch biosynthesis during pollen maturation is associated with altered patterns of gene expression in maize. *Plant Physiol.* **2002**, *130*, 1645–1656. [CrossRef]
37. Wu, Y.; Fox, T.W.; Trimnell, M.R.; Wang, L.; Xu, R.J.; Cigan, A.M.; Huffman, G.A.; Garnaat, C.W.; Hershey, H.; Albertsen, M.C. Development of a novel recessive genetic male sterility system for hybrid seed production in maize and other cross-pollinating crops. *Plant Biotechnol. J.* **2016**, *14*, 1046–1054. [CrossRef]
38. Zhu, J.K. Active DNA demethylation mediated by DNA glycosylases. *Annu. Rev. Genet.* **2009**, *43*, 143–166. [CrossRef]
39. Xu, J.H.; Wang, R.; Li, X.; Miclaus, M.; Messing, J. Locus- and Site-Specific DNA Methylation of 19 kDa Zein Genes in Maize. *PLoS ONE* **2016**, *11*, e0146416. [CrossRef]
40. Iyer, L.M.; Abhiman, S.; Aravind, L. Natural history of eukaryotic DNA methylation systems. *Prog. Mol. Biol. Transl. Sci.* **2011**, *101*, 25–104.
41. Gehring, M.; Bubb, K.L.; Henikoff, S. Extensive demethylation of repetitive elements during seed development underlies gene imprinting. *Science* **2009**, *324*, 1447–1451. [CrossRef] [PubMed]
42. Zhang, C.; Hung, Y.H.; Rim, H.J.; Zhang, D.; Frost, J.M.; Shin, H.; Jang, H.; Liu, F.; Xiao, W.; Iyer, L.M.; et al. The catalytic core of DEMETER guides active DNA demethylation in *Arabidopsis*. *Proc. Natl. Acad. Sci. USA* **2019**, *116*, 17563–17571. [CrossRef] [PubMed]
43. Song, J.; Rechkoblit, O.; Bestor, T.H.; Patel, D.J. Structure of DNMT1-DNA complex reveals a role for autoinhibition in maintenance DNA methylation. *Science* **2011**, *331*, 1036–1040. [CrossRef] [PubMed]
44. Irshad, F.; Li, C.; Wu, H.Y.; Yan, Y.; Xu, J.H. The function of DNA demethylase gene ROS1a null mutant on seed development in rice (*Oryza Sativa*) using the CRISPR/CAS9 system. *Int. J. Mol. Sci.* **2022**, *23*, 6357. [CrossRef] [PubMed]
45. Kumar, S.; Stecher, G.; Li, M.; Knyaz, C.; Tamura, K. MEGA-X: Molecular Evolutionary Genetics Analysis across Computing Platforms. *Mol. Biol. Evol.* **2018**, *35*, 1547–1549. [CrossRef]

46. Ma, X.; Zhang, Q.; Zhu, Q.; Liu, W.; Chen, Y.; Qiu, R.; Wang, B.; Yang, Z.; Li, H.; Lin, Y.; et al. A Robust CRISPR/Cas9 System for Convenient, High-Efficiency Multiplex Genome Editing in Monocot and Dicot Plants. *Mol. Plant* **2015**, *8*, 1274–1284. [CrossRef]
47. Sinay, H.; Karuwa, R.L. Proline and total soluble sugar content at the vegetative phase of six corn cultivars from Kisar Island Maluku, grown under drought stress conditions. *Int. J. Adv. Agric. Res.* **2014**, *2*, 77–82.

International Journal of *Molecular Sciences*

MDPI

Review

mRNA Localization to the Endoplasmic Reticulum in Plant Endosperm Cells

Laining Zhang [1], Qidong Si [1], Kejie Yang [1], Wenwei Zhang [1], Thomas W. Okita [2,*] and Li Tian [1,*]

[1] Collaborative Innovation Center for Efficient and Green Production of Agriculture in Mountainous Areas of Zhejiang Province, College of Horticulture Science, Zhejiang A&F University, Hangzhou 310007, China
[2] Institute of Biological Chemistry, Washington State University, Pullman, WA 99164, USA
* Correspondence: authors: okita@wsu.edu (T.W.O.); li.tian@zafu.edu.cn (L.T.)

Abstract: Subcellular mRNA localization is an evolutionarily conserved mechanism to spatially and temporally drive local translation and, in turn, protein targeting. Hence, this mechanism achieves precise control of gene expression and establishes functional and structural networks during cell growth and development as well as during stimuli response. Since its discovery in ascidian eggs, mRNA localization has been extensively studied in animal and yeast cells. Although our knowledge of subcellular mRNA localization in plant cells lags considerably behind other biological systems, mRNA localization to the endoplasmic reticulum (ER) has also been well established since its discovery in cereal endosperm cells in the early 1990s. Storage protein mRNA targeting to distinct subdomains of the ER determines efficient accumulation of the corresponding proteins in different endosomal storage sites and, in turn, underlies storage organelle biogenesis in cereal grains. The targeting process requires the presence of RNA localization elements, also called zipcodes, and specific RNA-binding proteins that recognize and bind these zipcodes and recruit other factors to mediate active transport. Here, we review the current knowledge of the mechanisms and functions of mRNA localization to the ER in plant cells and address directions for future research.

Keywords: mRNA localization; RNA-binding proteins; zipcode RNA; storage proteins

check for updates

Citation: Zhang, L.; Si, Q.; Yang, K.; Zhang, W.; Okita, T.W.; Tian, L. mRNA Localization to the Endoplasmic Reticulum in Plant Endosperm Cells. *Int. J. Mol. Sci.* **2022**, *23*, 13511. https://doi.org/10.3390/ijms232113511

Academic Editors: Jinsong Bao and Jianhong Xu

Received: 6 September 2022
Accepted: 31 October 2022
Published: 4 November 2022

Publisher's Note: MDPI stays neutral with regard to jurisdictional claims in published maps and institutional affiliations.

Copyright: © 2022 by the authors. Licensee MDPI, Basel, Switzerland. This article is an open access article distributed under the terms and conditions of the Creative Commons Attribution (CC BY) license (https://creativecommons.org/licenses/by/4.0/).

1. Introduction

Asymmetric distribution of mRNAs was first discovered in ascidian eggs and embryos in 1983 [1], where β-actin mRNA was further observed to be specifically localized at the site where muscle-forming cells reside [2]. This observation, followed by investigations of maternal mRNA distributions in *Xenopus* and *Drosophila* oocytes [3–5], supported the previous proposal of prelocalized RNAs during early development [6,7] and raised the hypothesis that specific mRNA pools localized to particular subcellular areas todetermine cell fate and tissue differentiation. The concept of subcellular mRNA localization was then proposed in 1986 by Lawrence and Singer [8], who applied the in situ hybridization technique to locate mRNAs in chicken fibroblasts. This phenomenon was subsequently observed in neurons [9], rice endosperm cells [10], and oligodendrocytes [11]. With the continual findings of mRNA localization in animal, plants, yeast, algae, and even bacteria [12], mRNA localization is proposed to be an ancient, prevalent, universal, and highly conserved mechanism.

Today, mRNA localization is referred to as a mechanism where mRNAs are specifically localized to discrete subcellular compartments. By creating local translation hotspots, mRNA localization provides a highly efficient process to concentrate newly synthesized proteins within a defined intracellular region. mRNA localization enables cells to fine-tune cell polarization, differentiation, and migration as well as quickly respond to intracellular and environmental stimuli.

In eukaryotic cells, localization of mRNAs involves multiple events in both the nucleus and cytoplasm. The process is initiated in the nucleus where cis-acting elements within the RNAs, also called zipcodes RNA elements, are recognized and bound by specific trans-acting factors, typically RNA-binding proteins (RBPs), to form an initial ribonucleoprotein (RNP) complex. The association of mRNAs with some of these RBPs may be maintained during RNA processing and maturation to form a primary messenger ribonucleoprotein (mRNP) complex competent for export from the nucleus. Once transported to the cytoplasm, the primary mRNP complex is remodeled by removing and recruiting one or more protein factors and linking to a cytoskeleton motor protein to initiate mRNA transport. When the mRNP complex arrives at the target location, further remodeling is required to activate local translation, turnover or storage. Thus, mRNA localization is a highly dynamic process. While zipcode elements are an essential prerequisite to determine the localization of mRNAs, multiple RBPs play an extremely important roles throughout the process.

In the past four decades, subcellular mRNA localization has been extensively studied in yeast and metazoan cells, and much of our current understanding on mRNA targeting emanates from research on these organisms. Although mRNA localization in plants was-discovered in the early 1990s, information underlying its basis is relatively poor compared to that generated from other systems. The main reason lies in the structural characteristics of plant cells, which usually contain one or more large central vacuoles, which squeezes the cytoplasm to the periphery of the cell and thus impedes effective observation of mRNA localization in a defined area. To date, the phenomenon in plant cells has only been described in the process of mRNA targeting to the cortical endoplasmic reticulum (ER) in cytoplasmic-rich rice endosperm cells [13], mRNAs enrichment on the mitochondrial surface in *Arabidopsis* [14–16] and potato [17], and localization of viral RNAs on the chloroplast envelope [18–20]. Among these different plant systems, localization of storage protein mRNAs to distinct ER subdomains in rice endosperm cells is relatively well-established, when compared to the bulk mRNA localization to mitochondria and viral mRNA localization to chloroplast. To assist further research on mRNA localization in plant cells, we summarize our current studies and knowledge of mRNA localization to the ER, using rice endosperm cell as a model system.

2. mRNA Targeting to the ER Subdomains Is Driven by Specific RNA Zipcodes

The early view of protein translation in eukaryotes assumed that after nuclear export to the cytoplasm, mRNAs were translated at random locations within the cytosol. Proteins were localized to specific cellular compartments or organelles by peptide-based determinants. N-terminal transit peptides (TP) served as the targeting signals to the chloroplast and mitochondria, while signal peptides directed the growing nascent polypeptide chain during protein synthesis to the ER. In this latter instance, a signal recognition particle (SRP) recognizes and binds to the signal peptide of the nascent polypeptide to direct the association of a complex of mRNA, ribosome and nascent polypeptide chain to the ER membrane. Signal peptides were considered to be necessary and sufficient information to target proteins to the ER in both plant and animal cells [21–23].

In rice endosperm cells, large amount of storage proteins, prolamine, glutelin and α-globulin are synthesized on the ER where these proteins are translocated to the lumen. Prolamines are retained in the lumen, where they co-assemble as intracisternal granules, which matures into an organelle labeled as ER-derived protein body-I (PB-I, prolamine) [24]. By contrast, glutelins and α-globulins are exported from the ER lumen to the Golgi and then transported to protein storage vacuoles (PSV, containing glutelin and α-globulin) to form PB-II (Figure 1A) [24]. Due to the presence of signal peptide elements in these storage proteins, their synthesis on the ER and packaging of the storage proteins into PB-I and PB-II were initially thought to be dependent on their signal peptides. In 1993, using in situ hybridization at the electron microscopy level, the Okita laboratory reported that *prolamine* and *glutelin* mRNAs were localized on two distinct subdomains of the ER [10]. *Prolamine* mRNAs were localized on the ER (PB-ER) that delimit PB-I, while *glutelin* mRNAs were distributed to the adjoining cisternal ER (cis-ER) (Figure 1A). Later, using optimized in situ RT-PCR technique, the group further discovered that when removing the translation initiation codon or signal peptide sequences, prolamine mRNAs remained targeted to the PB-ER [25]. These studies indicate that storage protein mRNAs are not randomly localized on the ER and that the localization process is RNA-based. This hypothesis was substantiated by expressing exogenous reporter genes containing *prolamine* or *glutelin* RNA sequences positioned at their 3′ UTR. The *β-glucuronidase (GUS)* mRNA, which by itself is normally targeted to the cis-ER, was redirected by *prolamine* RNA sequences to the PB-ER [25], while *GFP* mRNAs containing *prolamine* or *glutelin* RNA sequences were localized to the PB-ER and cis-ER, respectively [26].

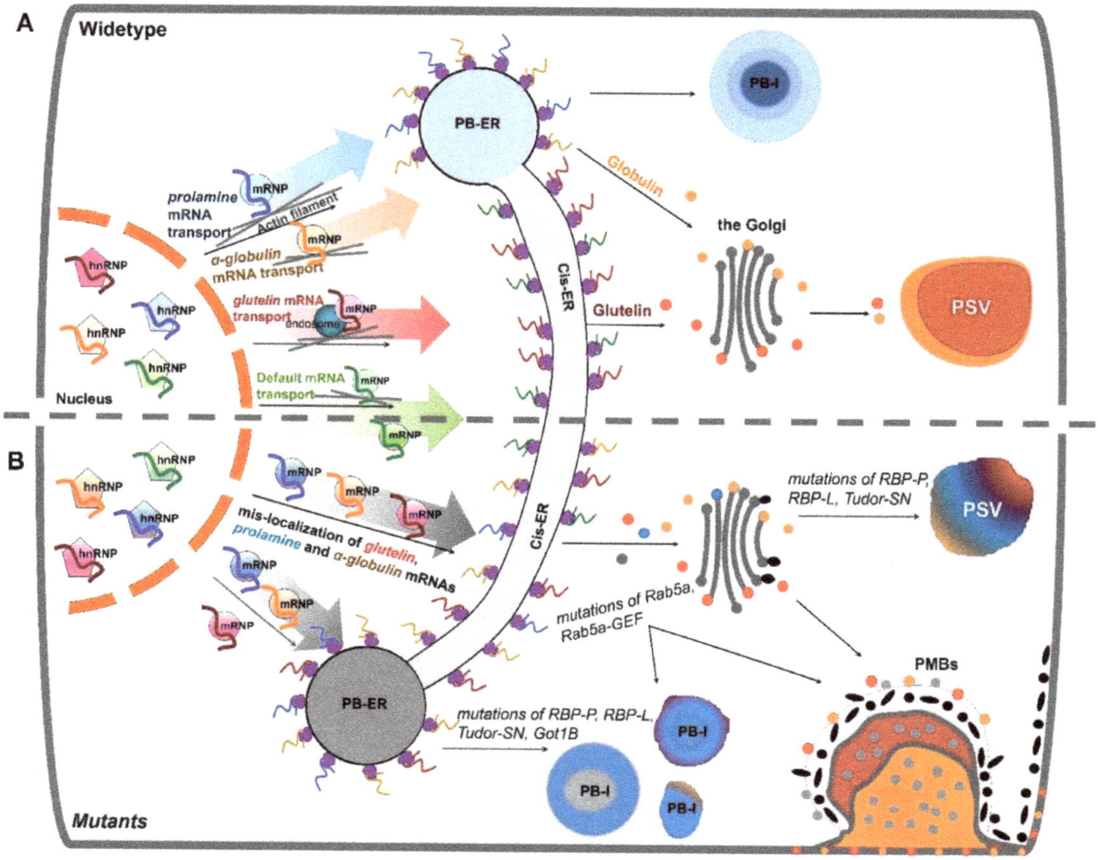

Figure 1. Schematic model of mRNA transport to the cortical ER in developing rice endosperm cells. (**A**) Working model of three pathways to transport mRNAs to distinct ER subdomains in wildtype. The localization of storage protein mRNAs initiates in the nucleus, where newly transcribed mRNAs are recognized and bound by various sets of specific RBPs, forming heterogenous nuclear ribonucleoprotein (hnRNP) complexes. After export from the nucleus to cytoplasm, the complexes undergo dynamic remodeling to form mRNPs that recruit molecular motor proteins for actively transporting to distinct ER subdomains likely on actin filaments (grey lines). *Glutelin* mRNAs (red curve line) are targeted to the cis-ER via endosomal trafficking. Following translation, proglutelins are transported via the Golgi complex to the irregularly shaped PSV, where they are proteolytically processed to acidic and basic subunits and accumulated in the crystalline regions of the PSV (shown in red). *Prolamine* (blue curve line) and *α-globulin* (orange curve line) mRNAs are transported to the PB-ER, where the mRNAs are translated and newly synthesized prolamine polypeptides are assembled as an intracisternal granule to form multi-layered PB-I. The synthesized α-globulins are

rapidly exported to the Golgi for subsequent transport, via dense vesicles, to the peripheral area of the PSVs. An additional default pathway transports zipcode-less mRNAs (green curve line) to the cis-ER. (**B**) Mutations of key RBPs or factors induce mistargeting of storage protein mRNAs and, in turn, their proteins. Mutants carrying mutations in *RBP-P*, *RBP-L* or *Tudor-SN* mistarget both *prolamine* and *glutelin* mRNAs and cause changes in the shape and/or protein components of PB-I and PSVs. *Got1B* mutations disrupts targeting of *prolamine* mRNAs to the PB-ER. Due to the dysfunction of endosomal factor Rab5a and its effector Rab5a-GEF in *glup6* and *glup4* mutants, respectively, mRNPs (light grey dots) carrying *glutelin* mRNAs are partially mislocalized from the cis-ER to the PB-ER and paramural bodies (PMBs). As endosomes are also involved in glutelin and α-globulin protein trafficking from the Golgi to PSVs, both α-globulin (red dots) and proglutelin (orange dots) proteins are found to partially mislocalize in PMBs.

The direct targeting of mRNAs to the ER is best established in yeast. A set of mRNAs, including *ASH1* mRNA, are co-transported on tubular ER that move to the emerging bud or daughter cell [27–30]. The process is found to be driven by multiple cis-acting elements within the mRNA sequences [29]. To investigate the cis-acting elements within the storage protein mRNAs, a series of transgenic rice lines carrying a reporter gene supplemented with partial storage protein mRNA sequences was constructed [26,31,32]. Exogenous GFP whose mRNA was localized on the cis-ER by a default pathway was used as a reporter gene to investigate *prolamine* localization elements [26]. The addition of various 5′ and 3′ deletions of the *prolamine* sequences led to the identification of two apparent cis-acting elements, one located downstream of the signal peptide coding sequence (zipcode 1) and the other in the 3′ UTR (zipcode 2, Figure 2A) [26]. The presence of a single zipcode resulted in only partial localization of prolamine mRNA to the PB-ER. Thus, two cis-acting elements, which shared a conserved U-rich motif element (Figure 2A), are required for restricted *prolamine* mRNA localization.

A similar strategy was applied to identify *glutelin* mRNA localization elements. When the maize *δ-zein* mRNA, a member of the cereal prolamine superfamily [33] was used as a reporter gene, two short sequences located at the 5′ and 3′ ends of the coding region as well as the 3′ UTR of *glutelin* mRNA were sufficient to redirect *δ-zein* mRNA from the PB-ER to the cis-ER (Figure 2B) [31]. These observations suggest that *glutelin* mRNA contains three cis-localization elements and that the glutelin zipcodes are dominant over the *δ-zein* mRNA zipcodes. Further sequence analysis suggests that these glutelin zipcode RNAs contain two conserved motifs (Figure 2B) with the U-rich motif 2 showing some homology to the *prolamine* zipcode [31].

In addition to prolamines and glutelins, rice endosperm cells accumulate small amounts of α-globulins. The saline-soluble proteins are synthesized on the ER-membrane, processed by the Golgi and ultimately deposited together with glutelin in the PSV. Interestingly, *α-globulin* mRNAs are distributed on the PB-ER (Figure 1A), and not on the cis-ER based on the analysis of in situ RT-PCR [32]. Its presence in the 3′ UTR of the *GFP* RNA redirects the hybrid RNA from the cis-ER to the PB-ER as well [32], suggesting that *α-globulin* mRNA sequence contains cis-acting elements for localization on the PB-ER. Sequence analysis revealed that *α-globulin* mRNA sequences possess three candidate zipcodes located at both the coding and non-coding regions, sharing high similarity to *prolamine* zipcode RNAs [32]. Collectively, these findings suggest that both *prolamine* and *α-globulin* mRNAs are directed to the cortical ER by specific zipcode RNAs.

Based on the results of these studies, three mRNA targeting pathways to the cortical ER exist in rice endosperm cells (Figure 1A). While *prolamine* and *α-globulin* mRNAs are localized to the PB-ER, *glutelin* mRNAs are targeted to the cis-ER. Both pathways are zipcode RNA-dependent. The third pathway is a default zipcode-independent pathway, which mediates RNAs, including *GFP* and GUS mRNAs, to the cis-ER. The three pathways are not independent but instead hierarchal and inter-related (Figure 3). The glutelin pathway is dominant, as its zipcodes can redirect the transport of *prolamine* and *α-globulin* mRNAs from the PB-ER to the cis-ER [27,28] (Figure 3). In turn, *prolamine* and *α-globulin* mRNAs are able to redirect *GFP* RNA from the default cis-ER to the PB-ER [26,32].

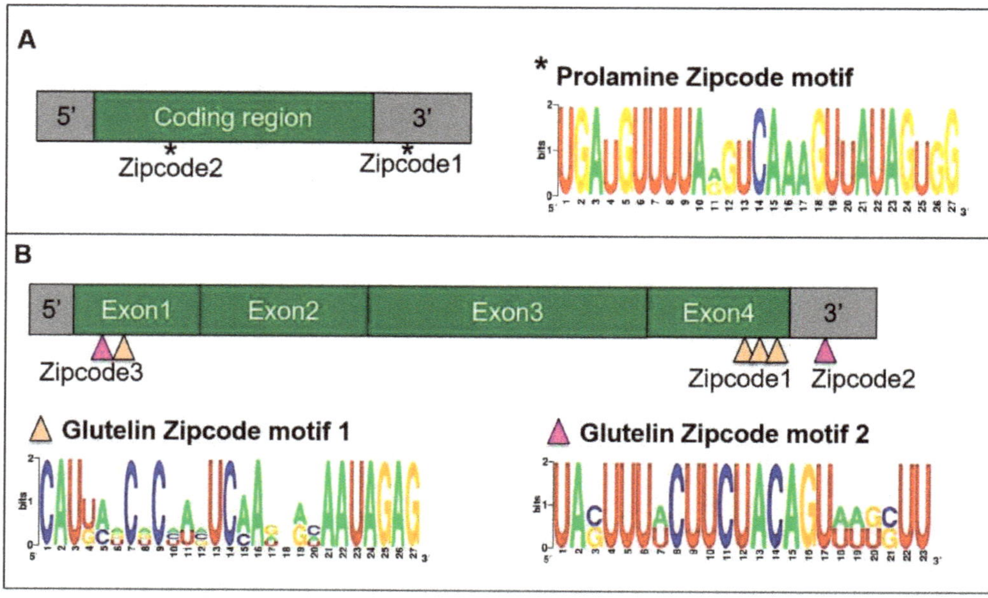

Figure 2. RNA zipcodes identified in *prolamine* (**A**) and *glutelin* (**B**) mRNAs (adapted from [34]). (**A**) The proximate locations (**left**) and the consensus motif sequence (**right**) of the two zipcode elements in *prolamine* mRNA. *Prolamine* zipcode elements are located in the coding region and 3′ UTR and consists of only a single zipcode motif (*). (**B**) The proximate locations (**top**) and the consensus motif sequences (**bottom**) of the zipcode elements in *glutelin* mRNA. Glutelin mRNAs possess three zipcodes (zipcode 1, 2, 3), which consist of two motifs, zipcode motif 1 (orange triangles) and zipcode motif 2 (magenta triangles). 5′ and 3′ denote 5′ UTR and 3′ UTR, respectively.

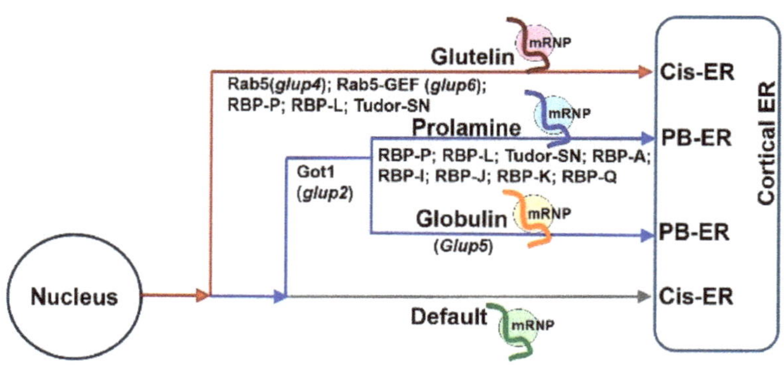

Figure 3. The hierarchal relationship among the three mRNA transport pathways in developing rice endosperm (adapted from [35]). The *glutelin* mRNA localization pathway to the cis-ER (red) is dominant over the pathway targeting *prolamine/α-globulin* mRNAs to the PB-ER (blue), which, in turn, is dominant over the default pathway (gray). Studies from mutant rice lines expressing defective Got1 (*glup2*), Rab5 (*glup4*) and Rab5-GEF (*glup6*), indicate that membrane trafficking mediates the mRNA transport to the PB-ER or cis-ER. Another rice mutant, *glup5*, possessing an undefined genetic defect, misdirects *α-globulin* mRNAs to the cis-ER without affecting the localization of *prolamine* mRNAs to the PB-ER. The key RBPs responsible for *glutelin* and *prolamine* mRNA localization, are marked under each pathway. The three RNA-transport pathways may share some common RBPs or factors for mRNA targeting to the cortical ER membrane, while additional specific factors are required for selective transport during each pathway.

3. mRNA Targeting to the ER Subdomains Requires a Set of Trans-Acting RBPs

Following the identification of cis-acting elements in RNAs, identification of transacting factors, mainly RNA-binding proteins (RBPs), was pursued. Two strategies were used to identify the RBPs required for *prolamine* and *glutelin* mRNA localization, respectively. Affinity chromatography using *prolamine* zipcode RNA as bait to pull-down interacting RBPs was initially applied [36]. Fifteen unique RBPs with specific binding affinity to the *prolamine* zipcode were selectively captured under highly stringent washing and elution conditions. Five of these RBPs, A, I, J, K, and Q, were heterogeneous nuclear ribonucleoproteins (hnRNPs) containing two RNA recognition motifs (RRMs) and were selected for further functional analysis [37]. All five RBPs have binding capability to *prolamine* zipcode RNAs. They form multiple complexes in the nucleus and cytoplasm, suggesting that they mediate various steps during *prolamine* mRNA transport and localization (Figures 4 and 5). RBPs A-J-K and I-J-K assemble into two complexes associated with *prolamine* zipcodes in both the nucleus and cytoplasm, while RBP-Q is involved in the formation of an undefined third complex in the nucleus that is released in the cytoplasm.

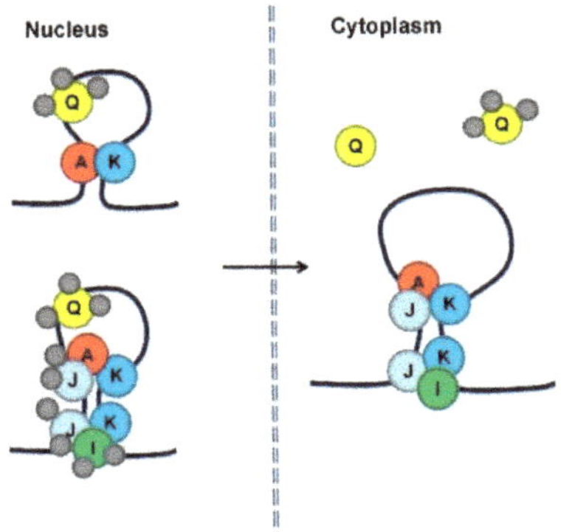

Figure 4. A proposed working model of five RBPs involved in *prolamine* mRNA localization (adapted from [37]). Five RBPs, A, I, J, K, and Q assemble into at least three different RBP complexes that recognize and bind to *prolamine* zipcodes. RBPs A, I, J and K form two cytoplasmic complexes, A-J-K and I-J-K. These two RBP complexes together with a third complex containing RBP-Q may also be present in the nucleus. In the nucleus, RBPs I and J may be associated with other proteins preventing their recognition by antibodies. When associated with RBP-A and RBP-K, RBP-Q is not accessible to its antibody, suggesting that it is bound by other proteins that comprise a third multiprotein family. Alternatively, the nucleus contains a simpler complex consisting of RBPs A and K as well as the RBP-Q complex.

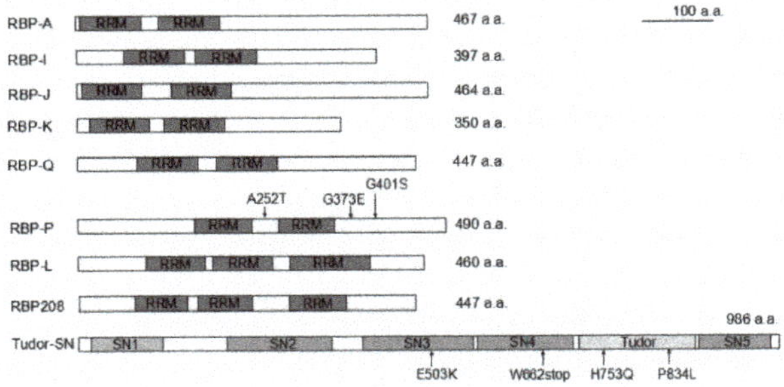

Figure 5. Schematic structure of the available RBPs responsible for *prolamine* and/or *glutelin* mRNA localization. While RBP-A, I, J, K, Q, and P contain two RRM motifs, there are three RRMs in RBP-L

and RBP208. Tudor-SN consists of four SN-like domains (SN1 to 4) followed by a Tudor domain and a fifth abbreviated SN-like domain (SN5). The mutations sites in RBP-P and Tudor-SN that cause mislocalization of *prolamine* and *glutelin* mRNAs are indicated by arrows followed by labeling amino acid substitutions. a.a., amino acids.

While affinity chromatography using prolamine zipcode RNA was successful in capturing specific *trans*-RBPs, it failed to identify specific *glutelin* zipcode binding proteins. An alternative North-western blot approach was later undertaken using *glutelin* zipcode and non-zipcode RNA as a comparison group, which identified RBP-P as a key *glutelin* zipcode-binding protein [38]. RBP-P was also captured by *prolamine* zipcode RNA chromatography [36], suggesting a dual functional role of RBP-P in localization of both glutelin and prolamine mRNAs. Indeed, the two RRMs containing RBP-P (Figure 5) showed highly specific binding to both *glutelin* and *prolamine* mRNAs, especially their zipcodes. Mutations in RBP-P led to a loss of RNA binding activity and caused partial mislocalization of both *glutelin* and *prolamine* mRNAs [39]. Collectively, these results indicate RBP-P plays essential roles in mediating specific targeting of both *glutelin* and *prolamine* mRNAs.

Similar to the case of prolamine, multi-protein complexes are also required to determine the localization of *glutelin* mRNAs. Protein–protein interaction revealed that RBP-P, RBP-L and RBP208 interact with each other, forming multiple complexes in the nucleus and/or cytoplasm [39]. RBP-L exhibits similar features as RBP-P (Figure 5), including binding to *glutelin* and *prolamine* mRNAs in vivo and in vitro [34]. When RBP-L expression is knocked down by a DNA segmental mutation in the 3′ UTR region, both *glutelin* and *prolamine* mRNAs [34] are partially mislocalized, suggesting that RBP-L is also a key RBP in regulating the localization of *glutelin* and *prolamine* mRNAs. Given that the complexes formed by RBP-P and RBP-L are not RNA-dependent and are located in both the nucleus and cytoplasm, the two RBPs may form a primary complex to serve as a scaffold to bind other RBPs to form a multi-protein complex that selectively targets *prolamine* and *glutelin* mRNAs to the cortical ER. In addition to RBP-P and RBP-L, a third candidate, RBP208, may also be involved. RBP208 interacts with RBP-P, and this interaction is weakened with several RBP-P mutant proteins. Hence, RBP208 may also be required for precise control of *glutelin* and *prolamine* mRNA localization. Unlike the RBP-P/RBP-L complex, RBP-P/RBP208 complexes are found in both the nucleus and cytoplasm and their interaction is RNA-dependent [39]. Hence, RBP208 may be specially recruited by RBP-P to the complex during mRNA localization. As RBP208 only interacts with RBP-L in the cytoplasm, the RBP group P/L/208 may also form multiple complexes to co-regulate *glutelin* and *prolamine* mRNA localization. However, the exact detailed function of RBP208 in the process deserves further investigation.

The requirement of RBP-P, RBP-L and RBP208 for both *glutelin* and *prolamine* mRNA localization indicates that the two transport pathways are inter-related by sharing common trans-factors to assemble the required mRNP complexes. This feature further contributes to the close relationship between the *glutelin* and *prolamine* mRNA transport pathways. The current results suggest a scenario that while the primary scaffold formed by RBP-P and RBP-L, with the undefined role from RBP208, selects *glutelin* and *prolamine* mRNAs for specific targeting to the cortical ER, the complexes formed by RBPs A, I, J, K and Q are involved to target *prolamine* mRNAs to the PB-ER. How the RBP group A/I/J/K/Q links to RBP-P/L scaffold and what specific factors control *glutelin* mRNA transport pathway require further study.

4. A Possible Role for Myosin Motor Protein Driving mRNA Transport to the ER Subdomains on Actin Filaments

How an mRNP complex carrying target mRNAs and a large collection of RBPs is transported to specific ER subdomains remains a mystery. Although passive mRNA diffusion and anchoring along with cytoplasmic streaming has been reported for *Nos* RNA during *Drosophila* oogenesis [40–42] and for most mRNAs in bacteria [12,43], active transport driven by cytoskeletal-associated motor proteins is the most common mode of mRNA localization in eukaryotic cells. In yeast, the well-known bud-localized *ASH1* mRNA is transported on actin filaments through its association with the RBPs, She2 and She3, with the type V myosin motor Myo4 [44–47]. The She2–She3 complex also recognizes *CLB2, TCB2, TCB3,* and *IST2* mRNAs, and drives their active transport on actin filaments [48,49]. In mammalian cells, mRNAs are actively transported on both actin and microtubules [50–52]. For example, *Actb* mRNAs are transported to the leading edge of migrating fibroblasts on both actin and microtubules, where the zipcode binding protein 1 (ZBP1) mediates the transport via its interaction with motor proteins [53–55].

In rice endosperm cells, a preliminary study using a GFP-based RNA movement system suggested that the storage protein mRNAs were transported on actin filaments. In this modified 2-hybrid system, *prolamine* RNA particles were visualized by co-expressing a GFP-MS2 fusion protein and a hybrid *prolamine* RNA containing tandem MS2 binding sites. Microscopic analysis showed that the *prolamine* RNA/GFP-MS2 particles moved in a stop-and-go manner [56,57]. The movement was overall unidirectional but with occasional bidirectional, random, and oscillatory movements, a movement behavior consistent with transport along cytoskeletal elements. Drug treatment known to disrupt the integrity of actin filaments using cytochalasin D and latrunculin B was found to efficiently suppress particle movement [56], indicating that *prolamine* mRNAs are likely transported via myosin along actin filaments.

5. mRNA Transport to the ER Subdomains Meets Membrane Trafficking

A mutant rice line, *glup2*, carrying a mutation in *Golgi Transport 1(Got1B)* gene, exhibits mislocalization of *prolamine* and *α-globulin* mRNAs [58] (Figure 3). Got1B is usually found on coat protein complex II (COPII) vesicles and functions as a membrane trafficking-related protein to mediate anterograde transport from the ER to the Golgi [58–60]. In rice, Got1B was reported to interact with COPII component Sec23 and thus mediate COPII vesicle formation at Golgi-associated ER exit sites [60], which may influence transport of glutelin polypeptides and cause abnormal accumulation of proglutelins in the relevant mutants [58,60]. It seems that in addition to protein transport, Got1B mediated COPII trafficking may also be involved in mRNA localization. Although it is not clear how *prolamine* and *α-globulin* mRNA transport is linked to COPII trafficking, it is obvious that the process of mRNA localization has a relationship with membrane trafficking as demonstrated by analysis of rice mutant lines.

The *glup4* and *glup6* rice lines carry mutations in the small GTPase Rab5 and its cognate guanine nucleotide exchange factor (Rab5-GEF), respectively. Rab5 plays multiple roles in membrane transport, ranging from early endosome formation to trafficking from the Golgi to the protein storage vacuole in rice endosperm [61]. Due to the critical function of Rab5 in endosomal transport, both *glup4* and *glup6* lines exhibit a pronounced aborted endocytosis phenotype, where extracellular paramural bodies (PMBs) are formed adjacent to the cytoplasmic membrane (Figure 1B). These paramural bodies contain protein markers for the ER, Golgi, prevacuolar compartment and plasma membrane [61] as well as glutelin, suggesting that normal membrane trafficking is disrupted. In addition, *glutelin* mRNAs were also detected in PMBs [61,62], suggesting a potential connection of endosomal trafficking and *glutelin* mRNA transport.

Long-distance endosomal transport of mRNAs is well studied in *Ustilago maydis*, a fungus that causes corn smut disease. A highly polarized growth of infectious hyphae is heavily dependent on motor-mediated endosomal transport along microtubules (Figure 6A) [27]. During the process, higher-order septin filaments are generated with gradients to set up polarized growth via endosomal transport of *septin* mRNAs, including *cdc3, cdc10, cdc11,* and *cdc12*. The endosomal transport of *septin* mRNAs requires their binding to the RNA binding protein Rrm4 [28,63,64], which interacts with a membrane-associated linker protein Upa1 through its FYVE domain to link mRNPs on endosomes. Thus, the complex of Rrm4 and Upa1 work together with the other two RBPs, Pab1 and Upa2 [63,64] to co-mediate endosomal transport of *cdc* mRNAs (Figure 6A). Apparently, specific adaptor proteins are required to hitch mRNPs onto endosomes for active transport of mRNAs.

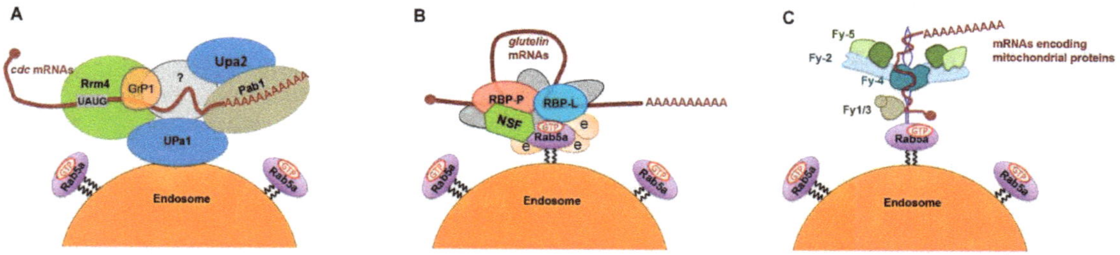

Figure 6. Proposed models of endosome-coupled trafficking of mRNAs in different organisms. (**A**) Working model of *cdc* mRNA transport via trafficking endosomes in the fungus *U. maydis* (adapted from [64]). The mRNAs encoding four septins, Cdc3, 10, 11 and 12, are bound by Rrm4 and Pab1, which recruit GrP1, UPa1, and UPa2 as well as an unknown linker protein (grey color). The resulting mRNP complex is associated with a trafficking endosome through membrane protein UPa1. (**B**) Working model of *glutelin* mRNA transport via trafficking endosomes in rice endosperm cells (adapted from [65]). *Glutelin* mRNAs are recognized and bound by RBP-P and RBP-L, forming an mRNP complex. Through direct or NSF-mediated interaction of RBP-L and RBP-P, respectively, with key endosomal factor, Rab5a, the quaternary complex links the mRNP complex onto endosomes for active transport via the cytoskeleton. Other unknown RBPs or factors (light gray) may also be involved to stabilize the mRNP complex and define the connection to endosomes. (**C**) Working model of the transport of mRNAs encoding mitochondrial proteins via trafficking endosomes in rat neuron cell (adapted from preprint studies of [66,67]). A FERRY (Five-subunit Endosomal Rab5 and RNA/ribosome intermediary) complex, composed of Fy-1 to 5, is required for transport. Fy-2 directly binds to mRNAs through its coiled-coils and recruits Fy4 and Fy-5 to form a clamp-like structure. Although Fy-1 and Fy-3 are not required for RNA binding, they assist Fy-2 to adopt the correct folding and conformation. In addition to functioning as the main binding protein to mRNAs, Fy-2 also interacts with the GTP-bound form of Rab5 via its C-terminal region, linking the mRNP complex onto trafficking endosomes for active mRNA transport.

In contrast to the working model in *U. maydis*, higher plants may employ a distinct set of proteins to accomplish endosomal trafficking of mRNAs. The abovementioned key RBPs, RBP-P, and RBP-L were found to directly interact with the membrane fusion factor N-ethylmaleimide-sensitive factor (NSF) and Rab5a [65], respectively, thus forming a quaternary complex linked to endosomes (Figure 6B). The complexes carry *glutelin* mRNAs through specific binding activities from RBP-P and RBP-L, while NSF and Rab5a are recruited for active transport of *glutelin* mRNAs on endosomes to the cortical ER membrane. Mistargeting of *glutelin* mRNAs, along with presence of the quaternary complex, to the PMBs in the *rab5a* mutant supports the endosomal transport of *glutelin* mRNAs.

Prior to these findings, the direct binding of RBPs with NSF or Rab5a had not been reported in any other organisms. Such protein-protein interaction may profit from a gain in binding properties by Rab5a and NSF, as well as RBP-P and RBP-L, which allow for the highly selective recognition of an RBP with the NSF membrane fusion factor or the molecular switch Rab5. Due to the high conservation of RRM domains between RBP-P and RBP-L, the N- and C-terminal regions of RBP-P and RBP-L may directly contribute to their protein recognition. In the case of RBP-P/NSF interaction, the N-terminal regions of RBP-P and NSF are likely responsible for their interaction. NSF is a soluble hexameric ATPase predominantly involved in membrane fusion events through its interaction with the soluble NSF attachment protein (SNAP) [68–71]. Due to the absence of SNAP in the assembly of RBP-P/NSF complex [65], NSF may gain a special function in mRNA metabolism by its interaction with RBP-P. Of the quaternary complex, Rab5a interacts with both NSF and RBP-L, but not RBP-P. Given that active GTPase activity of Rab5a is required for the transport of *glutelin* mRNAs on endosomes, NSF and RBP-L may act as Rab5a effectors to regulate endosomal transport of mRNAs.

The identification of these key linker proteins that enable hitchhiking of mRNPs on trafficking endosomes in rice endosperm cells provides new insights on membrane trafficking-mediated mRNA transport in eukaryotes. However, whether the requirement of NSF-Rab5a-RBP machinery in endosomal mRNA transport commonly exists in other eukaryotic organisms or is unique to higher plants needs further investigation.

A recent preprint study in rat neuron cells reported a novel FERRY (Five-subunit Endosomal Rab5 and RNA/ribosome intermediary) complex that directly interacts with mRNAs and Rab5, and functions as a Rab5 effector to mediate mRNA transport to mitochondria via early endosomes [67,72] (Figure 6C). The FERRY complex is assembled by five subunits, Fy-1 to Fy-5, in which the flexible Fy-2 serves as a binding hub to interact with Rab5 and connect all five subunits to mediate the binding to specific mitochondrial mRNAs [66]. While the FERRY complex is developed in some fungi and commonly exists as a full extent of five-subunit assembly in the Chordata [67], Fy proteins share very low sequence similarity with putative GTPase activator proteins in plants (lower than 28% when compared to *Arabidopsis*). Whether a similar mechanism exists in plant cells requires future investigation as well.

6. mRNA Localization Plays a Determinant Role in Storage Organelle Biogenesis in Cereal Grains

A prominent role of intracellular localization of mRNAs is to target the distribution of the encoded protein and, thereby, generate a high concentration at specific locales [73]. Specific localization of *prolamine* mRNAs on the PB-ER results in intensive local translation to generate a high concentration of prolamine polypeptides within the ER lumen, an environment conducive for self-assembly of prolamine polypeptides to form intracisternal granules, which eventually develop into ER-derived mature PB-I [74] (Figure 1A). Targeting of *glutelin* mRNAs to spatially separate cis-ER prevents the potential interaction of the newly synthesized glutelin with prolamine and α-globulin polypeptides in the ER lumen. This enables the proglutelin polypeptide to fold correctly and assemble into a quaternary structure competent for export from the ER lumen to the Golgi and subsequently to the PSVs.

Mislocalization of *glutelin* and *prolamine* mRNAs caused by mutations of key RBPs and other relevant factors results in mistargeting of the encoded proteins and the formation of abnormal storage organelles [32,34,39] (Figure 1B). For instance, the typical PB-I structure is spherical and contains an electron-dense core of cysteine-rich 10 kDa prolamine (CysR10), which is surrounded by an electron lucent layer of cysteine-poor 13 kDa prolamine (CysP13) [75]. When mRNA mislocalization occurs, the mistargeted glutelin or loss of target prolamine impairs the tight packaging of prolamine in PB-I, resulting in the loss of the electron-dense central core and appearance of irregular shaped PB-I [34,76] (Figure 1B).

Although α-globulin proteins are transported to the PSV similar to glutelins, *α-globulin* mRNAs are localized on the PB-ER. The separate transport pathways of *α-globulin* and *glutelin* mRNAs are likely responsible for the asymmetric distribution of α-globulin and glutelin proteins in the matrix and crystalloid regions, respectively, of the PSVs. Mislocalization of *α-globulin* mRNAs to the cis-ER disrupts the normal transport of α-globulin and the distinct, separate distribution of these storage proteins in PB-II. Taken together, the segregation of storage mRNAs to distinct ER subdomains plays a determinant role in precisely controlling the sorting, transport, and deposition of the encoded storage proteins within the storage organelles.

The asymmetric distribution of specific mRNAs on distinct ER subdomain is likely a prevalent and conserved mechanism in plant cells. In maize endosperm cells, mRNAs encoding prolamine family protein zeins and 11S globulin type protein legumin-1 display a similar pattern seen in rice endosperm cells, i.e., they are localized on the ER-bounded zein protein bodies and cis-ER, respectively [77]. The asymmetric distribution of *zein* and *legumin-1* mRNAs directly contributes to the deposition of zein proteins in the ER-derived protein body and legumin-1 in PSV [77]. Given that cereal grains usually accumulate a large amount of storage proteins, mRNA localization serves as a major mechanism to drive storage proteins targeting in cereal grains.

7. Accessory RBPs Is Required for mRNA Localization as well as Other Functions

An earlier study isolated a cytoskeleton-PB-enriched fraction from developing rice seed lysate by fractionation by sucrose density gradient centrifugation, which was then treated with high salt to solubilized cytoskeletal-associated proteins. This protein fraction was then subjected to a poly(U)-Sepharose chromatography, which resulted in the identification of a cytoskeleton-associated 120 kD protein with a prominent RNA-binding activity [78,79]. The protein was later named OsTudor-SN, an ortholog of the human transcriptional co-activator p100 [80]. OsTudor-SN possesses a multi-domain structure, consisting of four tandem staphylococcus nuclease (SN) domains (4SN module) and a Tudor domain followed by an abbreviated C-terminal SN (Tsn module). OsTudor-SN was found to specifically bind to the 3′ UTR regions of prolamine and glutelin mRNAs [78]. Involvement of OsTudor-SN in mRNA localization was evidenced by partial disruption of both *prolamine* and *glutelin* mRNA localization in mutant OsTudor-SN endosperm cells [81]. The N-terminal 4SN module is responsible for RNA binding activity to storage protein mRNAs, while the C-terminal Tsn module acts as a scaffold for protein–protein interaction with other RBPs [81]. Thus, the two modular regions of OsTudor-SN cooperate in *prolamine* and *glutelin* mRNA localization. Although OsTudor-SN functions as a non-zipcode trans-factor [35], its association with both cytoskeleton and storage protein mRNAs reveals that OsTudor-SN likely participates in the transport of target mRNAs.

In addition, OsTudor-SN is also required for storage protein expression, storage organelle biogenesis, and seed development. While a decrease in glutelin expression was observed in rice lines carrying mutations in the Tudor domain or loss of the Tsn module, mutation in the 4SN module caused elevated accumulation of the glutelin precursor [76]. The latter mutation also resulted in strong reduction in prolamine expression and, in turn, abnormal formation of protein bodies [76]. The phenotype caused by the 4SN mutation is partially restored by complementation with the wild-type OsTudor-SN gene, suggesting that the 4SN module plays a crucial role in seed development. Data from transcriptome analysis indicate that OsTudor-SN also has functions in regulating the expression of transcription factors and genes involved in seed development and stress response [76]. Collectively, the modular structure confers multiple functions of OsTudor-SN in mRNA localization as well as in other cellular properties.

8. Transport of mRNAs Is Highly Selective and as Large Regulons

In situ analysis of rice *glup4* and *glup6* mutants carrying loss-of-function Rab5 and its effector GEF, respectively, reveals that these mutations impact mistargeting of *glutelin* mRNAs to the PB-ER and PMBs while having no significant impact on the normal localization of *prolamine* and *α-globulin* mRNAs on the PB-ER. The study of these mutants demonstrates that mRNA transport is highly selective. Analysis of RNA transcripts isolated from purified PMBs from *glup6* mutant reveals that in addition to *glutelin* mRNAs, other mRNAs were also located in the purified PMBs [62]. The composition of PMB-associated mRNAs was found to be not random but selective. Sets of mRNAs encoding cell wall, respiration, photosynthesis, and ribosome-related proteins were highly enriched in the PMB-associated mRNA pool, suggesting that specific types of mRNAs are transported together with *glutelin* mRNAs via the endosomal-mediated transport pathway. This co-transport phenomenon is consistent with the "RNA regulon" hypothesis [82] where sets of mRNAs that encode similar intracellular location or functionally related proteins are co-transported and coordinately regulated.

Additional evidence of co-transport of mRNAs is from the study of *calreticulin* and *zein* mRNAs in maize calli cells [83]. The mRNAs encoding calcium binding protein calreticulin, an ER-resident chaperone protein, were found to be selectively targeted to the PB-ER subdomain where *zein* mRNAs are located. The distribution pattern is distinct from the diffuse distribution of the control mRNAs that encodes actin monomer binding protein profilin [83], further confirming that mRNA localization is a highly selective process. Calreticulin proteins were further found to localize in the ER-derived PBs containing zein.

Therefore, *calreticulin* mRNAs were thought to be co-transported with *zein* mRNAs to the PB-ER subdomain, possibly directly determining the enrichment of calreticulin in the PBs and assisting zein retention and assembly of PBs within the ER. Further investigations on the co-transport of these mRNAs may help to identify the factors that regulate protein synthesis and localization.

9. Future Perspectives

The localization of mRNAs plays an essential role in governing gene expression, protein targeting and thus determines cell fate, development, and polar growth. Although the phenomenon of mRNA localization in plant cells was discovered three decades ago, the underlying mechanism remains largely undefined. Studies from storage protein mRNA localization onto distinct subdomains of the ER in endosperm cells have identified the determinant zipcode RNA sequences and many key RBPs responsible for specific targeting pathways. Although a preliminary model of mRNA localization to the cortical ER in plant cells has been proposed, more questions are waiting to be addressed. Except for a few examples discovered in endosperm or callus cells, are there similar mRNA localization processes in other plant cells and plant species? While endosomal-mediated transport supports *glutelin* mRNA targeting to the cis-ER, what transport pathway regulates the localization of *prolamine* and *α-globulin* mRNAs to the PB-ER? Following the discovery of non-storage protein RNAs targeting to a specific subdomain of the ER, are these mRNAs co-transported as regulons to the same subcellular destination based on their similar function or intracellular location?

To address these questions, a combination of biochemical, cell biological, and high-throughput methods will allow for investigation of mRNA localization at a large-scale profiling level. Employment of RNA-protein immunoprecipitation combined with high-throughput sequencing and high-resolution mass spectrometry will help to further identify cis- and trans-factors responsible for specific mRNA localization. A more detailed network of co-transported mRNAs and the mechanism of assembly and remodeling of multi-RBP complexes to recognize and bind target mRNAs deserve further investigation. Application of fluorescence-based mRNA-labeling systems, such as boxB RNA stem-loop [84], bacteriophage PP7 aptamer tagging [85], GFP-MS2 [56,86], and spinach (*Spinacia oleracea*) tracking systems [13,87–89] in combination with high-resolution microscopy will enable visualization and simultaneously monitor the transport of mRNAs in live cells. All these efforts will help to identify the machinery involved in mRNA targeting and transport and to address the general significance of mRNA localization in plant cells.

Author Contributions: Writing—original draft, L.Z., Q.S., K.Y., W.Z. and L.T.; review and editing, T.W.O. and L.T. All authors have read and agreed to the published version of the manuscript.

Funding: This work was supported by Zhejiang A&F University Starting Funds of Scientific Research and Development (203402000101 and 203402001301) and the National Science Foundation (NSF) EAGER Grant (2029933).

Institutional Review Board Statement: Not applicable.

Informed Consent Statement: Not applicable.

Data Availability Statement: Not applicable.

Conflicts of Interest: The authors declare no conflict of interest.

References

1. Jeffery, W.R.; Tomlinson, C.R.; Brodeur, R.D. Localization of actin messenger RNA during early ascidian development. *Dev. Biol.* **1983**, *99*, 408–417. [CrossRef]
2. Jeffery, W.R. Requirement of cell division for muscle actin expression in the primary muscle cell lineage of ascidian embryos. *Development* **1989**, *105*, 75–84. [CrossRef]
3. Berleth, T.; Burri, M.; Thoma, G.; Bopp, D.; Richstein, S.; Frigerio, G.; Noll, M.; Nusslein-Volhard, C. The role of localization of bicoid RNA in organizing the anterior pattern of the Drosophila embryo. *EMBO J.* **1988**, *7*, 1749–1756. [CrossRef]

4. Frigerio, G.; Burri, M.; Bopp, D.; Baumgartner, S.; Noll, M. Structure of the segmentation gene paired and the Drosophila PRD gene set as part of a gene network. *Cell* **1986**, *47*, 735–746. [CrossRef]
5. Rebagliati, M.R.; Weeks, D.L.; Harvey, R.P.; Melton, D.A. Identification and cloning of localized maternal RNAs from Xenopus eggs. *Cell* **1985**, *42*, 769–777. [CrossRef]
6. Kandler-Singer, I.; Kalthoff, K. RNase sensitivity of an anterior morphogenetic determinant in an insect egg (Smittia sp., Chironomidae, Diptera). *Proc. Natl. Acad. Sci. USA* **1976**, *73*, 3739–3743. [CrossRef]
7. Davidson, E.H. Note on the control of gene expression during development. *J. Biol.* **1971**, *32*, 123–130. [CrossRef]
8. Lawrence, J.B.; Singer, R.H. Intracellular localization of messenger RNAs for cytoskeletal proteins. *Cell* **1986**, *45*, 407–415. [CrossRef]
9. Garner, C.C.; Tucker, R.P.; Matus, A. Selective localization of messenger RNA for cytoskeletal protein MAP2 in dendrites. *Nature* **1988**, *336*, 674–677. [CrossRef]
10. Li, X.; Franceschi, V.R.; Okita, T.W. Segregation of storage protein mRNAs on the rough endoplasmic reticulum membranes of rice endosperm cells. *Cell* **1993**, *72*, 869–879. [CrossRef]
11. Trapp, B.D.; Nishiyama, A.; Cheng, D.; Macklin, W. Differentiation and death of premyelinating oligodendrocytes in developing rodent brain. *J. Cell Biol.* **1997**, *137*, 459–468. [CrossRef] [PubMed]
12. Buskila, A.A.; Kannaiah, S.; Amster-Choder, O. RNA localization in bacteria. *RNA Biol.* **2014**, *11*, 1051–1060. [CrossRef] [PubMed]
13. Tian, L.; Okita, T.W. mRNA-based protein targeting to the endoplasmic reticulum and chloroplasts in plant cells. *Curr. Opin. Plant. Biol.* **2014**, *22*, 77–85. [CrossRef]
14. Margeot, A.; Garcia, M.; Wang, W.; Tetaud, E.; di Rago, J.P.; Jacq, C. Why are many mRNAs translated to the vicinity of mitochondria: A role in protein complex assembly? *Gene* **2005**, *354*, 64–71. [CrossRef] [PubMed]
15. Michaud, M.; Ubrig, E.; Filleur, S.; Erhardt, M.; Ephritikhine, G.; Marechal-Drouard, L.; Duchene, A.M. Differential targeting of VDAC3 mRNA isoforms influences mitochondria morphology. *Proc. Natl. Acad. Sci. USA* **2014**, *111*, 8991–8996. [CrossRef]
16. Ahmed, A.U.; Fisher, P.R. Import of nuclear-encoded mitochondrial proteins: A cotranslational perspective. *Int. Rev. Cell Mol. Biol.* **2009**, *273*, 49–68.
17. Vincent, T.; Vingadassalon, A.; Ubrig, E.; Azeredo, K.; Srour, O.; Cognat, V.; Graindorge, S.; Salinas, T.; Marechal-Drouard, L.; Duchene, A.M. A genome-scale analysis of mRNAs targeting to plant mitochondria: Upstream AUGs in 5′ untranslated regions reduce mitochondrial association. *Plant J.* **2017**, *92*, 1132–1142. [CrossRef]
18. Moriceau, L.; Jomat, L.; Bressanelli, S.; Alcaide-Loridan, C.; Jupin, I. Identification and Molecular Characterization of the Chloroplast Targeting Domain of Turnip yellow mosaic virus Replication Proteins. *Front. Plant Sci.* **2017**, *8*, 2138. [CrossRef]
19. Gomez, G.; Pallas, V. Noncoding RNA mediated traffic of foreign mRNA into chloroplasts reveals a novel signaling mechanism in plants. *PLoS ONE* **2010**, *5*, e12269. [CrossRef]
20. Budziszewska, M.; Obrepalska-Steplowska, A. The Role of the Chloroplast in the Replication of Positive-Sense Single-Stranded Plant RNA Viruses. *Front. Plant Sci.* **2018**, *9*, 1776. [CrossRef]
21. Walter, P.; Blobel, G. Translocation of proteins across the endoplasmic reticulum. *Oncodev. Biol. Med.* **1982**, *4*, 137–146. [CrossRef]
22. Meyer, D.I.; Krause, E.; Dobberstein, B. Secretory protein translocation across membranes-the role of the 'docking protein'. *Nature* **1982**, *297*, 647–650. [CrossRef]
23. Gilmore, R.; Kellaris, K.V. Translocation of proteins across and integration of membrane proteins into the rough endoplasmic reticulum. *Ann. N. Y. Acad. Sci.* **1992**, *674*, 27–37. [CrossRef]
24. Yamagata, H.; Tanaka, K. The Site of Synthesis and Accumulation of Rice Storage Proteins. *Plant Cell Physiol.* **1986**, *27*, 135–145.
25. Choi, S.B.; Wang, C.; Muench, D.G.; Ozawa, K.; Franceschi, V.R.; Wu, Y.; Okita, T.W. Messenger RNA targeting of rice seed storage proteins to specific ER subdomains. *Nature* **2000**, *407*, 765–767. [CrossRef]
26. Hamada, S.; Ishiyama, K.; Sakulsingharoj, C.; Choi, S.B.; Wu, Y.; Wang, C.; Singh, S.; Kawai, N.; Messing, J.; Okita, T.W. Dual regulated RNA transport pathways to the cortical region in developing rice endosperm. *Plant Cell* **2003**, *15*, 2265–2272. [CrossRef]
27. Haag, C.; Steuten, B.; Feldbrugge, M. Membrane-Coupled mRNA Trafficking in Fungi. *Annu. Rev. Microbiol.* **2015**, *69*, 265–281. [CrossRef]
28. Niessing, D.; Jansen, R.P.; Pohlmann, T.; Feldbrugge, M. mRNA transport in fungal top models. *Wiley Interdiscip. Rev. RNA* **2018**, *9*, e1453. [CrossRef]
29. Schmid, M.; Jaedicke, A.; Du, T.G.; Jansen, R.P. Coordination of endoplasmic reticulum and mRNA localization to the yeast bud. *Curr. Biol.* **2006**, *16*, 1538–1543. [CrossRef]
30. Jansen, R.P. mRNA localization: Message on the move. *Nat. Rev. Mol. Cell Biol.* **2001**, *2*, 247–256. [CrossRef]
31. Washida, H.; Kaneko, S.; Crofts, N.; Sugino, A.; Wang, C.; Okita, T.W. Identification of cis-localization elements that target glutelin RNAs to a specific subdomain of the cortical endoplasmic reticulum in rice endosperm cells. *Plant Cell Physiol.* **2009**, *50*, 1710–1714. [CrossRef]
32. Washida, H.; Sugino, A.; Doroshenk, K.A.; Satoh-Cruz, M.; Nagamine, A.; Katsube-Tanaka, T.; Ogawa, M.; Kumamaru, T.; Satoh, H.; Okita, T.W. RNA targeting to a specific ER sub-domain is required for efficient transport and packaging of alpha-globulins to the protein storage vacuole in developing rice endosperm. *Plant J.* **2012**, *70*, 471–479. [CrossRef]
33. Washida, H.; Sugino, A.; Kaneko, S.; Crofts, N.; Sakulsingharoj, C.; Kim, D.; Choi, S.B.; Hamada, S.; Ogawa, M.; Wang, C.; et al. Identification of cis-localization elements of the maize 10-kDa delta-zein and their use in targeting RNAs to specific cortical endoplasmic reticulum subdomains. *Plant J.* **2009**, *60*, 146–155. [CrossRef]

34. Tian, L.; Chou, H.L.; Zhang, L.; Okita, T.W. Targeted Endoplasmic Reticulum Localization of Storage Protein mRNAs Requires the RNA-Binding Protein RBP-L. *Plant Physiol.* **2019**, *179*, 1111–1131. [CrossRef]
35. Tian, L.; Chou, H.L.; Fukuda, M.; Kumamaru, T.; Okita, T.W. mRNA Localization in Plant Cells. *Plant Physiol.* **2020**, *182*, 97–109. [CrossRef]
36. Crofts, A.J.; Crofts, N.; Whitelegge, J.P.; Okita, T.W. Isolation and identification of cytoskeleton-associated prolamine mRNA binding proteins from developing rice seeds. *Planta* **2010**, *231*, 1261–1276. [CrossRef]
37. Yang, Y.; Crofts, A.J.; Crofts, N.; Okita, T.W. Multiple RNA binding protein complexes interact with the rice prolamine RNA cis-localization zipcode sequences. *Plant Physiol.* **2014**, *164*, 1271–1282. [CrossRef]
38. Doroshenk, K.A.; Tian, L.; Crofts, A.J.; Kumamaru, T.; Okita, T.W. Characterization of RNA binding protein RBP-P reveals a possible role in rice glutelin gene expression and RNA localization. *Plant Mol. Biol.* **2014**, *85*, 381–394. [CrossRef]
39. Tian, L.; Chou, H.L.; Zhang, L.; Hwang, S.K.; Starkenburg, S.R.; Doroshenk, K.A.; Kumamaru, T.; Okita, T.W. RNA-Binding Protein RBP-P Is Required for Glutelin and Prolamine mRNA Localization in Rice Endosperm Cells. *Plant Cell* **2018**, *30*, 2529–2552. [CrossRef]
40. Trcek, T.; Lehmann, R. Germ granules in Drosophila. *Traffic* **2019**, *20*, 650–660. [CrossRef]
41. Lecuyer, E.; Yoshida, H.; Parthasarathy, N.; Alm, C.; Babak, T.; Cerovina, T.; Hughes, T.R.; Tomancak, P.; Krause, H.M. Global analysis of mRNA localization reveals a prominent role in organizing cellular architecture and function. *Cell* **2007**, *131*, 174–187. [CrossRef]
42. Forrest, K.M.; Gavis, E.R. Live imaging of endogenous RNA reveals a diffusion and entrapment mechanism for nanos mRNA localization in Drosophila. *Curr. Biol.* **2003**, *13*, 1159–1168. [CrossRef]
43. Fei, J.; Sharma, C.M. RNA Localization in Bacteria. *Microbiol. Spectr.* **2018**, *6*, RWR-0024-2018. [CrossRef]
44. Edelmann, F.T.; Schlundt, A.; Heym, R.G.; Jenner, A.; Niedner-Boblenz, A.; Syed, M.I.; Paillart, J.C.; Stehle, R.; Janowski, R.; Sattler, M.; et al. Molecular architecture and dynamics of ASH1 mRNA recognition by its mRNA-transport complex. *Nat. Struct. Mol. Biol.* **2017**, *24*, 152–161. [CrossRef]
45. Bohl, F.; Kruse, C.; Frank, A.; Ferring, D.; Jansen, R.P. She2p, a novel RNA-binding protein tethers ASH1 mRNA to the Myo4p myosin motor via She3p. *EMBO J.* **2000**, *19*, 5514–5524. [CrossRef]
46. Gonsalvez, G.B.; Little, J.L.; Long, R.M. ASH1 mRNA anchoring requires reorganization of the Myo4p-She3p-She2p transport complex. *J. Biol. Chem.* **2004**, *279*, 46286–46294. [CrossRef]
47. Takizawa, P.A.; Vale, R.D. The myosin motor, Myo4p, binds Ash1 mRNA via the adapter protein, She3p. *Proc. Natl. Acad. Sci. USA* **2000**, *97*, 5273–5278. [CrossRef]
48. Takizawa, P.A.; DeRisi, J.L.; Wilhelm, J.E.; Vale, R.D. Plasma membrane compartmentalization in yeast by messenger RNA transport and a septin diffusion barrier. *Science* **2000**, *290*, 341–344. [CrossRef]
49. Shepard, K.A.; Gerber, A.P.; Jambhekar, A.; Takizawa, P.A.; Brown, P.O.; Herschlag, D.; DeRisi, J.L.; Vale, R.D. Widespread cytoplasmic mRNA transport in yeast: Identification of 22 bud-localized transcripts using DNA microarray analysis. *Proc. Natl. Acad. Sci. USA* **2003**, *100*, 11429–11434. [CrossRef]
50. Latham, V.M.; Yu, E.H.; Tullio, A.N.; Adelstein, R.S.; Singer, R.H. A Rho-dependent signaling pathway operating through myosin localizes beta-actin mRNA in fibroblasts. *Curr. Biol.* **2001**, *11*, 1010–1016. [CrossRef]
51. Denes, L.T.; Kelley, C.P.; Wang, E.T. Microtubule-based transport is essential to distribute RNA and nascent protein in skeletal muscle. *Nat. Commun.* **2021**, *12*, 6079. [CrossRef]
52. Pichon, X.; Moissoglu, K.; Coleno, E.; Wang, T.; Imbert, A.; Robert, M.C.; Peter, M.; Chouaib, R.; Walter, T.; Mueller, F.; et al. The kinesin KIF1C transports APC-dependent mRNAs to cell protrusions. *RNA* **2021**, *27*, 1528–1544. [CrossRef]
53. Katz, Z.B.; Wells, A.L.; Park, H.Y.; Wu, B.; Shenoy, S.M.; Singer, R.H. beta-Actin mRNA compartmentalization enhances focal adhesion stability and directs cell migration. *Genes Dev.* **2012**, *26*, 1885–1890. [CrossRef]
54. Oleynikov, Y.; Singer, R.H. Real-time visualization of ZBP1 association with beta-actin mRNA during transcription and localization. *Curr. Biol.* **2003**, *13*, 199–207. [CrossRef]
55. Song, T.; Zheng, Y.; Wang, Y.; Katz, Z.; Liu, X.; Chen, S.; Singer, R.H.; Gu, W. Specific interaction of KIF11 with ZBP1 regulates the transport of beta-actin mRNA and cell motility. *J. Cell Sci.* **2015**, *128*, 1001–1010.
56. Hamada, S.; Ishiyama, K.; Choi, S.B.; Wang, C.; Singh, S.; Kawai, N.; Franceschi, V.R.; Okita, T.W. The transport of prolamine RNAs to prolamine protein bodies in living rice endosperm cells. *Plant Cell* **2003**, *15*, 2253–2264. [CrossRef]
57. Okita, T.W.; Choi, S.B. mRNA localization in plants: Targeting to the cell's cortical region and beyond. *Curr. Opin. Plant Biol.* **2002**, *5*, 553–559. [CrossRef]
58. Fukuda, M.; Kawagoe, Y.; Murakami, T.; Washida, H.; Sugino, A.; Nagamine, A.; Okita, T.W.; Ogawa, M.; Kumamaru, T. The Dual Roles of the Golgi Transport 1 (GOT1B): RNA Localization to the Cortical Endoplasmic Reticulum and the Export of Proglutelin and alpha-Globulin from the Cortical ER to the Golgi. *Plant Cell Physiol.* **2016**, *57*, 2380–2391. [CrossRef]
59. Lorente-Rodriguez, A.; Heidtman, M.; Barlowe, C. Multicopy suppressor analysis of thermosensitive YIP1 alleles implicates GOT1 in transport from the ER. *J. Cell Sci.* **2009**, *122 Pt 10*, 1540–1550. [CrossRef]
60. Wang, Y.; Liu, F.; Ren, Y.; Wang, Y.; Liu, X.; Long, W.; Wang, D.; Zhu, J.; Zhu, X.; Jing, R.; et al. GOLGI TRANSPORT 1B Regulates Protein Export from the Endoplasmic Reticulum in Rice Endosperm Cells. *Plant Cell* **2016**, *28*, 2850–2865. [CrossRef]
61. Fukuda, M.; Satoh-Cruz, M.; Wen, L.; Crofts, A.J.; Sugino, A.; Washida, H.; Okita, T.W.; Ogawa, M.; Kawagoe, Y.; Maeshima, M.; et al. The small GTPase Rab5a is essential for intracellular transport of proglutelin from the Golgi apparatus to the protein storage vacuole and endosomal membrane organization in developing rice endosperm. *Plant Physiol.* **2011**, *157*, 632–644. [CrossRef]

62. Yang, Y.; Chou, H.L.; Crofts, A.J.; Zhang, L.; Tian, L.; Washida, H.; Fukuda, M.; Kumamaru, T.; Oviedo, O.J.; Starkenburg, S.R.; et al. Selective sets of mRNAs localize to extracellular paramural bodies in a rice glup6 mutant. *J. Exp. Bot.* **2018**, *69*, 5045–5058. [CrossRef]
63. Pohlmann, T.; Baumann, S.; Haag, C.; Albrecht, M.; Feldbrugge, M. A FYVE zinc finger domain protein specifically links mRNA transport to endosome trafficking. *elife* **2015**, *4*, e06041. [CrossRef] [PubMed]
64. Muntjes, K.; Devan, S.K.; Reichert, A.S.; Feldbrugge, M. Linking transport and translation of mRNAs with endosomes and mitochondria. *EMBO Rep.* **2021**, *22*, e52445. [CrossRef] [PubMed]
65. Tian, L.; Doroshenk, K.A.; Zhang, L.; Fukuda, M.; Washida, H.; Kumamaru, T.; Okita, T. Zipcode RNA-Binding Proteins and Membrane Trafficking Proteins Cooperate to Transport Glutelin mRNAs in Rice Endosperm. *Plant Cell* **2020**, *32*, 2566–2581. [CrossRef]
66. Quentin, D.; Schuhmacher, J.S.; Klink, B.U.; Lauer, J.; Shaikh, T.R.; Huis, P.J.; Welp, L.M.; Urlaub, H.; Zerial, M.; Raunser, S. Structure of the human FERRY Rab5 effector complex. *bioRxiv* **2021**. [CrossRef]
67. Schuhmacher, J.S.; Dieck, S.T.; Christoforidis, S.; Landerer, C.; Hersemann, L.; Seifert, S.; Giner, A.; Toth-Petroczy, A.; Kalaidzidis, Y.; Schumann, E.M.; et al. The novel Rab5 effector FERRY links early endosomes with the translation machinery. *bioRxiv* **2021**. [CrossRef]
68. Guo, Y.; Yue, Q.; Gao, J.; Wang, Z.; Chen, Y.R.; Blissard, G.W.; Liu, T.X.; Li, Z. Roles of Cellular NSF Protein in Entry and Nuclear Egress of Budded Virions of Autographa californica Multiple Nucleopolyhedrovirus. *J. Virol.* **2017**, *91*, e01111-17. [CrossRef]
69. Mastick, C.C.; Falick, A.L. Association of N-ethylmaleimide sensitive fusion (NSF) protein and soluble NSF attachment proteins-alpha and -gamma with glucose transporter-4-containing vesicles in primary rat adipocytes. *Endocrinology* **1997**, *138*, 2391–2397. [CrossRef]
70. Zhao, C.; Matveeva, E.A.; Ren, Q.; Whiteheart, S.W. Dissecting the N-ethylmaleimide-sensitive factor: Required elements of the N and D1 domains. *J. Biol. Chem.* **2010**, *285*, 761–772. [CrossRef]
71. Zhao, C.; Slevin, J.T.; Whiteheart, S.W. Cellular functions of NSF: Not just SNAPs and SNAREs. *FEBS Lett.* **2007**, *581*, 2140–2149. [CrossRef] [PubMed]
72. Mantegazza, A. Preprint Highlight: The novel Rab5 effector FERRY links early endosomes with the translation machinery. *Mol. Biol. Cell* **2022**, *33*, mbcP22041001. [CrossRef] [PubMed]
73. Kloc, M.; Zearfoss, N.R.; Etkin, L.D. Mechanisms of subcellular mRNA localization. *Cell* **2002**, *108*, 533–544. [CrossRef]
74. Hara-Nishimura, I.; Matsushima, R.; Shimada, T.; Nishimura, M. Diversity and formation of endoplasmic reticulum-derived compartments in plants. Are these compartments specific to plant cells? *Plant Physiol.* **2004**, *136*, 3435–3439. [CrossRef]
75. Nagamine, A.; Matsusaka, H.; Ushijima, T.; Kawagoe, Y.; Ogawa, M.; Okita, T.W.; Kumamaru, T. A role for the cysteine-rich 10 kDa prolamin in protein body I formation in rice. *Plant Cell Physiol.* **2011**, *52*, 1003–1016. [CrossRef] [PubMed]
76. Chou, H.L.; Tian, L.; Fukuda, M.; Kumamaru, T.; Okita, T.W. The Role of RNA-Binding Protein OsTudor-SN in Post-Transcriptional Regulation of Seed Storage Proteins and Endosperm Development. *Plant Cell Physiol.* **2019**, *60*, 2193–2205. [CrossRef]
77. Washida, H.; Sugino, A.; Messing, J.; Esen, A.; Okita, T.W. Asymmetric localization of seed storage protein RNAs to distinct subdomains of the endoplasmic reticulum in developing maize endosperm cells. *Plant Cell Physiol.* **2004**, *45*, 1830–1837. [CrossRef]
78. Sami-Subbu, R.; Choi, S.B.; Wu, Y.; Wang, C.; Okita, T.W. Identification of a cytoskeleton-associated 120 kDa RNA-binding protein in developing rice seeds. *Plant Mol. Biol.* **2001**, *46*, 79–88. [CrossRef]
79. Sami-Subbu, R.; Muench, D.G.; Okita, T.W. A cytoskeleton-associated RNA-binding protein binds to the untranslated regions of prolamine mRNA and to poly(A). *Plant Sci.* **2000**, *152*, 115–122. [CrossRef]
80. Ponting, C.P. P100, a transcriptional coactivator, is a human homologue of staphylococcal nuclease. *Protein Sci.* **1997**, *6*, 459–463. [CrossRef]
81. Chou, H.L.; Tian, L.; Kumamaru, T.; Hamada, S.; Okita, T.W. Multifunctional RNA Binding Protein OsTudor-SN in Storage Protein mRNA Transport and Localization. *Plant Physiol.* **2017**, *175*, 1608–1623. [CrossRef] [PubMed]
82. Keene, J.D. RNA regulons: Coordination of post-transcriptional events. *Nat. Rev. Genet.* **2007**, *8*, 533–543. [CrossRef] [PubMed]
83. Samaj, J.; Salaj, J.; Obert, B.; Baluska, F.; Menzel, D.; Volkmann, D. Calreticulin mRNA and protein are localized to protein bodies in storage maize callus cells. *Plant Cell Rep.* **2008**, *27*, 231–239. [CrossRef]
84. Urbinati, C.R.; Long, R.M. Techniques for following the movement of single RNAs in living cells. *Wiley Interdiscip. Rev. RNA* **2011**, *2*, 601–609. [CrossRef] [PubMed]
85. Wu, B.; Chao, J.A.; Singer, R.H. Fluorescence fluctuation spectroscopy enables quantitative imaging of single mRNAs in living cells. *Biophys. J.* **2012**, *102*, 2936–2944. [CrossRef]
86. Luo, K.R.; Huang, N.C.; Yu, T.S. Selective Targeting of Mobile mRNAs to Plasmodesmata for Cell-to-Cell Movement. *Plant Physiol.* **2018**, *177*, 604–614. [CrossRef]
87. Bann, D.V.; Parent, L.J. Application of live-cell RNA imaging techniques to the study of retroviral RNA trafficking. *Viruses* **2012**, *4*, 963–979. [CrossRef]
88. Dean, K.M.; Palmer, A.E. Advances in fluorescence labeling strategies for dynamic cellular imaging. *Nat. Chem Biol.* **2014**, *10*, 512–523. [CrossRef]
89. Paige, J.S.; Wu, K.Y.; Jaffrey, S.R. RNA mimics of green fluorescent protein. *Science* **2011**, *333*, 642–646. [CrossRef]

International Journal of *Molecular Sciences*

MDPI

Article

Effects of Storage Temperature on Indica-Japonica Hybrid Rice Metabolites, Analyzed Using Liquid Chromatography and Mass Spectrometry

Lin Zhu [1,†], Yu Tian [2,†], Jiangang Ling [1], Xue Gong [2], Jing Sun [1,*] and Litao Tong [1,*]

1 Key Laboratory of Preservation Engineering of Agricultural Products, Ningbo Academy of Agricultural Sciences, Institute of Agricultural Products Processing, Ningbo 315040, China; zhulin0822@163.com (L.Z.); nbnjg@163.com (J.L.)

2 Key Laboratory of Agro-Products Processing Ministry of Agriculture, Institute of Food Science and Technology, Chinese Academy of Agricultural Sciences, Beijing 100193, China; tianyu152302@163.com (Y.T.); xuegong1214@163.com (X.G.)

* Correspondence: ycsunjing2008@126.com (J.S.); tonglitao@caas.cn (L.T.); Tel./Fax: +86-10-6281-0295 (J.S.); +86-10-6281-7417 (L.T.)

† These authors contributed equally to this work.

Abstract: The Yongyou series of indica-japonica hybrid rice has excellent production potential and storage performance. However, little is known about the underlying mechanism of its storage resistance. In this study, Yongyou 1540 rice (*Oryza sativa cv. yongyou 1540*) was stored at different temperatures, and the storability was validated though measuring nutritional components and apparent change. In addition, a broad-targeted metabolomic approach coupled with liquid chromatography-mass spectrometry was applied to analyze the metabolite changes. The study found that under high temperature storage conditions (35 °C), Yongyou 1540 was not significantly worse in terms of fatty acid value, whiteness value, and changes in electron microscope profile. A total of 19 key differential metabolites were screened, and lipid metabolites related to palmitoleic acid were found to affect the aging of rice. At the same time, two substances, guanosine 3′,5′-cyclophosphate and pipecolic acid, were beneficial to enhance the resistance of rice under harsh storage conditions, thereby delaying the deterioration of its quality and maintaining its quality. Significant regulation of galactose metabolism, alanine, aspartate and glutamate metabolism, butyrate metabolism, and arginine and proline metabolism pathways were probably responsible for the good storage capacity of Yongyou 1540.

Keywords: indica/japonica hybrid rice; liquid chromatography–mass spectrometry; wide-targeted metabolomics; storage temperature; storage performance

check for updates

Citation: Zhu, L.; Tian, Y.; Ling, J.; Gong, X.; Sun, J.; Tong, L. Effects of Storage Temperature on Indica-Japonica Hybrid Rice Metabolites, Analyzed Using Liquid Chromatography and Mass Spectrometry. *Int. J. Mol. Sci.* **2022**, *23*, 7421. https://doi.org/10.3390/ijms23137421

Academic Editors: Jinsong Bao and Jianhong Xu

Received: 7 May 2022
Accepted: 30 June 2022
Published: 4 July 2022

Publisher's Note: MDPI stays neutral with regard to jurisdictional claims in published maps and institutional affiliations.

Copyright: © 2022 by the authors. Licensee MDPI, Basel, Switzerland. This article is an open access article distributed under the terms and conditions of the Creative Commons Attribution (CC BY) license (https://creativecommons.org/licenses/by/4.0/).

1. Introduction

Rice (*Oryza sativa* L.) is one of the major staple foods consumed all over the world, notably in China, India, Indonesia, Bangladesh, Vietnam, Thailand, Myanmar, and the Philippines. These countries contribute most highly to rice production, corresponding to 82% of global rice production and 69% of global rice consumption [1]. At the same time, rice also provides 35–60% of dietary calories for most people in the world [2]. However, in the next few decades, with the increase of population, the demand for food in China and the world will remain a serious challenge. Breeding high-yielding rice varieties and developing high-yielding cultivation techniques are considered two key approaches to address this challenge [3,4]. Indica-japonica hybrid rice is a new type of rice cultivar developed by hybridizing indica as the male parent and japonica as the female parent. Moreover, the hybrid rice plant structure has the ability to support super high yield, and the Yongyou series have a positive effect [5].

The storage of rice is also an important part of its production and transactions, and only with good storage performance can its commercial value be maintained in the commodity transaction link. Starch, protein, lipid, mineral elements and vitamins, are the main nutritional components of rice [6–9], which is susceptible to microbial contamination and oxidative deterioration during storage, leading to reduced physiological activity, fat oxidation, and protein degradation [10]. The rice quality reduction was usually manifested in the weakening of grain respiration, the reduction of vitality, the change of physicochemical properties and the change of protoplasmic colloid structure [11]. There are many factors affecting the aging of rice, such as the variety, water content, temperature, humidity, storage time, storage conditions, and local climatic conditions.

Yongyou 1540 rice is a new variety jointly cultivated by the research group, and it has been proved that Yongyou 1540 has good storability. However, the underlying mechanisms are still largely unknown, which limits the promotion and application of the products. Metabolomics is the systemic study of the metabolites, that is, of all small molecules in a biological sample, to provide a snap-shot of the ongoing biochemical processes [12,13]. At the same time, it is also the science of qualitative and quantitative analysis of all low molecular weight (<1000) metabolites of biological cells or organisms in a specific physiological period [14].

In this study, in order to reveal the storage mechanism of Yongyou 1540, and to provide theoretical data support for the production and promotion of this series of indica-japonica hybrid rice, Yongyou 1540 was stored at different temperatures. The fatty acid and whiteness values of each sample were determined, and the rice surface was observed with an electron microscope. In addition, the changes of Yongyou 1540 metabolites under different storage conditions were investigated using a broad-targeted metabolomic technology based on LC-MS/MS. After multivariate statistical analysis such as principal component analysis, cluster analysis, and path analysis, significant differential metabolites were screened out. Finally, through enrichment analysis of metabolic pathways, the most relevant metabolic pathways were found. The results showed that the key differential metabolites of Yongyou 1540 that were significantly up-regulated were guanosine 3′,5′-cyclic phosphate, pipecolic acid and GABA. Based on the properties and effects of these substances, it is speculated that these substances increase the resistance of rice to relatively harsh storage conditions, thereby delaying the aging process and maintaining its quality.

2. Result

2.1. Measurement Results of Fatty Acid Value and Whiteness Value of Rice

As shown in Figure 1A, the higher the storage temperature, the higher was the fatty acid value (from 0.085 to 0.350 g kg^{-1}). According to the Chinese national standard GB/T20569-2006, the fatty acid value of fresh rice should be less than 0.250 g kg^{-1} and that of stored rice should be less than 0.350 g kg^{-1} [15]. Even under the storage condition of 35 °C, the fatty acid value met the requirements, and the degree of quality deterioration of the Yongyou rice was not significant.

The test results of whiteness value are shown in Figure 1B. The whiteness value represents the level of whiteness of the rice surface. Obviously, with the prolongation of storage time, the whiteness value of rice at different storage temperatures showed a significant downward trend, and then entered a stable stage and continued to decrease slowly. Due to the influence of storage temperature, the higher the temperature, the greater is the decrease of the whiteness value. However, the whiteness values (from 68% to 76%) are still in the reasonable whiteness value range [16].

2.2. Electron Microscope Section Observation Results

The section images of rice at different storage temperatures were observed under different magnifications (left: 1 mm, right: 200 μm) (Figure 2). Through observation, it can be seen that with the continuous increase of the ambient temperature of the rice, the width and depth of the cracks on the cut surface of the rice samples increased significantly by the

90th day of storage. This may be due to the influence of temperature in the early storage period. The loss of nutrients was accelerated, as well as the rate of surface water loss and the degree of deterioration. On the 180th day of storage, there was no obvious further deterioration compared with the previous period. This may be due to the fact that when kept under unsuitable storage conditions for a long time, a certain metabolic system was activated by the rice itself, which enhanced its resistance to stress and storage performance.

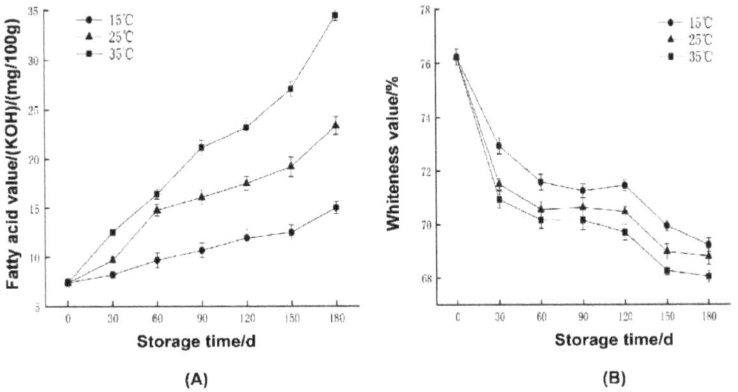

Figure 1. Changes in fatty acid value (**A**) and whiteness value (**B**) at different storage temperatures.

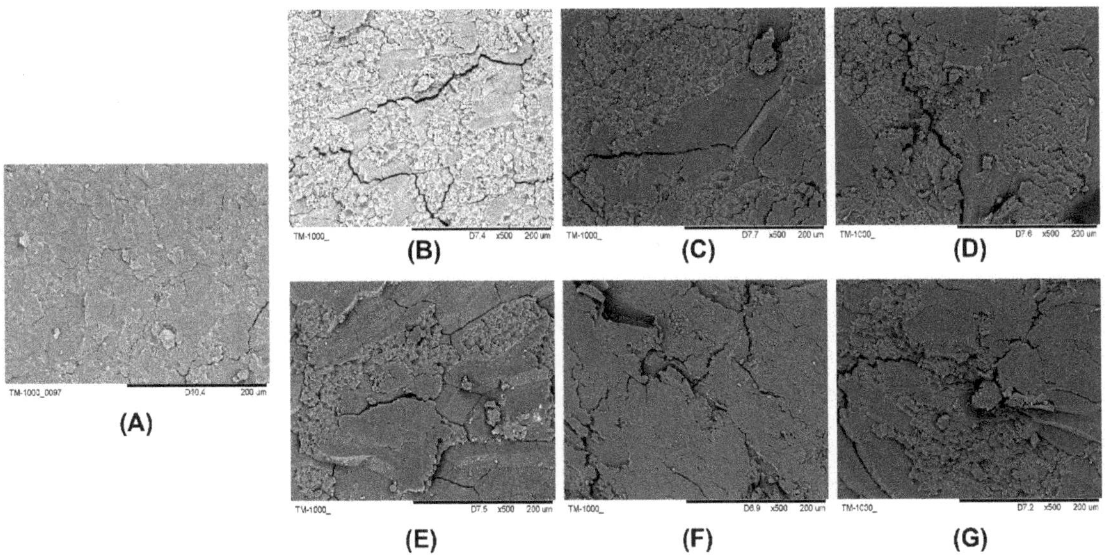

Figure 2. Electron microscope profiles of rice (200 μm) at different storage temperatures. (**A**) Morphological images of rice stored on day 0. (**B–D**) are the morphological images of rice on the 90th day of storage at 15 °C, 25 °C, and 35 °C storage temperature, respectively. (**E–G**) are the morphological images of rice on the 180th day of storage at 15 °C, 25 °C, and 35 °C storage temperature, respectively.

2.3. Metabolite Characterization of Samples

The LC-MS-based wide-targeted metabolomics approach was employed in the study. By matching the substance database with conditions such as retention time and mass-to-charge ratio, the substance could be qualitatively identified [17]. A total of 177 metabolites were identified, as shown in Supplementary Table S1. In terms of quantity, alkaloids account for 14.12%, amino acids and their derivatives account for 9.60%, phenols and their derivatives account for 8.74%, organic acids and their derivatives account for 8.47%, fatty acyl groups account for 6.7%, and flavonoids account for 7.3%, carbohydrates and their derivatives accounted for 6.21%, terpenoids accounted for 6.21%, and other small amounts of substances such as steroids and their derivatives, plant hormones, coumarin, purines, pyridines, etc. accounted for 12.99%. Alkaloids accounted for the largest proportion, followed by amino acids and their derivatives, while steroids and their derivatives and terpenes were also present.

2.4. Principal Component Analysis

The PCA score scatter plot of all samples (including QC samples) is shown in Figure 3. The value of R^2X in the scatter plot of the PCA model was 0.527, which further indicated that the experimental model was not over-fitting, and the experimental results were reliable. It could be clearly observed that there was a significant difference between HT_6, LT_6, NT_6 groups and the CK group, while the difference between the HT_6 group and CK group was the most significant. The comparisons between the LT_6 group and the CK group were close in degree of difference, suggesting that storage temperature was an important factor affecting the metabolic activity of rice. However, there are a certain number of samples with no obvious changes in the principal component differences. This may indicate that the rice of this variety produces a certain substance that is self-regulating even under inappropriate storage conditions.

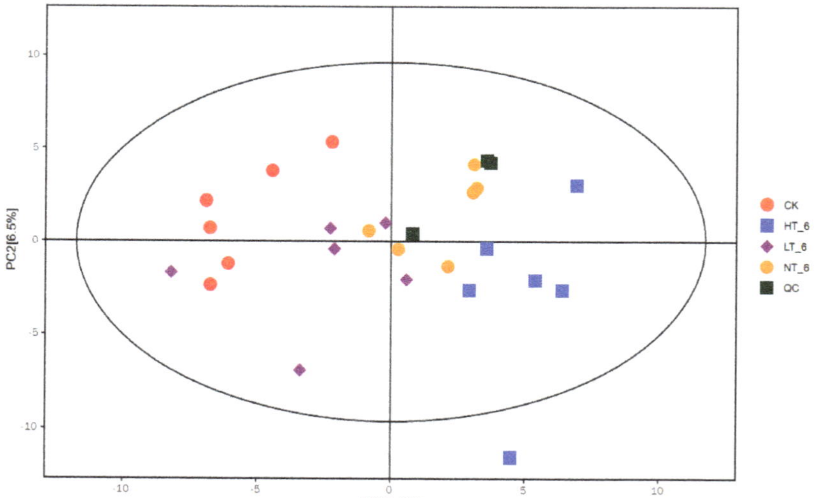

Figure 3. Scatter plot from PCA model for CK_6, LT_6, NT_6, HT_6, and QC groups. The abscissa PC[1] and ordinate PC[2] in the figure represent the scores of the first and second principal components, respectively. Each scatter represents a sample, and the color and shape of the scatter represent different groups. The samples were basically within the 95% confidence interval, and the QC group was tightly gathered and close to the middle of all samples, suggesting great system stability within the entire measurement queue. R^2X represents the model's interpretation of the X variable and Q^2 represents the predictability of the model. The closer the two metrics are to 1, the better the model performs and the higher the interpretability.

2.5. OPLS-DA Analysis

The abscissa t[1]P in the OPLS-DA scored graph represented the predicted principal component score of the first principal component, showing the difference between sample groups, and the ordinate t[1]O represented the orthogonal principal component score, showing the difference within the sample group. In the present study, OPLS-DA was modeled for the classification between the HT_6 and CK_6 groups, and the HT_6 and LT_6 groups, respectively. From the results of the score plot, it can be seen that the HT_6 group was very significantly distinguished from the CK group and the LT_6 group, and the samples were all within the 95% confidence interval (Hotelling's T-squared ellipse), indicating that the storage temperature changed the metabolites of the rice samples. According to the statistics, HT_6 vs. CK groups Q^2 = 0.701, HT_6 group vs. LT_6 group Q^2 = 0.654, indicating that if new samples are added to the model, a relatively approximate distribution will be obtained. The permutation test randomly changed the arrangement order of the categorical variable Y, and established the corresponding OPLS-DA, modeled multiple times (n = 200 times) to obtain the R^2Y and Q^2 values of the random model. The intercept of the regression line of Q^2 and the vertical axis was less than zero. At the same time, with the gradual decrease of the permutation retention, the proportion of the permuted Y variable increased, and the Q^2 of the random model gradually decreased. The results above showed that the original models of HT_6 vs. CK groups and HT_6 vs. LT_6 groups could meet the acceptable requirements.

The Q^2 values of NT_6 vs. CK groups, NT_6 vs. LT_6 groups, HT_6 vs. NT_6 groups were close to 0.5, indicating that there were differences between sample groups, but the degree of difference was not significant. The Q^2 < 0.5 of the LT_6 group and the CK group indicated that the metabolic difference between the two samples was not significant within a certain range. This phenomenon shows that the metabolic activity of Yongyou 1540 rice is slow under the storage condition of 15 °C (see Figure 4).

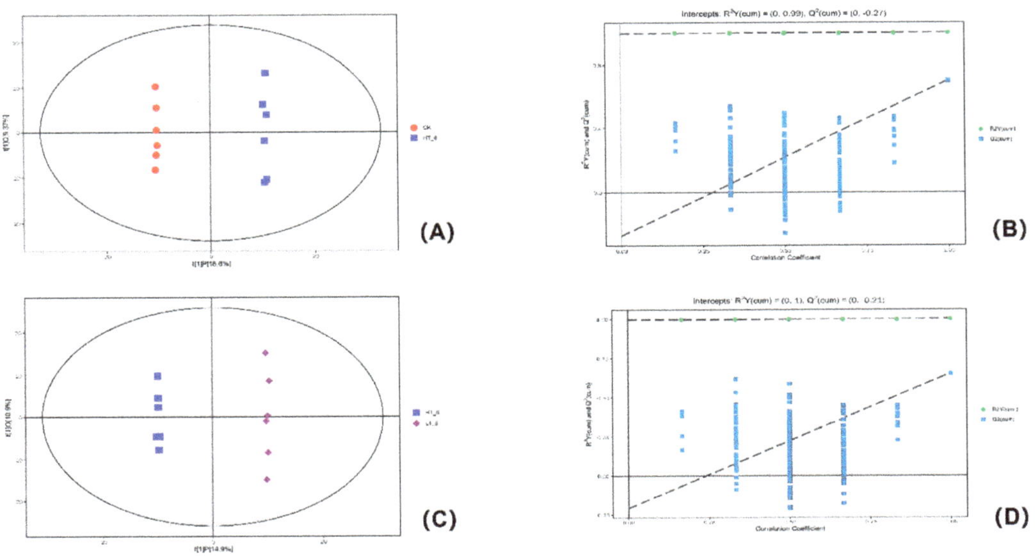

Figure 4. The scatter plot and permutation tested results of the OPLS-DA model for the HT_6 vs. CK groups, and the HT_6 vs. LT_6 groups. (**A**) The scatter plot of the OPLS-DA modeled on the HT_6 vs. CK groups; (**B**) The permutation test of the OPLS-DA modeled on the HT_6 vs. CK groups; (**C**) The scatter plot of the OPLS-DA modeled on the HT_6 vs. LT_6 groups; (**D**) The permutation test of the OPLS-DA modeled on the HT_6 vs. LT_6 groups.

2.6. Screening and Analysis of Differential Metabolites with Significant Changes in Content

In order to explore the effects of different storage temperatures on rice metabolites and quality, the VIP value of the OPLS-DA model (threshold > 1) and the p-value of Student's t-test (threshold < 0.05) were used to screen for differential metabolites. The results showed that there were 105 differential metabolites in HT_6 vs. CK group, 99 differential metabolites in the HT_6 vs. LT_6 group, 81 differential metabolites in HT_6 vs. NT_6 group, 93 differential metabolites in NT_6 vs. CK group, 87 differential metabolites in NT_6 vs. LT_6 group and 49 differential metabolites in LT_6 vs. CK group. The differential metabolites obtained through the above analysis often have the results and functional similarity/complementarity in biology, or are positively regulated/negatively regulated by the same metabolic pathway, showing similar or opposite expression characteristics between different experimental groups. Hierarchical cluster analysis was used to help explore similarities and differences between different groups. Figure 5 illustrated the thermodynamic diagram of the hierarchical cluster analysis of experimental groups at different storage temperatures [18]. The differential metabolites of the above experimental groups were also subjected to matchstick analysis and chord analysis. Matchstick analysis and chord analysis help to further screen the most closely related differential metabolites. The identification of differential metabolites can help to find the relevant metabolic pathways. HT_6 vs. CK groups were used as the core of the analysis. The research groups compared the results of their matchstick analysis and chord analysis, and screened for 19 key differential metabolites. These 19 substances are shown in Table 1. Among the above 19 differential metabolites, 8 substances were distinct differential metabolites in the HT_6 vs. NT_6 groups, and 14 substances were distinct differential metabolites in the HT_6 vs. LT_6 groups. The metabolites of LT_6 and CK were so similar that it was difficult to find the differential metabolites.

This result indicates that some reactions are only excited at the higher temperature conditions where HT_6 and NT_6 are located. The two groups had 8 identical differential metabolites, indicating that when the storage temperature was higher, the corresponding metabolic pathways would be activated to produce corresponding metabolites. The only other 11 differential metabolites expressed in the HT_6 vs. CK groups were those produced by the corresponding stimulated metabolic pathways in this rice variety under relatively harsher storage conditions. Among these 11 differential metabolites, adenosine 2′,3′−cyclic phosphate, guanosine 3′,5′−cyclic monophosphate, (S) −2−aceto−2−hydroxybutanoate, streptozotocin, palmitoylethanolamide, pimelic acid, pipecolic acid, these 7 substances are up-regulated. Stachyose, maltotetraose, kojibiose, and denudatine were down-regulated. The first three substances that were down-regulated are carbohydrates, and the down-regulation of carbohydrates indicated that the metabolism of energy substances was significant under high temperature storage conditions. HT_6 vs. LT_6 groups had the highest number of overlapping differential metabolites with HT_6 vs. CK groups. The HT_6 vs. CK group was the focus of the analysis, with a larger temperature span and wider coverage.

2.7. Effects of Storage on Metabolic Pathways of Rice

The research group mapped 19 key differential metabolites screened by the HT_6 vs. CK groups to authoritative metabolite databases such as KEGG and PubChem, and found all the pathways involved in the regulation of these differential metabolites (Figures 6 and 7). Through a comprehensive analysis of metabolic pathways (including enrichment analysis and topology analysis), the key pathways most associated with metabolite differences were found. As shown in Figure 6A, five of the most relevant pathways were located, including valine, leucine and isoleucine biosynthesis, galactose metabolism, lysine biosynthesis and degradation, and glutathione metabolism. Meanwhile, the differential metabolites with the highest correlation between the HT_6 and LT_6 groups were also located, and the most closely related metabolic pathways were identified, as shown in Figure 6B. It can be seen that the difference in metabolic pathways between the two experimental groups is only in the high expression of starch and sucrose metabolism in the HT_6 and LT_6 groups. The metabolic activity of rice stored at 15 °C was higher than that of rice stored at −80 °C. Energy metabolism is the normal physiological activity of rice.

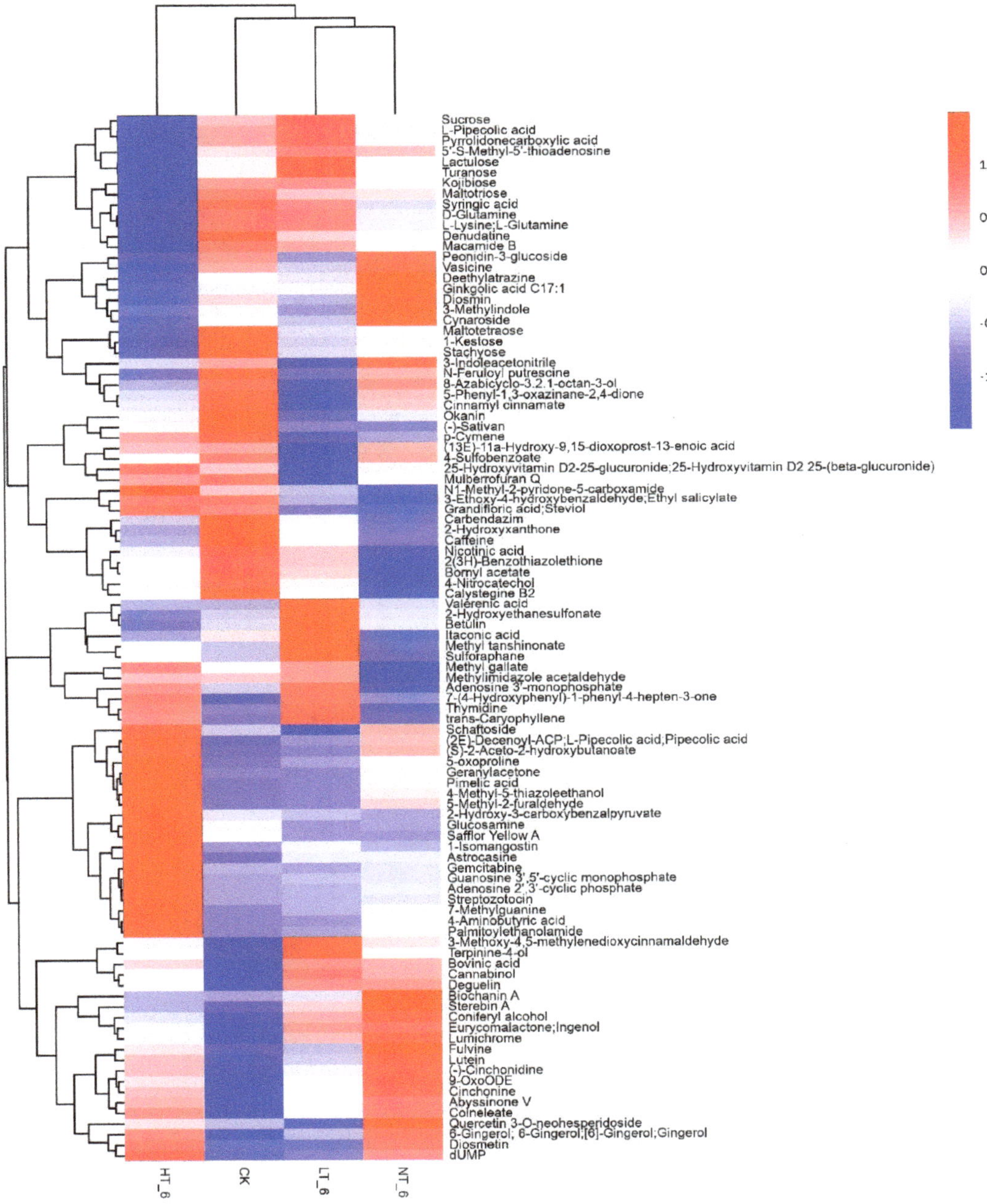

Figure 5. Hierarchical cluster analysis heatmap for all the experimental groups. The abscissa in the figure represents different experimental groups, the ordinate represents the differential metabolites compared in the group, the color blocks of different positions represent the relative expression levels of the metabolites at the corresponding positions, with the red indicating high expression of the substance, and the blue indicating low expression of the substance.

Table 1. Key differential metabolites in HT_6 group.

No.	Compound Name	Rt(min)	KEGG_ID	EXACT_MASS	VIP	*p*-Value	Fold Change	LOG_ Foldchange
1	Adenosine 2′,3′−cyclic phosphate	1.86	C02353	329.0525	2.192	0.0001337	30.64	4.937
2	(S)−2−Aceto−2−hydroxybutanoate	0.66	C06006	146.0579	2.258	0.00002167	3.301	1.723
3	Pimelic acid	0.55	C02656	160.0736	2.131	0.00008361	3.421	1.774
4	Pipecolic acid	1.47	C00408	129.0790	2.303	0.0002440	3.809	1.930
5	9−OxoODE	11.98	C14766	294.2195	1.504	0.0005610	3.886	1.958
6	Palmitoylethanolamide	12.61	C16512	299.2824	2.259	0.0008559	4.063	2.023
7	4−Methyl−5−thiazoleethanol	2.66	C04294	143.0405	2.419	0.000003395	5.152	2.365
8	5−Methyl−2−furaldehyde	0.76	C11115	110.0368	2.329	0.00000001875	5.440	2.444
9	5−Oxoproline	0.56	C01879	129.0426	2.238	0.0002517	6.411	2.681
10	7−Methylguanine	1.56	C02242	165.0651	2.425	0.000007288	10.393	3.378
11	Streptozotocin	0.72	C07313	265.0910	2.025	0.0002545	16.726	4.064
12	Guanosine 3′,5′−cyclic monophosphate	2.26	C00942	345.0474	2.135	0.00002303	18.115	4.179
13	4−Aminobutyric acid	0.66	C00334	103.0633	2.062	0.000000003593	21.814	4.447
14	D−Glutamine	0.66	C00819	146.0691	2.373	0.0005950	0.2028	−2.302
15	L−Glutamine	0.64	C00047; C00064	146.1055	2.319	0.0005209	0.2270	−2.139
16	Denudatine	1.59	C08680	343.2511	2.327	0.000001120	0.3601	−1.474
17	Maltotetraose	0.64	C02052	666.2219	1.865	0.0026925	0.3738	−1.420
18	Stachyose	0.73	C01613	666.2219	1.865	0.0026925	0.3738	−1.420
19	Kojibiose	0.72	C19632	342.1162	2.316	0.0002408	0.4181	−1.258

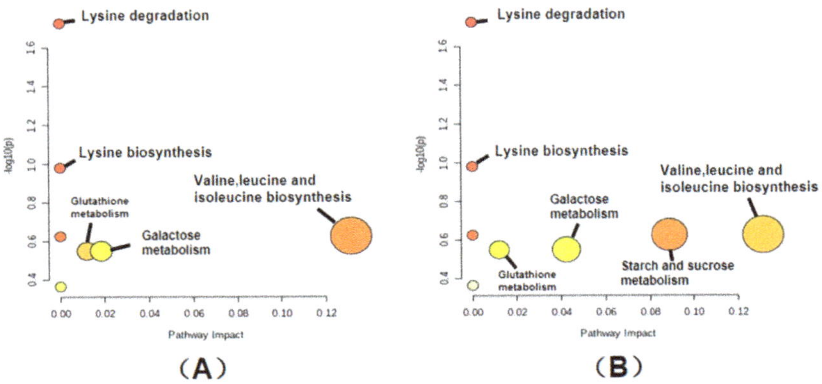

Figure 6. The key differential metabolites of the HT_6 group were mapped to metabolic pathways to analysis, and the results of the metabolic pathway analysis are shown in the bubble chart. (**A**) represents HT_6 vs. ck groups, (**B**) represents HT_6 vs. LT_6 groups. Each bubble in the bubble diagram represented a metabolic pathway. The abscissa of the bubble and the size of the bubble represent the size of the pathway influencing factor in the topology analysis. The larger the scale, the greater the influencing factors. The ordinate where the bubble is located and the color of the bubble represented the *p* value of the enrichment analysis (taking the negative natural logarithm, −ln(p), the darker the color, the smaller the *p* value, and the more significant the enrichment).

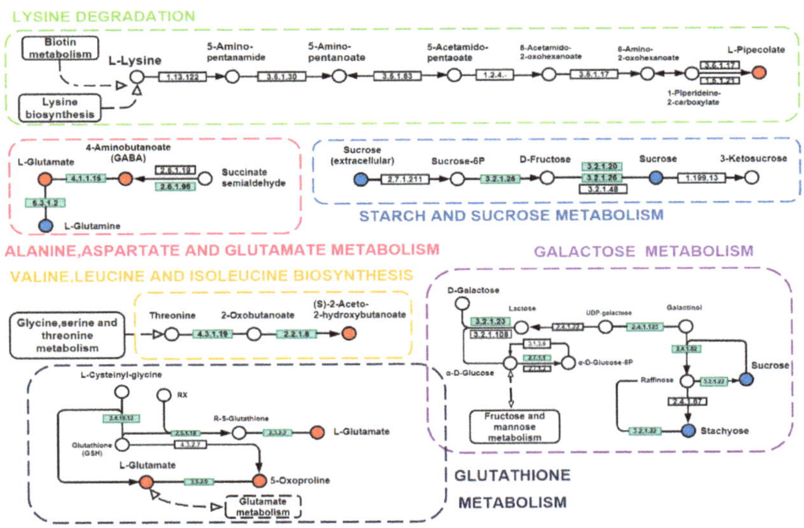

Figure 7. The key differential metabolites of HT_6 vs. CK groups and HT_6 vs. LT_6 groups were mapped to metabolic pathways, and the results of metabolic pathway analysis are shown in the KEGG pathway results.

3. Discussion

The basic quality of You 1540 was evaluated from three aspects: fatty acid value, whiteness value, and electron microscope section observation. According to the experimental results, the following conclusions can be drawn: At different storage temperatures, the changing trend of fatty acid value is different. The higher the storage temperature, the more obvious was the increase of fatty acid value. However, under the storage condition of 35 °C, the fatty acid value of rice was also within 0.350 g kg^{-1}, indicating that the rice was slightly unsuitable for storage at this temperature. However, the degree of quality deterioration of

Yongyou rice was not significant. Similarly, for the whiteness value of Yongyou 1540, the research team believes that the change of whiteness value was within an acceptable range after evaluation, and there was no obvious deterioration. On the other hand, the section of Yongyou 1540 was observed under an electron microscope with a magnification of 200 μm. On the 90th day, the depth and width of surface cracks of rice stored at 35 °C were indeed more obvious than those of rice at the other two storage temperatures. However, on the 180th day, the change of rice cracks was not obvious under different storage temperatures, and the degree was lower than that on the 90th day. Based on the above conclusions, it can be speculated that Yongyou 1540 performed certain metabolic activities and regulated certain pathways under unsuitable storage conditions, and this activity was related to its storage resistance and stress resistance.

To further analyze and explore the effects of metabolites on the storage performance of Yongyou 1540, the research group used liquid chromatography and mass spectrometry-based broad-target metabolomics techniques to isolate a total of 719 metabolites. Of these, 615 were recorded in public databases. Principal component analysis (PCA) and OPLS-DA analysis were used in this study. PCA analysis is a statistical method that transforms a set of observed, possibly correlated variables into linearly uncorrelated variables (i.e., principal components) through an orthogonal transformation. The PCA method can reveal the internal structure of the data and thus better explained the data variables between different groups [19]. In fact, LC-MS-based metabolomics data are characterized by high dimensionality (many types of metabolites detected) and small sample size (too few samples detected). These variables contain not only differential variables related to categorical variables, but also a large number of interrelated indifference variables [20]. Therefore, the OPLS-DA method, a supervised pattern recognition technique that outperforms PCA in class discrimination, was used to find distinct metabolites between each pair of groups. Through OPLS-DA analysis, orthogonal variables unrelated to categorical variables in metabolites were filtered out, and non-orthogonal variables and orthogonal variables were analyzed separately, so as to obtain more reliable information on the difference and influence degree between metabolite groups. PCA analysis can provide support in the acquisition of related metabolites, and OPLS-DA analysis can support the screening and identification of key differential metabolites. At the same time, the 19 most important differential metabolites were finally selected by using matchstick analysis and chord analysis to assist the screening.

Different storage temperatures were set in the experiments. The maximum storage temperature of 35 °C was used as the core of the exploration, and the key differential metabolites were analyzed. A certain number of differential metabolites overlapped with differential metabolites screened at other storage temperatures. In order to explore the metabolic activities and reactions of Yongyou 1540 at higher storage temperatures, the research team chose to analyze 11 non-overlapping differential metabolites one by one. Palmitoylethanolamide (PEA) is a natural amide of ethanolamine and palmitic acid. Pimelic acid exists in the free acid form and is synthetically assembled from fatty acids. These two substances are metabolites produced by the aging phenomenon caused by lipid metabolism in rice during storage. Guanosine 3′,5′−cyclic monophosphate (cGMP), a second messenger discovered in the 1960s, is found in both prokaryotes and eukaryotes [21]. The molecule cGMP is synthesized from guanosine triphosphate (GTP) by guanylate cyclase enzymes (GCs) and is involved in various cellular responses, such as protein kinase activity, cyclic nucleotide gated ion channels and cGMP regulated cyclic nucleotide phosphodiesterases [22,23]. There is a positive correlation between the accumulation of cGMP in plants and various developmental processes and responses to abiotic and pathogenic stresses. Multiple groups have demonstrated that both NO-dependent and NO-independent cGMP signaling pathways are important in the activation of defense responses during biological stress [24–28]. Furthermore, NO-cGMP-dependent signaling pathways have been reported to be involved in adventitious root development [29,30], stomatal closure during abiotic and biotic stresses [31,32], protein phosphorylation [33,34] and transcriptional regulation [35]. Internal cGMP production in rice involves various signaling processes in the plant, especially in control of stomatal pore size. In a high-temperature storage environment, this is important for surviving water shortages, and this key metabolite helps rice maintain water content in the body and delay quality deterioration. Pipecolic acid is L-Pip (hereinafter referred to as Pip); the widespread occurrence of the non-protein amino acid L-Pip in plants, animals, fungi, and microorganisms and its biosynthetic origin from Lys in plants and animals was realized in the 1950s [36,37]. Pip, a common lysine catabolite in plants and animals, is a key regulator of induced plant immunity. Pip is one of several key metabolic mediators for the induction of resistance [38–40]. Based on the above analysis of differential metabolites with significant effects, it can be speculated that under the storage condition of higher temperature, Yongyou 1540 significantly accumulated pipecolic acid, which induced rice to improve its own stress resistance to relatively harsh conditions and maintain its own quality.

In conclusion, a broad target metabolomics technique based on liquid chromatography and mass spectrometry was used in this study. The changes of metabolites of Yongyou 1540 at different storage temperatures were fully analyzed. Through the screening analysis and metabolic pathway mapping of key differential metabolites, Yongyou 1540 was shown to produce palmitoleic acid-related lipid metabolites that are associated with rice aging when stored at 35 °C. At the same time, the key differential metabolites of Yongyou 1540 were significantly up-regulated: guanosine 3′,5′-cyclic phosphate, pipecolic acid and GABA.

Based on the properties and effects of these substances, it is speculated that these substances increase the resistance of rice to relatively harsh storage conditions, thereby delaying the aging process and maintaining its own quality [41,42]. The results show that Yongyou 1540 has self-regulating ability under unsuitable storage conditions, and has good storage performance, which is beneficial to maintain the quality of production and transaction links. Under the increasingly heavy pressure for global food staples, this research can provide a theoretical and scientific basis for the popularization and production of Yongyou 1540.

4. Materials and Methods

4.1. Rice Materials and Sampling

Yongyou 1540 rice was harvested in Xiangshan County, Ningbo City, Zhejiang Province. The harvested rice was dried (RH 13–14%), and then ground into first-class rice according to GB/T1354 standard. All samples were divided into four groups. The control (CK) was frozen at −80 °C immediately after grinding. After the other three groups were milled, the rice was put into PE bags (2.5 kg per bag) and stored in a constant temperature and humidity box at 15 °C (LT_6), 25 °C (NT_6), or 35 °C (HT_6) (RH 60%), which were used as the experimental groups. There were six samples in parallel for each group. All samples were sent for testing after being stored for 6 months.

4.2. Materials and Reagents

Methanol (CAS: 67-56-1) and acetonitrile (CAS: 75-05-8) were supplied by CNW Technologies, and SIGMA brand formic acid reagent (CAS: 64-18-6) was used. The purity of the above reagents was all LC-MS grade. Ultrapure water was prepared in-house by a Milli-Q Integral water purification system (Millipore, Bedford, MA, USA). During the experiment, Sciex's model ExionLC AD ultra-high performance liquid phase and QTrap 6500+ high-sensitivity mass spectrometer were used. In addition, a Thermo centrifuge (Heraeus Fresco17), a Merck Millipore water purifier (Clear D24 UV) and a Waters chromatographic column (ACQUITY UPLC HSS T3 1.8 µm 2.1 ∗ 100 mm) were also used in the experiments.

4.3. Metabolites Extraction

The freeze-dried samples were crushed with a mixer mill for 240 s at 60 Hz. 400 mg aliquots of individual samples were precision weighed and then transferred to an Eppendorf tube. After addition of 2000 µL of extract solution (methanol/water = 3:1, precooled at −40 °C, containing internal standard), the samples were vortexed for 30s, homogenized at

40 Hz for 4 min, and sonicated for 5 min in an ice-water bath. After repeated homogenization and sonication for 3 times, the samples were extracted over night at 4 °C on a shaker. The resultant mixtures were centrifuged at 12,000× g rpm (RCF = 13,800× g, R = 8.6 cm) for 15 min at 4 °C, and the supernatant was carefully filtered through a 0.22 μm microporous membrane, and transferred to 2 mL glass vials. Moreover, 20 μL aliquots of each sample were pooled and prepared in parallel with other rice samples to yield quality control (QC) samples. All samples were stored at −80 °C until the UHPLC-MS analysis.

4.4. LC-MS Analysis

The UHPLC separation was carried out using an EXIONLC System (Sciex). The mobile phase A was 0.1% formic acid in water, and the mobile phase B was acetonitrile. The column temperature was set at 40 °C. The auto-sampler temperature was set at 4 °C and the injection volume was 2 μL. A Sciex QTrap 6500+ (Sciex Technologies, Framingham, MA, USA), was applied for assay development. Typical ion source parameters were: IonSpray Voltage: +5500/−4500 V, Curtain Gas: 35 psi, Temperature: 400 °C, Ion Source Gas 1:60 psi, Ion Source Gas 2: 60 psi, DP: ±100 V. In fact, due to the large sample size of this study, the detection task lasts for a long time, so it is very important to monitor the stability of the instrument and whether the signal is normal during the detection process in real time. To ensure the accuracy of the experimental results, the original data included 3 quality control (QC) samples and 24 experimental samples. QC samples were made by mixing 20 μL of each experimental sample. The retention times and peak areas of the QC samples TIC overlap well, indicating good instrument stability. In addition, the retention time and peak area of the internal standard 2-chlorophenylalanine response were very stable, indicating that the instrument's data acquisition stability was very good (Figures S1 and S2). The experimental equipment is reliable and the data is credible [43–51].

4.5. Data Analysis

SCIEX Analyst Work Station Software (Version 1.6.3, Framingham, MA, USA) was employed for MRM data acquisition and processing. MS raw data (.wiff) files were converted to the TXT format using MSconventer. In-house R program and database were applied to peak detection and annotation. After obtaining the collated data, the SIMCA software (V16.0.2, Sartorius Stedim Data Analytics AB, Umea, Sweden) was used to screen for differential metabolites using multivariate statistical analysis, such as principal component analysis (PCA), orthogonal partial least squares discriminant analysis (OPLS-DA), Student's t-test, and variable importance in projection (VIP) principal components of the OPLS-DA model. The card value standard used in this project is that the p-value of the Student's t-test is less than 0.05, and the VIP of the first principal component of the OPLS-DA model greater than 1. According to the expression profiles of differential metabolites, the changes of metabolites between groups can be summarized, and the useful information behind them can be mined in combination with the disciplinary background. For example, if some metabolites have the same or different variation trends among groups, combining metabolic pathways can help to mine important metabolic pathways and regulatory relationships. Differential metabolites were annotated in CAS and KEGG databases based on retention time and mass-to-charge ratio (m/z). Afterwards, through comprehensive analysis of differential metabolite pathways (including enrichment analysis and topology analysis), the pathways were further screened to identify the key pathways of the most relevant differential metabolites [52–54].

4.6. Determination of Rice Quality

4.6.1. Fatty Acid Value (Petroleum Ether Extraction Method)

Nutrients such as starch, protein, moisture, and lipids in rice itself will deteriorate due to the influence of the environment and internal factors during the storage process. Fatty acid value is an important indicator to measure the quality of rice. According to the Chinese national standard GB/T5510-2011 "Grain and Oil Inspection—Determination of

Fatty Acid Value of Grain and Oilseeds", the research team measured the fatty acid value of rice to judge the degree of deterioration of rice at different storage temperatures. Rice was crushed by grinder under different storage conditions (the groups needed cleaning and were passed through a 1.0 mm round hole sieve). A sample of 10 g (precision 0.01 g) was weighed and placed in a 250 mL conical flask. 50 mL petroleum ether was added to the pipette, and the plug was added before shaking for several seconds. The plug was opened and deflated, and then the bottle was closed and the oscillator was shaken for 10 min. Next steps: Take off the conical bottle, tilt and stand for 1–2 min, put filter paper into the short neck glass funnel to filter. Go to the first few drops of filtrate, use colorimetric tube to collect filtrate more than 25 mL, cover and save. (Tight timing is important: placed at 4 °C, use within 24 h). 25 mL filtrate was removed in 150 mL conical flask with a pipette, 75 mL 50% ethanol solution was added into the measuring cylinder, 4–5 drops of phenolphthalein indicator was added, shaken, titrated with potassium hydroxide solution until the lower ethanol solution was slightly red, 30 s did not fade, the titration solution volume (V_1) was recorded, and 25 mL petroleum ether was used as the blank control group, the titration solution volume (V_0) was recorded.

The acid value (A_K) formula is as follows:

$$A_K = (V_1 - V_0) * c * 56.1 * \frac{50}{25} * \frac{100}{m(100 - w)} * 100$$

c-potassium hydroxide concentration (mol/L)
m—Sample mass (g)
w—Sample moisture mass (per 100 g)
56.1—Potassium hydroxide molar mass (g/mol)
50—The volume of the extraction solution used to extract the sample(mL)
25—Volume of sample extract for titration (mL)
100—Converted to the mass of 100 g dry sample (g)

4.6.2. Whiteness Value

The whiteness of rice under different storage conditions was measured with a whiteness meter, and each sample was measured three times.

4.6.3. Observation of Appearance Features in Slices under Electron Microscope

Section observation (2 mm thick cross section), sampling at 90 d and 180 d, respectively. Instrument name of the microscope used in this project: Thermal Field Emission Scanning Electron Microscope (Zeiss G300); Model: GeminiSEM 300.

Supplementary Materials: The following supporting information can be downloaded at: https://www.mdpi.com/article/10.3390/ijms23137421/s1.

Author Contributions: Y.T.: Methodology, Validation, Writing—Original Draft. L.Z. and L.T.: Conceptualization, Methodology, Investigation, Resources, Writing—Review and Editing, Supervision. J.L.: Investigation, Resources and Supervision. X.G.: Formal analysis and Data Curation. J.S.: Investigation, Formal analysis, Data Curation, Writing—Review and Editing. All authors have read and agreed to the published version of the manuscript.

Funding: This work was supported by Funding: Science and Technology Cooperative Innovation Project between Ningbo Academy of Agricultural Sciences and Chinese Academy of Agricultural Sciences (2019CXGC005).

Institutional Review Board Statement: Not applicable.

Informed Consent Statement: Not applicable.

Data Availability Statement: Not applicable.

Conflicts of Interest: The authors declare that they have no known competing financial interest or personal relationships that could have appeared to influence the work reported in this paper.

References

1. USDA. Grain: World Markets and Trade. Available online: https://www.fas.usda.gov/data/grain-world-markets-and-trade (accessed on 10 June 2022).
2. Fageria, N.K.; Baligar, V.C. Methodology for evaluation of lowland rice genotypes for nitrogen use efficiency. *J. Plant Nutr.* **2003**, *26*, 1315–1333. [CrossRef]
3. Normile, D. Reinventing rice to feed the world. *Science* **2008**, *321*, 330–333. [CrossRef] [PubMed]
4. Peng, S.; Tang, Q.; Zou, Y. Current status and challenges of rice production in China. *Plant Prod. Sci.* **2009**, *12*, 3–8. [CrossRef]
5. Yin, M.; Liu, S.; Zheng, X.; Chu, G.; Xu, C.; Zhang, X.; Wang, D.; Chen, S. Solar radiation-use characteristics of *indica/japonica* hybrid rice (*Oryza sativa* L.) in the late season in southeast China. *Crop J.* **2021**, *9*, 427–439. [CrossRef]
6. Zhou, Z.; Robards, K.; Helliwell, S.; Blanchard, C. Composition and functional properties of rice. *Int. J. Food Sci. Technol.* **2002**, *37*, 849–868. [CrossRef]
7. Yoshida, H.; Tomiyama, Y.; Mizushina, Y. Lipid components, fatty acids and triacylglycerol molecular species of black and red rices. *Food Chem.* **2010**, *123*, 210–215. [CrossRef]
8. Kim, N.H.; Kwak, J.; Baik, J.Y.; Yoon, M.R.; Lee, J.S.; Yoon, S.W.; Kim, I.H. Changes in lipid substances in rice during grain development. *Phytochemistry* **2015**, *116*, 170–179. [CrossRef]
9. Heinemann, R.J.B.; Fagundes, P.D.L.; Pinto, E.A.; Penteado, M.D.V.C.; Lanfer-Marquez, U.M. Comparative study of nutrient composition of commercial brown, parboiled and milled rice from Brazil. *J. Food Compos. Anal.* **2005**, *18*, 287–296. [CrossRef]
10. Chen, H.H.; Hung, C.L.; Lin, S.Y.; Liou, G.J. Effect of low-pressure plasma exposure on the storage characteristics of brown rice. *Food Bioprocess Technol.* **2015**, *8*, 471–477. [CrossRef]
11. Saikrishna, A.; Dutta, S.; Subramanian, V.; Moses, J.A.; Anandharamakrishnan, C. Ageing of rice: A review. *J. Cereal Sci.* **2018**, *81*, 161–170. [CrossRef]
12. Sun, Y.V.; Hu, Y.J. Integrative analysis of multi-omics data for discovery and functional studies of complex human diseases. *Adv. Genet.* **2016**, *93*, 147–190. [PubMed]
13. Link, H.; Fuhrer, T.; Gerosa, L.; Zamboni, N.; Sauer, U. Real-time metabolome profiling of the metabolic switch between starvation and growth. *Nat. Methods* **2015**, *12*, 1091–1097. [CrossRef] [PubMed]
14. Wang, C.; Feng, Y.; Zhang, S.; Fu, T.; Sheng, Y.; Zhang, Y.; Jiang, Y.; Yu, M.; Zhang, L.; Zhang, D. Effects of storage on brown rice (*Oryza sativa* L.) metabolites, analyzed using gas chromatography and mass spectrometry. *Food Sci. Nutr.* **2020**, *8*, 2882–2894. [CrossRef] [PubMed]
15. Yin, Q.; Jiang, Z.; Wei, M. Research on quality changing of stored rice in high temperature. *Agric. Sci. Technol. Equip.* **2015**, *8*, 50–53.
16. Fang, G.; Wang, X.; Zhao, Y. A measuring method of rice whitness. *J. Harbin Univ. Sci. Technol.* **2008**, *4*, 14–16+20.
17. Saccenti, E.; Hoefsloot, H.C.; Smilde, A.K.; Westerhuis, J.A.; Hendriks, M.M. Reflections on univariate and multivariate analysis of metabolomics data. *Metabolomics* **2014**, *10*, 361–374. [CrossRef]
18. Kolde, R.; Kolde, M.R. Package 'pheatmap'. R Package Version 1.0.8.
19. Jolliffe, I.T.; Cadima, J. Principal component analysis: A review and recent developments. *Philos. Trans. R. Soc. A* **2016**, *374*, 20150202. [CrossRef]
20. Trygg, J.; Wold, S. Orthogonal projections to latent structures (O-PLS). *J. Chemom.* **2002**, *16*, 119–128. [CrossRef]
21. Gross, I.; Durner, J. In search of enzymes with a role in 3′,5′-cyclic guanosine monophosphate metabolism in plants. *Front. Plant Sci.* **2016**, *7*, 576. [CrossRef]
22. Denninger, J.W.; Marletta, M.A. Guanylate cyclase and the NO/cGMP signaling pathway. *Biochimica et Biophysica Acta (BBA)-Bioenergetics* **1999**, *1411*, 334–350. [CrossRef]
23. Römling, U.; Galperin, M.Y.; Gomelsky, M. Cyclic di-GMP: The first 25 years of a universal bacterial second messenger. *Microbiol. Mol. Biol. R.* **2013**, *77*, 1–52. [CrossRef] [PubMed]
24. Klessig, D.F.; Durner, J.; Noad, R.; Navarre, D.A.; Wendehenne, D.; Kumar, D.; Zhou, J.; Shah, J.; Zhang, S.; Kachroo, P.; et al. Nitric oxide and salicylic acid signaling in plant defense. *Proc. Natl. Acad. Sci. USA* **2000**, *97*, 8849–8855. [CrossRef] [PubMed]
25. Pasqualini, S.; Meier, S.; Gehring, C.; Madeo, L.; Fornaciari, M.; Romano, B.; Ederli, L. Ozone and nitric oxide induce cGMP-dependent and -independent transcription of defence genes in tobacco. *New Phytol.* **2009**, *181*, 860–870. [CrossRef] [PubMed]
26. Liu, S.; Ma, X.; Gong, M.; Shi, L.; Lincoln, T.; Wang, S. Glucose down-regulation of cGMP-dependent protein kinase I expression in vascular smooth muscle cells involves NAD(P)H oxidase-derived reactive oxygen species. *Free Radic. Biol. Med.* **2007**, *42*, 852–863. [CrossRef]
27. Xu, J.; Morris, L.; Thapa, A.; Ma, H.; Michalakis, S.; Biel, M.; Baehr, W.; Peshenko, I.; Dizhoor, A.; Ding, X.Q. cGMP accumulation causes photoreceptor degeneration in CNG channel deficiency: Evidence of cGMP cytotoxicity independently of enhanced CNG channel function. *J. Neurosci.* **2013**, *33*, 14939–14948. [CrossRef]
28. Qi, Z.; Verma, R.; Gehring, C.; Yamaguchi, Y.; Zhao, Y.; Ryan, C.A.; Berkowitz, G.A. Ca^{2+} signaling by plant *Arabidopsis thaliana* Pep peptides depends on AtPepR1, a receptor with guanylyl cyclase activity, and cGMP-activated Ca^{2+} channels. *Proc. Natl. Acad. Sci. USA* **2010**, *107*, 21193–21198. [CrossRef]
29. Pagnussat, G.C.; Lanteri, M.L.; Lamattina, L. Nitric oxide and cyclic GMP are messengers in the indole acetic acid-induced adventitious rooting process. *Plant Physiol.* **2003**, *132*, 1241–1248. [CrossRef]
30. Xuan, J.; Xiao, W.J. Visible-light photoredox catalysis. *Angew. Chem. Int. Ed.* **2012**, *51*, 6828–6838. [CrossRef]

31. Neill, S.; Barros, R.; Bright, J.; Desikan, R.; Hancock, J.; Harrison, J.; Morris, P.; Ribeiro, D.; Wilson, I. Nitric oxide, stomatal closure, and abiotic stress. *J. Exp. Bot.* **2008**, *59*, 165–176. [CrossRef]
32. Joudoi, T.; Shichiri, Y.; Kamizono, N.; Akaike, T.; Sawa, T.; Yoshitake, J.; Yamada, N.; Iwai, S. Nitrated cyclic GMP modulates guard cell signaling in Arabidopsis. *Plant Cell* **2013**, *25*, 558–571. [CrossRef]
33. Isner, J.C.; Nühse, T.; Maathuis, F.J. The cyclic nucleotide cGMP is involved in plant hormone signalling and alters phosphorylation of Arabidopsis thaliana root proteins. *J. Exp. Bot.* **2012**, *63*, 3199–3205. [CrossRef] [PubMed]
34. Marondedze, C.; Wong, A.; Thomas, L.; Irving, H.; Gehring, C. Cyclic nucleotide monophosphates in plants and plant signaling. In *Non-Canonical Cyclic Nucleotides. Handbook of Experimental Pharmacology*; Seifert, R., Ed.; Springer: Cham, Switzerland, 2015; Volume 238, pp. 87–103.
35. Suita, K.; Kiryu, T.; Sawada, M.; Mitsui, M.; Nakagawa, M.; Kanamaru, K.; Yamagata, H. Cyclic GMP acts as a common regulator for the transcriptional activation of the flavonoid biosynthetic pathway in soybean. *Planta* **2009**, *229*, 403–413. [CrossRef]
36. Broquist, H.P. Lysine-pipecolic acid metabolic relationships in microbes and mammals. *Annu. Rev. Nutr.* **1991**, *11*, 435–448. [CrossRef]
37. Zacharius, R.M.; Thompson, J.F.; Steward, F.C. The Detection, Isolation and Identification of L(—)Pipecolic Acid in the Non-protein Fraction of Beans (*Phascolus vulgaris*). *J. Am. Chem. Soc.* **1954**, *76*, 2908–2912. [CrossRef]
38. Lugon-Moulin, N.; Martin, F.; Krauss, M.R.; Ramey, P.B.; Rossi, L. Cadmium concentration in tobacco (*Nicotiana tabacum* L.) from different countries and its relationship with other elements. *Chemosphere* **2006**, *63*, 1074–1086. [CrossRef] [PubMed]
39. Pálfi, G.; Dézsi, L. Pipecolic acid as an indicator of abnormal protein metabolism in diseased plants. *Plant Soil* **1968**, *29*, 285–291. [CrossRef]
40. Yatsu, L.; Boynton, D. Pipecolic acid in leaves of strawberry plant as influenced by treatments affecting growth. *Science* **1959**, *130*, 864–865. [CrossRef] [PubMed]
41. Xiong, H.; Guo, H.; Xie, Y.; Zhao, L.; Gu, J.; Zhao, S.; Li, J.; Liu, L. RNAseq analysis reveals pathways and candidate genes associated with salinity tolerance in a spaceflight-induced wheat mutant. *Sci. Rep.* **2017**, *7*, 2731. [CrossRef]
42. Obata, T.; Witt, S.; Lisec, J.; Palacios-Rojas, N.; Florez-Sarasa, I.; Yousfi, S.; Araus, J.L.; Cairns, J.E.; Fernie, A.R. Metabolite profiles of maize leaves in drought, heat, and combined stress field trials reveal the relationship between metabolism and grain yield. *Plant Physiol.* **2015**, *169*, 2665–2683. [CrossRef]
43. Sawada, Y.; Akiyama, K.; Sakata, A.; Kuwahara, A.; Otsuki, H.; Sakurai, T.; Saito, K.; Hirai, M.Y. Widely targeted metabolomics based on large-scale MS/MS data for elucidating metabolite accumulation patterns in plants. *Plant Cell Physiol.* **2009**, *50*, 37–47. [CrossRef]
44. Zha, H.; Cai, Y.; Yin, Y.; Wang, Z.; Li, K.; Zhu, Z.J. SWATHtoMRM: Development of high-coverage targeted metabolomics method using SWATH technology for biomarker discovery. *Anal. Chem.* **2018**, *90*, 4062–4070. [CrossRef] [PubMed]
45. Shi, X.; Wang, S.; Jasbi, P.; Turner, C.; Hrovat, J.; Wei, Y.; Liu, J.; Gu, H. Database-assisted globally optimized targeted mass spectrometry (dGOT-MS): Broad and reliable metabolomics analysis with enhanced identification. *Anal. Chem.* **2019**, *91*, 13737–13745. [CrossRef]
46. De Vos, R.C.; Moco, S.; Lommen, A.; Keurentjes, J.J.; Bino, R.J.; Hall, R.D. Untargeted large-scale plant metabolomics using liquid chromatography coupled to mass spectrometry. *Nat. Protoc.* **2007**, *2*, 778–791. [CrossRef] [PubMed]
47. Chen, W.; Gong, L.; Guo, Z.; Wang, W.; Zhang, H.; Liu, X.; Yu, S.; Xiong, L.; Luo, J. A novel integrated method for large-scale detection, identification, and quantification of widely targeted metabolites: Application in the study of rice metabolomics. *Mol. Plant* **2013**, *6*, 1769–1780. [CrossRef] [PubMed]
48. Doppler, M.; Kluger, B.; Bueschl, C.; Schneider, C.; Krska, R.; Delcambre, S.; Hiller, K.; Lemmens, M.; Schuhmacher, R. Stable isotope-assisted evaluation of different extraction solvents for untargeted metabolomics of plants. *Int. J. Mol. Sci.* **2016**, *17*, 1017. [CrossRef]
49. Cai, Y.; Weng, K.; Guo, Y.; Peng, J.; Zhu, Z.J. An integrated targeted metabolomic platform for high-throughput metabolite profiling and automated data processing. *Metabolomics* **2015**, *11*, 1575–1586. [CrossRef]
50. Dunn, W.B.; Broadhurst, D.; Begley, P.; Zelena, E.; Francis-McIntyre, S.; Anderson, N.; Brown, M.; Knowles, J.D.; Halsall, A.; Haselden, J.N.; et al. Procedures for large-scale metabolic profiling of serum and plasma using gas chromatography and liquid chromatography coupled to mass spectrometry. *Nat. Protoc.* **2011**, *6*, 1060–1083. [CrossRef]
51. Want, E.J.; Wilson, I.D.; Gika, H.; Theodoridis, G.; Plumb, R.S.; Shockcor, J.; Holmes, E.; Nicholson, J.K. Global metabolic profiling procedures for urine using UPLC–MS. *Nat. Protoc.* **2010**, *5*, 1005–1018. [CrossRef]
52. Smith, C.A.; Want, E.J.; O'Maille, G.; Abagyan, R.; Siuzdak, G. XCMS: Processing mass spectrometry data for metabolite profiling using nonlinear peak alignment, matching, and identification. *Anal. Chem.* **2006**, *78*, 779–787. [CrossRef]
53. Kuhl, C.; Tautenhahn, R.; Bottcher, C.; Larson, T.R.; Neumann, S. CAMERA: An integrated strategy for compound spectra extraction and annotation of liquid chromatography/mass spectrometry data sets. *Anal. Chem.* **2012**, *84*, 283–289. [CrossRef]
54. Zhang, Z.M.; Tong, X.; Peng, Y.; Ma, P.; Zhang, M.J.; Lu, H.M.; Chen, X.Q.; Liang, Y.Z. Multiscale peak detection in wavelet space. *Analyst* **2015**, *140*, 7955–7964. [CrossRef] [PubMed]

International Journal of *Molecular Sciences*

MDPI

Article

Molecular Characterization and SNP-Based Molecular Marker Development of Two Novel High Molecular Weight Glutenin Genes from *Triticum spelta* L.

Yuemei Cao †, Junwei Zhang †, Ruomei Wang, Haocheng Sun and Yueming Yan *

Beijing Key Laboratory of Plant Gene Resources and Biotechnology for Carbon Reduction and Environmental Improvement, College of Life Science, Capital Normal University, Beijing 100048, China

* Correspondence: yanym@cnu.edu.cn

† These authors contributed equally to this work.

Abstract: Spelt wheat (*Triticum spelta* L., 2n=6x=42, AABBDD) is a valuable source of new gene resources for wheat genetic improvement. In the present study, two novel high molecular weight glutenin subunits (HMW-GS) 1Ax2.1* at *Glu-A1* and 1By19* at *Glu-B1* from German spelt wheat were identified. The encoding genes of both subunits were amplified and cloned by allele-specific PCR (AS-PCR), and the complete sequences of open reading frames (ORF) were obtained. *1Ax2.1** with 2478 bp and *1By19** with 2163 bp encoded 824 and 720 amino acid residues, respectively. Molecular characterization showed that both subunits had a longer repetitive region, and high percentage of α-helices at the N- and C-termini, which are beneficial for forming superior gluten macropolymers. Protein modelling by AlphaFold2 revealed similar three-diamensional (3D) structure features of 1Ax2.1* with two x-type superior quality subunits (1Ax1 and 1Ax2*) and 1By19* with four y-type superior quality subunits (1By16, 1By9, 1By8 and 1By18). Four cysteine residues in the three x-type subunits (1Ax2.1*, 1Ax1 and 1Ax2*) and the cysteine in intermediate repeat region of y-type subunits were not expected to participate in intramolecular disulfide bond formation, but these cysteines might form intermolecular disulfide bonds with other glutenins and gliadins to enhance gluten macropolymer formation. The SNP-based molecular markers for *1Ax2.1** and *1By19** genes were developed, which were verified in different F2 populations and recombination inbred lines (RILs) derived from crossing between spelt wheat and bread wheat cultivars. This study provides data on new glutenin genes and molecular markers for wheat quality improvement.

Keywords: spelt wheat; HMW-GS; three-diamensional structure; SNP marker; quality improvement

check for updates

Citation: Cao, Y.; Zhang, J.; Wang, R.; Sun, H.; Yan, Y. Molecular Characterization and SNP-Based Molecular Marker Development of Two Novel High Molecular Weight Glutenin Genes from *Triticum spelta* L.. *Int. J. Mol. Sci.* **2022**, *23*, 11104. https://doi.org/10.3390/ijms231911104

Academic Editors: Jinsong Bao and Jianhong Xu

Received: 26 August 2022
Accepted: 11 September 2022
Published: 21 September 2022

Publisher's Note: MDPI stays neutral with regard to jurisdictional claims in published maps and institutional affiliations.

Copyright: © 2022 by the authors. Licensee MDPI, Basel, Switzerland. This article is an open access article distributed under the terms and conditions of the Creative Commons Attribution (CC BY) license (https://creativecommons.org/licenses/by/4.0/).

1. Introduction

Wheat (*Triticum aestivum* L., 2n=6x=42, AABBDD) is one of the three major food crops globally with a long history of cultivation, accounting for almost 35% of the human population's dietary needs [1]. Wheat is an allohexaploid species, and has a huge genome (up to 17 GB) containing A, B, and D subgenomes and a large amount of repetitive sequences [2]. Wheat flour can be processed into various foods such as bread, noodles, cookies as well as non-food products because of its versatile functional features and excellent textural properties, which are closely related to the structure of gluten and interactions within the protein complex in grains [3,4].

The gluten proteins include water-insoluble monomeric gliadins and polymeric glutenins, which are considered as the largest protein molecule in nature [5]. Gliadins act as a "plasticizer" to endow dough extensibility or viscous flow. The glutenin proteins include high and low molecular weight glutenin subunits (HMW-GS, LMW-GS), of which HMW-GS serve as "backbone" for interactions with other glutenin subunits and confer dough elasticity or strength [6,7]. HMW-GS are encoded by two tightly linked paralogous genes (larger x- and smaller y-type) located at the *Glu-A1*, *Glu-B1* and *Glu-D1* loci on the long arm of the chromosome 1A, 1B and 1D, respectively. Theoretically, individual hexaploid wheat cultivars can express 6 different HMW-GS, but usually 3~5 HMW-GS are present due to gene silence [8,9]. The allelic variations at *Glu-1* loci are closely associated with flour processing quality [10]. For example, the subunit pairs and subunits 1Dx5+1Dy10, 1Bx7+1By8, 1Bx13+1By16, 1Bx17+1By18, 1Ax1, 1Ax2* have positive effects on gluten quality whereas 1Dx2+1Dy12 and 1Bx20 are related to poor dough strength [11–13].

Typical HMW-GS have three structural domains: a long and variable central repeat region conferring elasticity to protein molecules, and flanked by the short and conservative N- and C-terminals, in which most or all cysteine residues are present [14,15]. The repeat region is rich in β-turns flanked by spherical conservative regions formed by α-helices [16]. Most of the cysteine residues form intra-chain disulfide bonds, some of them form inter-chain disulfide bonds and directly affect dough formation and rheological characteristics [17–19]. Generally, the β-turn structure confers protein molecules with significant deformation resistance [20]. In particular, the central repeat domain can be folded in the action of protein disulfide isomerase, which is the basis of dough extension [21]. However, although different methods such as homology modelling and fold identification algorithm were used for deciphering the structure features of HMW-GS [22,23], the detailed 3D structures of wheat glutenin subunits and gluten proteins are still not clear due to the complexity of the protein compositions and the difficulty of crystallization [24–26]. The recently developed protein predicting tool Alphafold has provided the possibility to dissect the 3D structure of glutenin subunits, which is considered as the first computational method that can regularly predict protein structures with atomic accuracy even in cases in which no similar structure is known [27–29]. The recent study has shown that AlphaFold2 using the transformer design can obtain a high accuracy and is easy for 3D structure prediction of wheat MATE proteins [30].

In the past decades, a lot of studies have showed that the allelic variations at *Glu-1* in bread wheat are limited. However, more extensive *Glu-1* allelic variations are present in wheat related species such as spelt wheat [31,32], *T. dicoccum* [33], *Ae. tauschii* [34,35], *Ae. longissima* [36,37], *Ae. speltoides* and *Ae. kotschyi* [38]. Spelt wheat (*T. spelta* L., 2n=6x=42, AABBDD) is closely related to common wheat and belongs to the same species with few differences among the subspecies [39–41]. As the first-class gene resource of wheat, spelt wheat is rich in nutritional value, various disease resistance factors, tolerance for various abiotic stresses and can be used for wheat genetic improvement through introducing the beneficial gene from spelt wheat [42–45]. In particular, some novel HMW-GS and subunit combinations in European spelt wheat varieties were identified such as 1Ax2.1*, 1Bx13*+1By19* and 1Bx6.1+1By22.1 [31]. These new allelic variations can be used as potential gene resources for improving wheat gluten quality. However, the molecular characterization of these novel glutenin genes and their application values for wheat quality improvement are still unknown.

To create novel germplasm for the improvement of wheat quality, we amplified and cloned the encoding genes of 1Ax2.1* and 1By19* subunits from European spelt wheat varieties by AS-PCR, and their molecular characterization and phylogentic relationships among HMW-GS genes were investigated. Meanwhile, AlphaFold modeling was used to reveal the 3D structural features of 1Ax2.1* and 1By19* subunits and previously characterized superior HMW subunits. On the basis of the sequence variations of promoter region, the specific SNP-based markers for *1Ax2.1** and *1By19** genes were developed and validated by using a wide range of wheat cultivars, F2 populations and RILs from different crossings between spelt wheat and bread wheat cultivars. The results are discussed in relation to provide new gene resources and molecular markers for the improvement of wheat's breadmaking quality.

2. Results and Discussion

2.1. Identification of Novel HMW-GS in Spelt Wheat

HMW-GS compositions at *Glu-1* loci in Spelt 137 and Spelt 6 were identified by SDS-PAGE (Figure 1). The results showed that Spelt 137 contained 1Ax2.1*, 1Bx13+1By22* and 1Dx2+1Dy12 subunits at the *Glu-A1*, *Glu-B1* and *Glu-D1* locus, respectively. The new x-type subunit 1Ax2.1* encoded by *Glu-A1* located between 1Ax1 and 1Ax2* and had a lower electrophoretic mobility and higher molecular weight than 1Ax2* (Figure 1A). In addition, one new y-type subunit 1By22* was present at *Glu-B1* locus, which had a similar electrophoretic mobility and molecular weight with 1By18 subunit. The HMW-GS compositions in Spelt 6 were null at *Glu-A1*, 1Bx13*+1By19* at *Glu-B1* and 1Dx2+1Dy12 at *Glu-D1*. The new subunits 1Bx13* and 1By19* encoded by *Glu-B1* showed a similar electrophoretic mobility and molecular weight with 1Bx13 and 1By16 subunit, respectively (Figure 1B). The above two new subunits in spelt wheat were firstly found in our previous study [31]. These new allelic variations at *Glu-1* loci provide potential gene resources for wheat quality improvement.

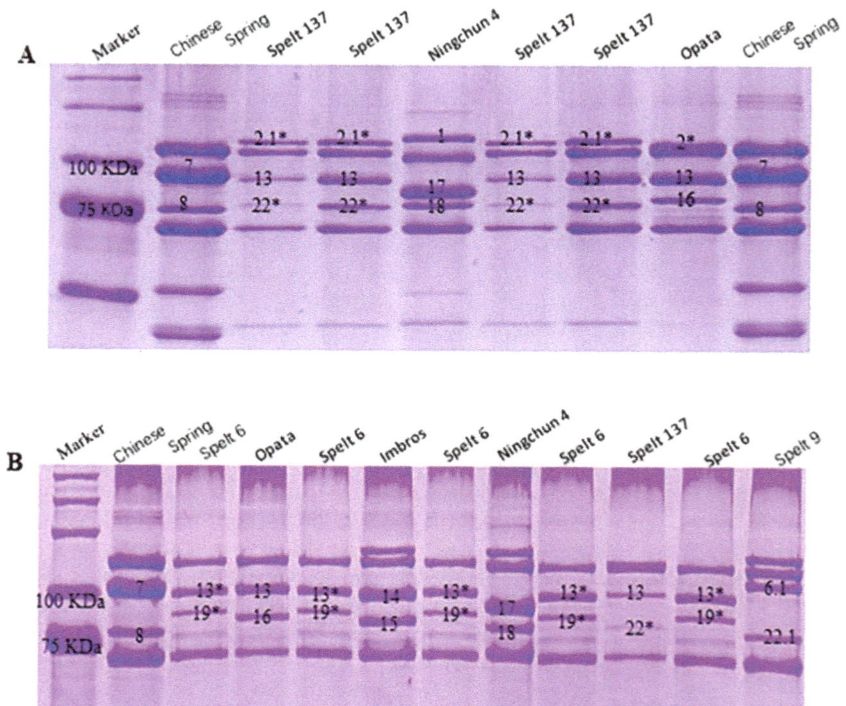

Figure 1. Identification of HMW-GS in European spelt wheat varieties Spelt 137 (**A**) and Spelt 6 (**B**) by SDS-PAGE. The HMW-GS in wheat is marked on the gel image with numbers or numbers with asterisks.

2.2. Molecular Cloning and Characterization of 1Ax2.1* and 1By19* Genes from Spelt Wheat

Two specific amplification fragments with about 2500 bp from Spelt 137 and 2200 bp from Spelt 6 were obtained by AS-PCR (Figure S1), which were corresponding to the sizes of x-type and y-type HMW-GS genes, respectively. Both PCR fragments were collected, cloned and sequenced, and their complete coding sequences with typical structural characteristics of previously characterized HMW-GS genes were obtained and designated as *1Ax2.1** and *1By19**. Both *1Ax2.1** and *1By19** gene sequences were deposited in GenBank with the accession numbers of MK395158 and MK395159, respectively. The nucleotide sequences of *1Ax2.1** (2478 bp) and *1By19** (2163 bp) genes encoded 824 and 720 amino acid residues, respectively.

The deduced amino acid sequence of *1Ax2.1** gene showed four distinct domains as other x-type HMW-GSs in Figure 2: a signal peptide of 21 amino acid residues, an N-terminal sequence of 86 amino acid residues followed by a central repetitive domain of 672 residues, a C-terminal domain of 42 residues. The repeat region of 1Ax2.1* subunit consisted of tandem and interspersed repeats of 33 hexapeptide motifs (consensus PGQGQQ), 10 nonapeptide motifs (consensus GYYPTSPQQ and GYYPTSLQQ) and 23 tripeptide motifs (consensus GQQ). The deduced amino acid sequences of both genes were aligned with other HMW-GSs from wheat and related species to compare their sequence characteristics. Compared with the superior subunit 1Ax2* [46], 1Ax2.1* had 6 and 9 amino acid insertions, 6 amino acid deletions and 22 single amino acid substitutions, which ultimately increased the length of the repeat region. Differences in subunit size were mainly caused by the sequences of repetitive polypeptide motifs caused by base deletion or insertion [15,47]. It is known that the longer and more regular repeating units are usually more conducive to superior gluten quality [15,36]. 1Ax2.1* contained a longer repetitive domain with additional two hexapeptides (consensus PGQGQQ) than the superior quality subunit 1Ax2*. Besides, 1Ax2.1* had 6 amino acids deletion in repeat domain compared with the superior quality 1Ax1, but the tripeptides (GQQ), hexapeptides (PGQGQQ) and nonapeptides (GYYPTSPQQ) were highly consistent. And just like 1Ax2.1*, the number and distribution of cysteine residues of these subunits were highly conserved between them, including sites 31, 46, 61 at the N-terminal and site 812 or 818 or 803 at the C-terminal.

Similarly, *1By19** gene encoded 720 amino acid residues, and showed a similar primary structures with y-type HMW-GSs in Figure 3, including a signal peptide of 21 amino acid residues, an N-terminal of 104 amino acid residues, a repeat region of 553 amino acid residues and a C-terminal of 42 amino acid residues. The repeat region contained 8 nona- and 34 hexapeptides. In addition, 1Ax2.1* subunit contained four conserved cysteine residues (three at the N-terminus and one at the C-terminus) while 1By19* subunits had seven conserved cysteine residues (five at the N-terminus, one at the repetitive domain, and one at the C-terminus). Likewise, the repeat unit in 1By19* accounted for about 50% of the total amino acids in the repeat region, in which the number of peptide segment PGQGQQ was equivalent to the y-type superior quality subunit 1Dy10. Overall, 1By19* had the same length of amino acid sequence as the superior quality subunits 1By8 and 1By18. But the amino acid length of 1By19* located between the superior quality subunits 1By9 and 1By16 due to the insertion and deletion of amino acids in the middle repeat region. Besides, these subunits all contained the same number of nonapeptides (consensus GYYPTSLQQ), but 1By19* contained 1~2 additional hexapeptides (consensus PGQGQQ). And each of these subunits contained seven cysteine residues, including site 606 or 591 or 624 near the C-terminus flanked by sites 31, 43, 65, 66 and 76 at N-terminus and site 708 or 693 or 726 at the C-terminus.

2.3. SNP and InDel Variations in 1Ax2.1 and 1By19* Genes*

Sequence variations resulting from point mutation and insertion/deletions (InDels) are the most common cause of wheat storage protein variation [48,49]. As listed in Table 1, 11 SNPs in *1Ax2.1** gene were present at different positions in N-terminal domain and in repeat region. Among them, 8 SNPs were nonsynonymous and resulted in the changes of corresponding amino acid residues of H-Y, L-S/P/G, E-G/T/Q, R-G, R-G/L, A-P/G, L-S/P. The remaining three SNPs at positions 198 bp (T-C), 288 bp (G-A) and 1302 bp (T-A/C) were synonymous and did not cause amino acid residue changes. One nonsynonymous SNP variation in *1By19** gene was detected: A/G-T transversion at the positions of 1137 bp, which led to the amino acid residue change (Q/E/G-H). No InDel variations were found in *1By19** gene.

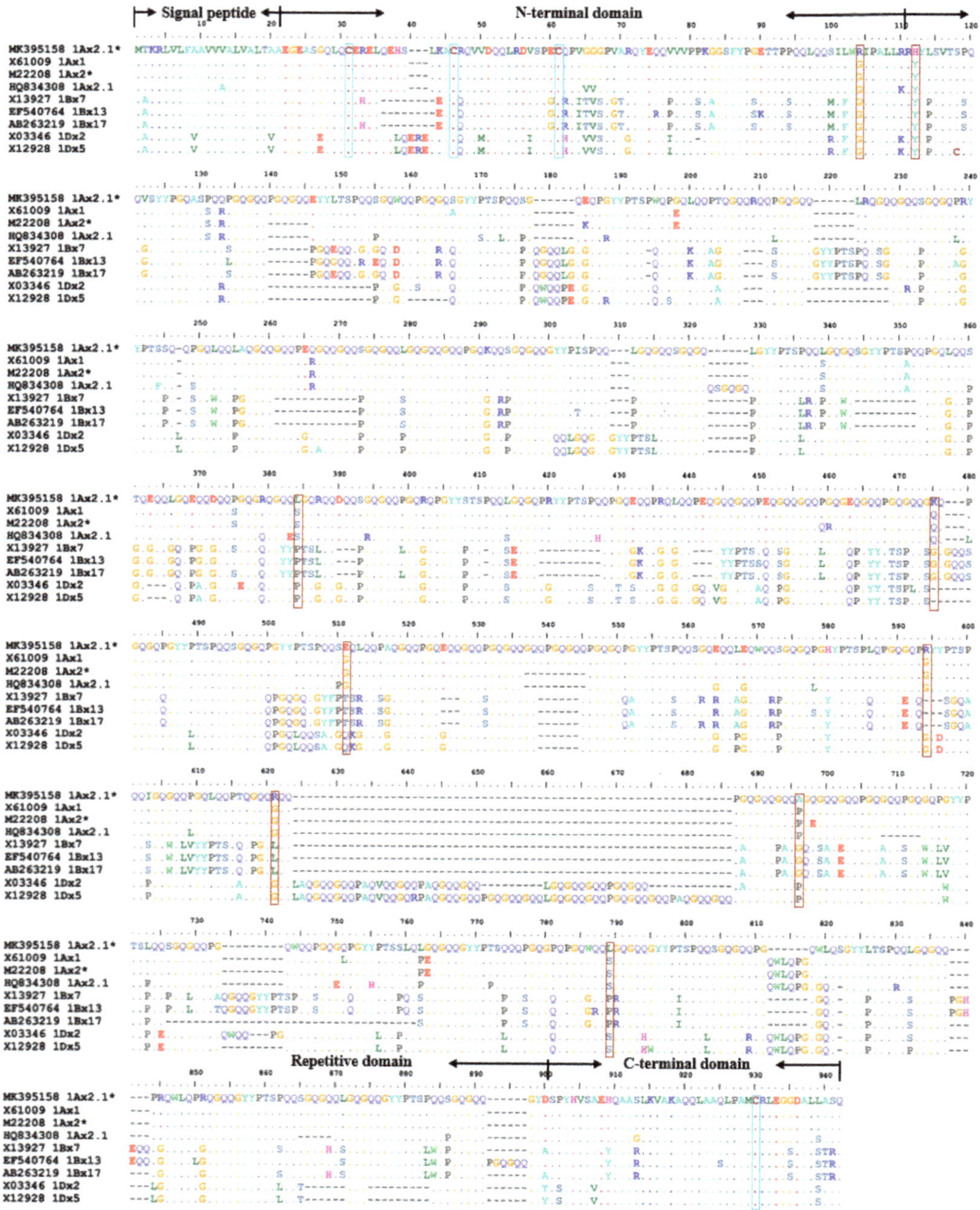

Figure 2. Comparison of the deduced amino acid sequences of *1Ax2.1** gene from spelt wheat with other eight x-subunit genes. The blue boxes indicate the cysteine residues, red boxes indicate differences of amino acid.

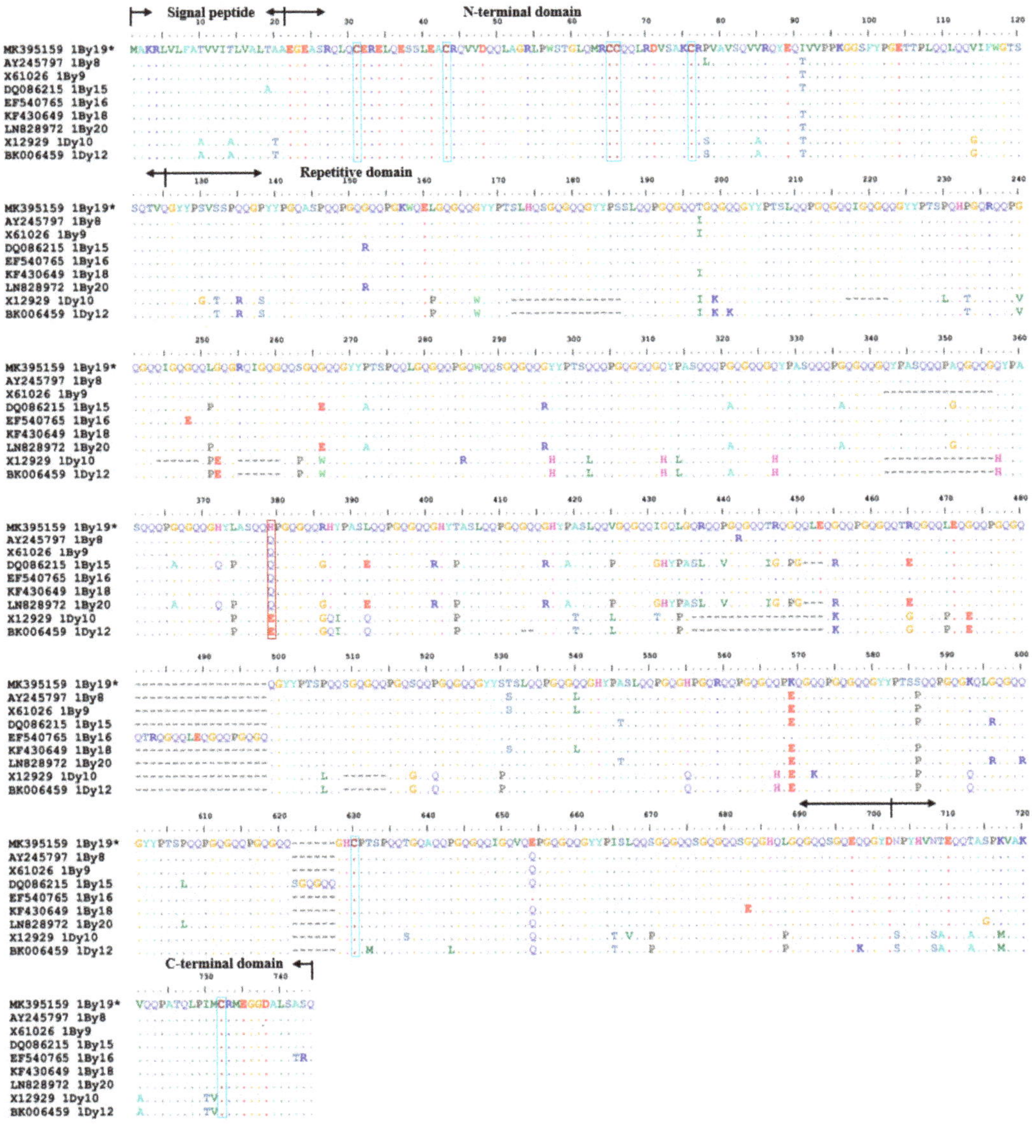

Figure 3. Comparison of the deduced amino acid sequences of *1By19** gene from spelt wheat with other eight y-type subunit genes. The blue boxes indicate the cysteine residues, red boxes indicate differences of amino acid.

Table 1. Positions of SNPs identified between *1Ax2.1** and other x-type HMW-GS genes.

Positions (bp)	198	288	325	1076	1302	1448	1696	1777	1813	1815	2066
*1Ax2.1**	T	G	C	T	T	A	A	A	G	G	T
15 other x-type HMW-GS genes	C	A	T	C/–	A/C	G	G	G/–	C/T	A	C

2.4. Verification of the Cloned 1Ax2.1* and 1By19* Genes from Spelt Wheat by Tandem Mass Spectrometry Analysis

The corresponding HMW-GS on the SDS-PAGE gel were collected and digested, and then identified by MALDI-TOF/TOF-MS (Table 2). The results showed that 1Ax2.1* and the x-type subunit MK395158 could well match in five peptide segments with 35 (45–79), 24 (195–218), 49 (357–405), 6 (725–730) and 24 (792–815) amino acid residues, respectively (Figure 2). 1By19* was well marched to the y-type subunit MK395159 in one peptide segment of 43 residues (QLQCERELQESSLEACRQVVDQQLAGRLPWSTGLQMRCCQQLR) at the position of 28-70 amino acid (Figure 3). All peptide segments identified by tandem mass spectrometry acquired a high protein score (C.I.%) of 100, indicating a high realiability of MS/MS analysis. These results further verified the validity of the cloned *1Ax2.1** and *1By19** genes.

Table 2. Identification of 1Ax2.1* and 1By19* subunits by MALDI-TOF/TOF-MS.

HMW-GS	Identified Protein	Accession No	Tryptic Fragments Identified by MS Data	Positions	Matched Peptides	Protein Score C. I. %
1Ax2.1*	x-type	MK395158	QVVDQQLRDVSPECQPVGGG PVARQYEQQVVVPPK QQPGQGQQLRQGQQGQQSGQGQPR QQDQQSGQGQQPGQRQPGY YSTSPQQLGQGQPRYYPTSPQQPGQEQQPR QWLQPR AQQLAAQLPAMCRLEGGDALLASQ	45–79 195–218 357–405 725–730 792–815	5	100
1By19*	y-type	MK395159	QLQCERELQESSLEACRQVVD QQLAGRLPWSTGLQMRCCQQLR	28–70	1	100

2.5. Phylogenetic Analysis of Ax2.1* and 1By19* Genes

The polymorphism of glutenin subunits is beneficial for understanding the origin and evolution of spelt wheat. The early study on the allelic variations at *Glu-1* and *Glu-3* in spelt wheat varieties supported the hypothesis of secondary origin of spelt wheat from hybridizing between cultivated emmer (*T. dicoccum*, AABB) and club wheat (*T. aestivum* subsp. *compactum*, AABBDD) [31]. Here, the complete encoding sequences of the cloned *1Ax2.1** and *1By19** genes from spelt wheat and other 27 HMW-GS genes from common wheat and related species were used to construct a neighbor-joining phylogenetic tree (Figure 4). The results showed that the x-type and y-type HMW-GS genes in the phylogenetic tree were clearly separated into two clades. In the x-type gene clade, 1Bx genes together with two Sx genes from S genome were classified into a subgrouop while the Dx and Ax genes were classified into another subgroup. In particular, *1Ax2.1** gene from spelt wheat showed a close phylogenetic relationship with *1Ax1*, *1Ax2** and *1Ax2.1* genes. In the y-type gene clade, the Dy and By genes were classified into different subgroups while *1By19** gene from spelt wheat was more closed to *1By16*, *1By9*, *1By8* and *1By18* genes.

To further explore the evolutionary relationships between *1Ax2.1** and *1By19** and other HMW-GS genes, their divergence times (million years ago, MYA) were estimated and the results are shown in Table S3. In general, the divergent times between x-type subunit genes between y-type subunit genes at each locus were about 0-13 MYA and 0.2-6.0 MYA while the divergence between x- and y-type subunit genes in each locus occurred about 14-17 MYA. The divergence among *1Ax2.1** and *1Ax1* and *1Ax2** genes occurred more recently, at about 0.92-1.00 MYA while earlier divergence between *1Ax2.1** and *1Ax2.1* occurred at 2.31 MYA. *1By19** displayed a high similarity with *1By16* and they were diverged more recently at 0.31 MYA. It is known that the hexaploid wheat is originated from the hybridation between tetraploid wheat and *T. tauschii* that occurred at about ten-thousand years ago. Thus, *1Ax2.1** and *1By19** genes should emerge before spelt wheat formation.

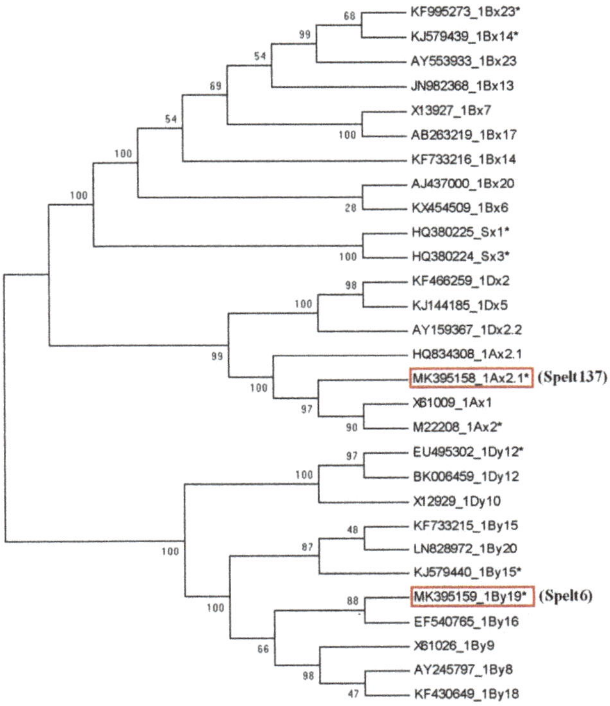

Figure 4. Phylogenetics of *1Ax2.1** and *1By19** genes and other 27 HMW-GS genes from A, B and D genomes of common wheat and related species. Bootstrap value ≥28% is indicated above or below the branches. The red frame indicates *1Ax2.1** and *1By19** genes, respectively. The 27 HMW-GS genes include X61009, M22208, KX454509, X13927, JN982368, KF733216, KJ579439, AB263219, AY553933, KF466259, AY159367, KJ144185, X61026, KF733215, KJ579440, EF540765, KF430649, X12929, BK006459 from *T. aestivum*; HQ380225 and HQ380224 from *Ae. speltoides*; KF995273 from *Triticum turgidum*; EU495302 from *T. aestivum* subsp. *Yunnanense*; AJ437000, LN828972 and AY245797 from *T. turgidum* subsp. *Durum*; HQ834308 from *T. monococcum* subsp. *Monococcum*.

2.6. Secondary Structure and 3D Structure Analysis of 1Ax2.1* and 1By19* Protein Subunits

The secondary structure characteristics of glutenin proteins can be used to uncover the molecular mechanisms of gluten quality formation [50]. In general, higher content of β-sheets and α-helix in glutenin subunits is helpful for forming good gluten structure and superior breadmaking quality [51,52]. In this study, the secondary structures of 1Ax2.1* and 1By19* subunits and other eight wheat HMW-GS (1Ax2*, 1Ax1, 1By8, 1By9, 1By15, 1By16, 1By18 and 1By20) were predicted by the PSIPRED server and a comparative analysis for their secondary structure features was performed. Among them, 1Ax2*, 1Ax1, 1By8, 1By15, 1By16 and 1By18 are considered as superior quality subunits whereas 1By20 is a poor quality subunit [11–13]. As shown in Table 3, 1Ax2.1* subunit had seven α-helixes (9.58%) and four β-strands (0.82%), better than the superior quality subunits 1Ax2* with the same seven α-helixes (9.33%) and two β-strands (0.98%), and 1Ax1 with six α-helixes (9.40%) and without β-strands. 1By19* subunit and other six y-type subunits generally contained eight α-helixes with different percentages, but no β-strands were present. 1By19* had 12.36% α-helixes, higher than 1By15 (12.03%), 1By16 (11.65%), 1By18 (12.22%), and 1By20 (11.85%), and lower than 1By8 (16.81%) and 1By9 (13.19%).

Table 3. Secondary structure prediction of 1Ax2.1* and 1By19* subunits and other eight wheat HMW-GS.

HMW-GS (Accession No.)	Type	Structure Motifs	Content (%)	Total	Amino Acid Length
1Ax2.1* (MK395158)	x-type	α-helix	9.58	7	824
		β-strand	0.82	4	
1Ax2* (M22208)	x-type	α-helix	9.33	7	815
		β-strand	0.98	2	
1Ax1 (X61009)	x-type	α-helix	9.40	6	830
		β-strand	0.00	0	
1By19* (MK395159)	y-type	α-helix	12.36	8	720
		β-strand	0.00	0	
1By8 (AY245797)	y-type	α-helix	16.81	8	720
		β-strand	0.00	0	
1By9 (X61026)	y-type	α-helix	13.19	8	705
		β-strand	0.00	0	
1By15 (DQ086215)	y-type	α-helix	12.03	6	723
		β-strand	0.00	0	
1By16 (EF540765)	y-type	α-helix	11.65	8	738
		β-strand	0.00	0	
1By18 (KF430649)	y-type	α-helix	12.22	8	720
		β-strand	0.00	0	
1By20 (LN828972)	y-type	α-helix	11.85	8	717
		β-strand	0.00	0	

The 3D structures of 1Ax2.1* and 1By19* subunits and other six superior quality subunits 1Ax2*, 1Ax1, 1By8, 1By9, 1By16 and 1By18 were further predicted by AlphaFold2 (Figure 5). As a deep learning algorithm, AlphaFold2 can visualize, analyze and interpret 3D protein structures basing on amino acid sequences [53]. The results showed that Alphafold2 predictions with five cycles for eight HMW-GS were almost identical, including three indicators of alignment errors, per-residue confidence scores, matching templates (Figure S2). Eight subunits had different confidence scores in 3D structure models after five cycles of prediction, which were divided into five different ranks according to their scores (Figure S3). The highest level for eight subunits is shown in Figure 5, and the confidence score for 1Ax2.1* was 33.9, slightly close to 1Ax1 (34.9) and 1Ax2* (34.4). The confidence score for 1By19* was 36.9, slightly close to 1By16 (37.4), 1By9 (37.7), 1By8 (37.2) and 1By18 (37.0). All subunits predicted had a typical 3D structure, including two non-repetitive structural domains (N- terminal and C-terminal) with a repetitive fragment in the middle and a large number of α-helices at the N- and C-termini. In general, three x-type subunits (1Ax2.1*, 1Ax1 and 1Ax2*), and five y-type subunits (1By19*, 1By16, 1By9, 1By18 and 1By8) showed a similar 3D structural characteristics, respectively. It is worth noting that the number and distribution of Cys residues play an important role in determining the formation of glutenin polymer and the subsequent rheological parameters of dough [54–56]. Particularly, three x-type subunits (1Ax2.1*, 1Ax1 and 1Ax2*) contained four cysteine residues (three in the N-terminal domain and one in the C-terminal domain), which did not form intramolecular disulfide bonds within the subunits. Five y-type subunits (1By19*, 1By16, 1By18, 1By8 and 1By9) contained seven cysteine residues (five in the N-terminal domain, one in the C-terminal domain and one in the intermediate repeating structural domain), and formed three disulfide bonds: Cys 31~76, 43~65 and 66~706 in 1By19*, Cys 31~76, 43~65 and 66~726 in 1By16, Cys 31~76, 43-65 and 66~693 in 1By9, Cys 31~76, 43~65 and 66~708 in 1By8, and Cys 31~76, 43~65 and 66~708 in 1By8. However, the cysteines at the intermediate repeat region were unable to form a disulfide bond (Figure 5). Meanwhile, 1Ax2.1* and 1By19* subunits contained a large number of α-helices at the N- and C-termini, consistent with the above mentioned secondary structure prediction results as well as previous report [57,58]. Although the cysteines in 1Ax2.1*, 1Ax1 and 1Ax2* subunits and three cysteines in 1By19*, 1By16, 1By9, 1By8 and 1By18 subunits did not form intramolecular disulfide bonds, they might form intermolecular disulfide bonds with other glutenins or gliadins to further form gluten macropolymers [59].

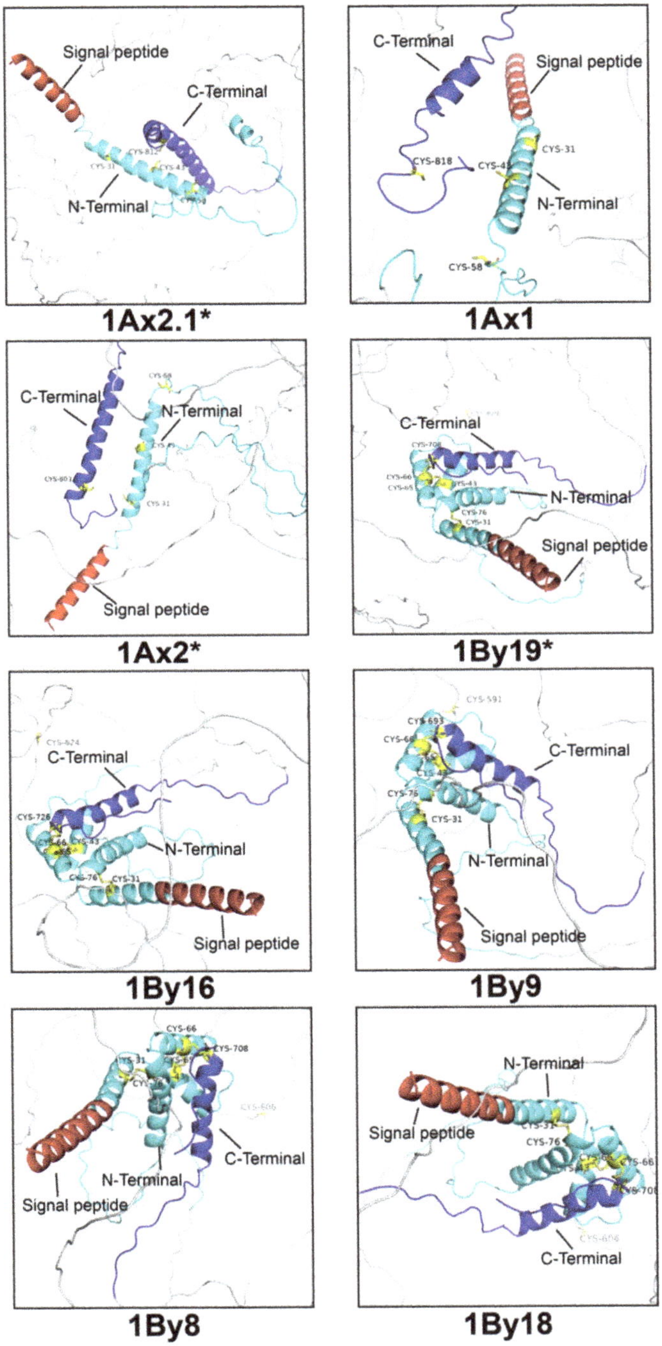

Figure 5. Localized view of the 3D structures of eight wheat HMW-GS by AlphaFold2.

2.7. Development and Validation of SNP-Based Molecular Markers for 1Ax2.1 * and 1By19* Genes

A new marker type named SNP, is only a bi-allelic type of marker in nature with low expectation of heterozygosity [60,61]. In the past years, a number of SNPs-based molecular markers for glutenin genes have been developed such as *1Dx2* and *1Dx5* alleles [62], *1By18* [49], *1Slx2.3** and *1Sly16** [36]. These markers can be for rapid improvement of wheat gluten quality. Herein, we amplified and cloned the upstream promoter sequences of *1Ax2.1** and *1By19** genes. Then BioEdit7.0 was used to compare their promoter sequence differences with other x- and y-type HMW-GS genes and the specific SNP variations were identified. Ultimately, we used these SNP variations to develop SNP-based molecular markers for *1Ax2.1** and *1By19** genes.

The specific primers were designed (Table S4), and then used to amplify the upstream promoter sequences of *1Ax2.1** and *1By19** genes, and two specific fragments 905 bp and 728 bp from *1Ax2.1** and *1By19** genes were obtained, respectively (Figure S4). After cloning and sequencing, the typical promoter sequences of *1Ax2.1** and *1By19** genes were obtained, which were used for sequence alignment with the promoter sequences of other 12 x- and y-type HMW-GS genes deposited in GenBank (Figure S5 and S6). Two specific SNP sites in the upstream promoter region of each gene were identified: −377 bp and −203 bp in *1Ax2.1** and −591 bp and −56 bp in *1By19**. According to these specific SNP sites, two pairs of specific primers (2.1*F/R and 19*F/R) were designed (Table S5) and used to develop SNP-based molecular markers of *1Ax2.1** and *1By19** genes as showed in Figures S5 and S6 via combining with SDS-PAGE identification. The results from 47 bread wheat and spelt wheat varieties with different *Glu-1* allelic variations (Table S1) showed that one 209 bp specific fragment was amplified by using 2.1*F/R primer pair from the varieties containing 1Ax2.1* subunit whereas no amplified products were obtained in the varieties without 1Ax2.1* subunit (Figure 6A,B). Similarly, one 570 bp specific fragment was amplified by the 19*F/R primer pair from varieties with 1By19* subunit, which was absent in the varieties without 1By19* subunit (Figure 6C,D). These results were well consistent with the SDS-PAGE identification. Further collection and sequencing of both amplified fragments also showed a consistence with their amplified regions as shown in Figures S5 and S6.

To further validate the developed SNP markers, two pairs of specific primers were used to amplify the promoter sequences of the target genes in different hybrid populations and RILs. Meanwhile, SDS-PAGE identification was conducted to verify the PCR results. For the marker validation, 200–250 grains from F2 generation populations and 50–80 grains from each RIL were detected. In the F2 populations of Spelt 137×Zhongmai 175, Spelt 137×Ningchun 4 and Spelt 137×Zhongmai 8601 crosses, 2.1*F/R primer could specifically amplify the 209 bp fragment in the F2 grains containing 1Ax2.1* subunit, and no amplified products were produced in the F2 grains without 1Ax2.1* subunit. At the same time, three F2 populations of Spelt 6×Zhongmai 175, Spelt 6×Ningchun 4 and Spelt 6×Zhongmai 8601 crosses were used to verify the molecular marker of *1By19** gene and the similar results were obtained. The F2 grains having 1By19* subunit showed the 570 bp specific fragment when amplified by 19*F/R primer pair (Figure 7). Furthermore, seven RILs from Spelt 137×Ningchun 4 crossing and six RILs from Spelt 6×Ningchun 4 crossing were further used to verify the developed molecular markers for *1Ax2.1** and *1By19** genes. As expected, all RILs containing 1Ax2.1* and 1By19* subunits could exhibit 209 bp and 570 bp specific fragments, respectively (Figure 8). These results were well consistent with SDS-PAGE identification, confirming that the developed SNP-based molecular markers have high specificity and accuracy for identifying *1Ax2.1** and *1By19** genes. Therefore, these SNP-based markers are expected to be used for breadmaking quality improvement via marker-assisted selection during wheat breeding program.

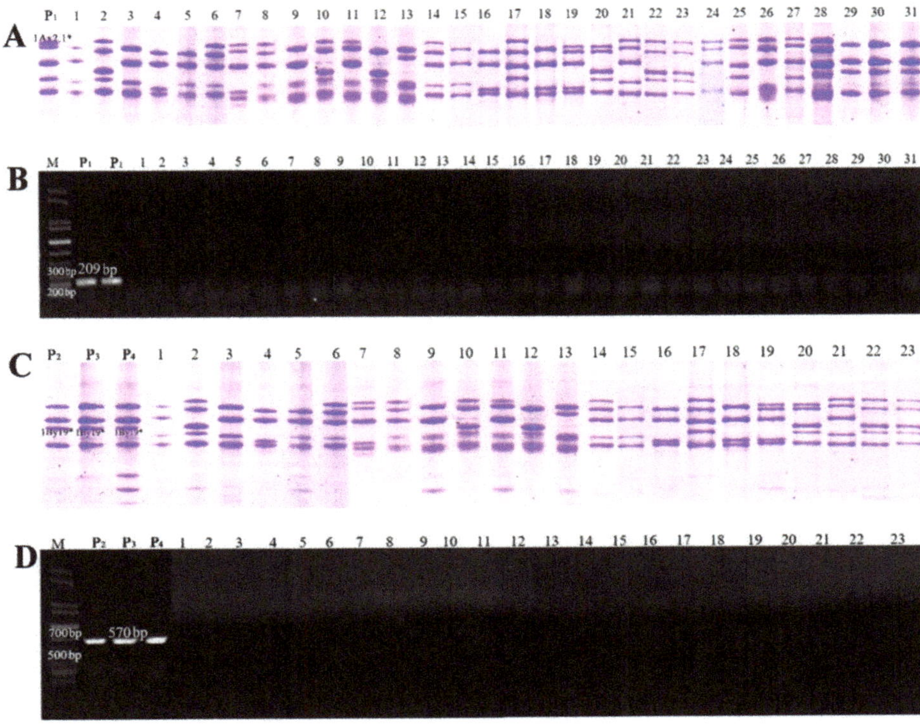

Figure 6. Development of the SNP-based molecular markers for *1Ax2.1** and *1By19** genes. (**A**) Identification of 1Ax2.1* subunit in different wheat varieties by SDS-PAGE. (**B**) PCR amplification of *1Ax2.1** gene from different wheat varieties by the primer 2.1*F/R. The 209 bp specific amplified fragment is indicated. (**C**) Identification of 1By19* subunit in different wheat varieties by SDS-PAGE. (**D**) PCR amplification of *1By19** gene from different wheat varieties by the primer 19*F/R. The 570 bp specific amplified fragment is indicated. The varied numbers in the figure are the same as those in Table S1. P_1: Spelt 137; P_2: Spelt 6; P_3: Spelt 20; P_4: Spelt 24.

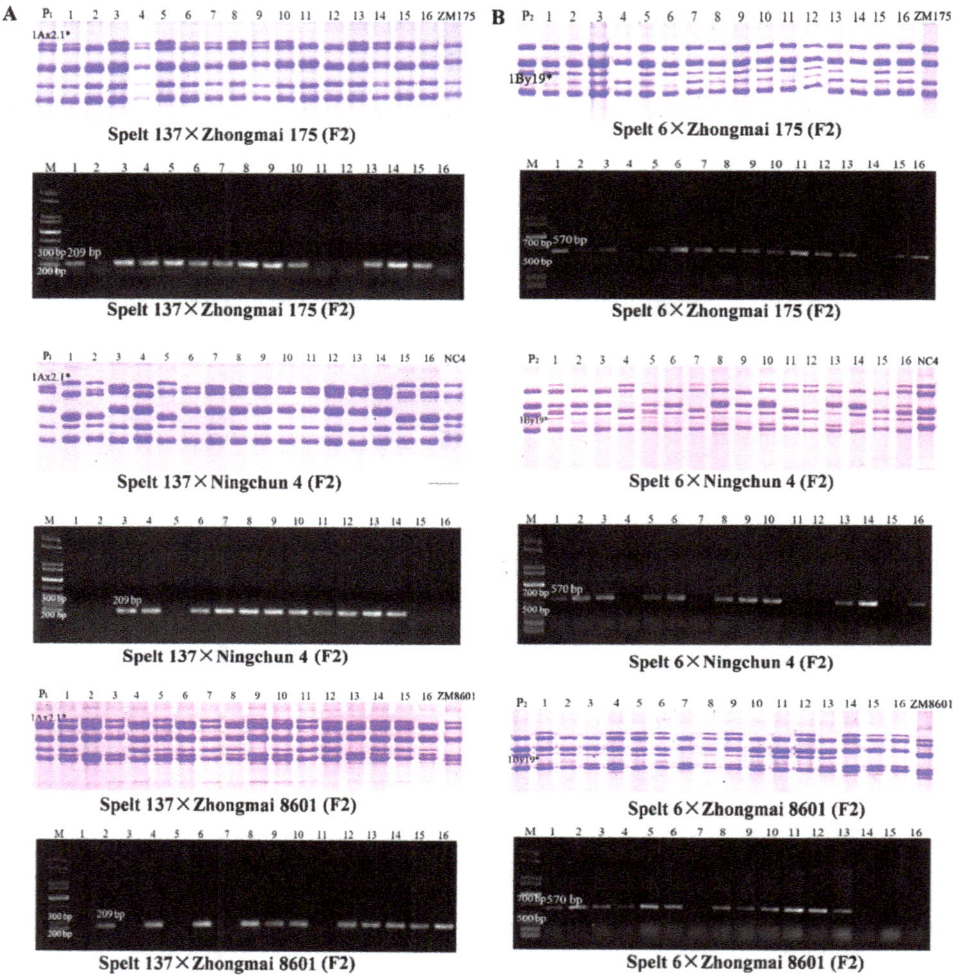

Figure 7. Validation of the SNP-based molecular marker for *1Ax2.1** (**A**) and *1By19** (**B**) genes in the F2 populations from crossing between spelt wheat and bread wheat varieties. 1Ax2.1* and 1By19* subunits and their special amplification fragments 209 bp and 570 bp are indicated. P_1: Spelt 137; P_2: Spelt 6; ZM175: Zhongmai 175; NC4: Ningchun 4; ZM8601: Zhongmai 8601.

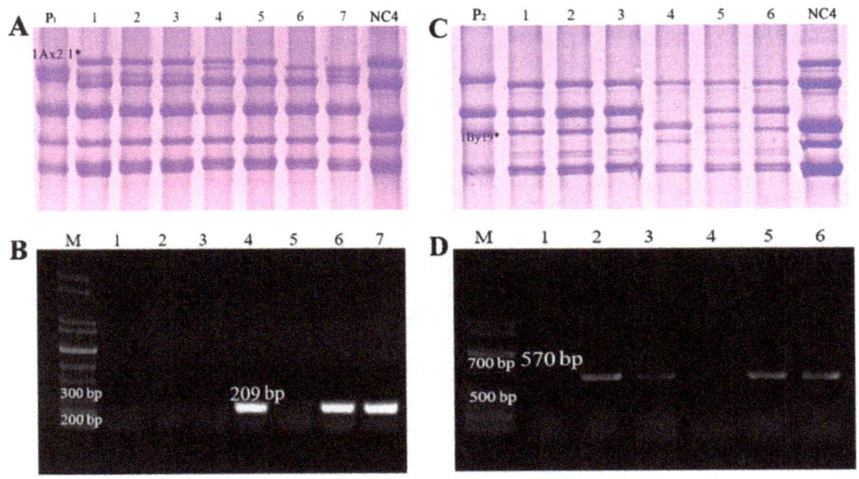

Figure 8. Validation of the SNP-based molecular markers for *1Ax2.1** and *1By19** genes in the RILs derived from Spelt 137×Ningchun 4 and Spelt 6×Ningchun 4 crosses. (**A**) HMW-GS from seven RILs (Lanes 1–7) from Spelt 137×Ningchun 4 identified by SDS-PAGE; (**B**) PCR amplification of *1Ax2.1** gene from seven RILs (Lanes 1–7) of Spelt 137×Ningchun 4; (**C**) HMW-GS from six RILs (lanes 1–6) from Spelt 6×Ningchun 4 identified by SDS-PAGE; (**D**) PCR amplification of *1By19** gene from six RILs (lanes 1–6) from Spelt 6×Ningchun 4. 1Ax2.1* and 1By19* subunits and the special amplification fragments 209 bp and 570 bp are indicated. P_1: Spelt 137; P_2: Spelt 6; NC4: Ningchun 4.

3. Materials and Methods

3.1. Plant Materials

The materials used in this work included two European spelt wheat varieties Spelt 6 and Spelt 137, 45 bread wheat and spelt wheat varieties with different allelic variations at *Glu-1* loci (Table S1), six F2 generation populations from Spelt 137×Zhongmai 175, Spelt 137×Ningchun 4, Spelt 137×Zhongmai 8601, Spelt 6×Zhongmai 175, Spelt 6×Ningchun 4, and Spelt 6×Zhongmai 8601, seven and six F6 RILs, respectively, derived from crossing between Spelt 137×Ningchun 4 and Spelt 6×Ningchun 4 via consecutive self-crossing combining with screening and identification. All spelt wheat varieties were collected from Plant Breeding Institute, Technical University of Munich, Germany.

3.2. HMW-GS Extraction and SDS-PAGE

The extraction of seed HMW-GS was carried out with referred to the previous study with minor modifications [63]. Firstly, 70% ethanol (*v/v*) and 55% isopropanol (*v/v*) were sequentially added to homogenized seeds to remove albumins, globulins and gliadins. Then, glutenins were extracted by a commonly used glutenin extraction buffer (50% isopropanol, 80 μL Tris-HCl, pH 8.0) with 1% dithiothreitol (*w/v*) and 1.4% 4-vinylpyridine (*w/v*). The final extracted glutenin proteins were used for SDS-PAGE based on the reported method [34].

3.3. DNA Isolation, AS-PCR Amplication and Sequencing

The wheat seedling leaves were used to extract genomic DNA according to the improved CTAB method [37]. Two pair of AS-PCR primers were designed according to the published HMW-GS gene sequences and listed in Table S2. The high-fidelity polymerases (Vazyme Biotech, Nanjing, China) were used for amplifying the complete coding sequence of HMW-GS. PCR reaction was performed by CFX96 Real Time system (Bio-Rad Laboratories) programmed at an initial denaturation at 95 °C for 3 min followed by 35 cycles of 95 °C for 15 s, 61 °C for 15 s, 72 °C for 150 s and finally extended at 72 °C for 5 min. The

PCR products were separated by 1% agrose gel in Tris-acetic acid-EDTA buffer and the fragments of expected size were collected and purified by using Gel Extraction Kit (Omega, Bienne, Switzerland), and then the purified products were ligated into pMD18-T vector (TaKaRa Biotechnology, Dalian, China) and transformed into Trans-2-blue competent cells. Three positive clones were randomly selected to reduce the sequencing error, and then sequenced by TaKaRa Biotechnology, Dalian, China.

3.4. Sequence Alignment and SNP/InDel Identification

Multiple sequences alignments of the cloned HMW-GS genes and previously characterized x-type and y-type subunit genes were conducted by using Bioedit (version 7.0, Tom Hall, Scotts Valley, USA) SNPs and InDels variations in HMW-GS genes were identified. 15 x-type and 12 y-type HMW-GS genes deposited in GenBank are as follows: *1Ax1* (X61009), *1Ax2** (M22208), *1Ax2.1* (HQ834308), *1Bx6* (KX454509), *1Bx7* (X13927), *1Bx13* (JN982368), *1Bx14* (KF733216), *1Bx17* (AB263219), *1Bx20* (AJ437000), *1Bx23* (AY553933), *1Dx2* (KF466259), *1Dx2.2* (AY159367), *1Dx5* (KJ144185), *Sx1** (HQ380225), and *Sx3** (HQ380224), *1Ay* (FJ404595), *1By8* (AY245797), *1By9* (X61026), *1By15* (KF733215), *1By15** (KJ579440), *1By16* (EF540765), *1By18* (KF430649), *1By20* (LN828972), *1Dy10* (X12929), *1Dy12* (BK006459), *Sy9** (HQ380223), and *Sy18** (HQ380222).

3.5. MALDI-TOF/TOF-MS

According to the previously described method [49,51], the corresponding HMW subunits bands on SDS-PAGE gel were excised, and then digested with trypsin. HMW-GS identification of tandem mass spectrometry was performed by using EASY-nLC 1000 (Thermo/Finnigan, San jose, CA, USA) equipped with orbitrap Q Exactive mass spectrometry (Thermo/Finnigan).

3.6. Construction of Phylogenetic Tree and Estimation of Divergence Time

The full-length homologous nucleotide sequences were aligned using the Clustal W program, and the alignment file was used to construct phylogenetic tree according to the complete coding regions of HMW-GS by using software MEGA (version 6.0, Koichiro Tamura at al., Auckland, New Zealand). An evolution rate of 6.5×10^{-9} substitution/site year was used to estimate the divergence times of HMW-GS genes based on the reported method [64].

3.7. Secondary Structure and 3D Structure Prediction of HMW-GS

Secondary structure prediction of the deduced amino acid sequences from the cloned HMW-GS genes were carried out by PSIPRED (http://bioinf.cs.ucl.ac.uk/psipred/psiform.html, accessed on 10 May 2022) [65,66]. The 3D structures of HMW-GS were predicted by AlphaFold2 [67,68], and then editing was carried out by Pymol software (version 1.7.4, Schrödinger, Warren Lyford DeLano, New York City, USA).

3.8. Development and Validation of SNP-Based Molecular Markers

According to the SNP variations in the upstream promoter sequences of the coding region in the cloned HMW-GS genes (Table S4), two pairs of AS-PCR primers were designed (Table S5). Genomic DNA was extracted using the Trans Fast Taq DNA Polymerase system (TransGen Biotech, Beijing, China). PCR cycles consisted of 94 °C for 3 min for activation, followed by 35 cycles of 94 °C for 5 s, 58 °C/55 °C (Table S6) for 15 s, 72 °C for 2/4 s, and finally ended at 72 °C for a 5 min extension step. The materials used for marker development and verification were described in the section of plant materials.

4. Conclusions

Two novel HMW-GS 1Ax2.1* and 1By19* were identified in European spelt wheat, and their complete encoding gene sequences of corresponding genes were cloned and sequenced. Molecular characterization showed that both subunits had longer and more regular repeating units, and a high percentage of α-helices and β-sheets. AlphaFold2 prediction showed that 1Ax2.1* and 1By19* showed similar 3D structural characteristics with three superior quality x-type subunits (1Ax2.1*, 1Ax1 and 1Ax2*) and four superior quality y-type subunits (1By16, 1By9, 1By8 and 1By18), respectively. Both subunits contained a large number of α-helices at the N- and C-termini. In particular, four cysteine residues in three x-type subunits (1Ax2.1*, 1Ax1 and 1Ax2*) and the cysteines in the intermediate repeat region were unable to form intramolecular disulfide bonds, but these cysteines might form intermolecular disulfide bonds with other glutenins and gliadins to enhance gluten macropolymer formation. The SNP-based molecular markers for *1Ax2.1** and *1By19** genes were developed according to the SNP variations in the promoter regions and verified in different F2 generation populations as well as different recombination inbred lines derived from crossing between spelt wheat and bread wheat cultivars. These molecular markers have shown a high reliability and have a good application prospect for improving bread making through marker-assisted selection.

Supplementary Materials: The supporting information can be downloaded at: https://www.mdpi.com/article/10.3390/ijms231911104/s1.

Author Contributions: Y.C., J.Z. and R.W. performed all of the experiments, data analysis and wrote the paper. H.S. performed protein identification and data analysis. Y.Y. designed and supervised experiments. All authors have read and agreed to the published version of the manuscript.

Funding: This research was financially supported by a grant from the National Natural Science Foundation of China (31971931).

Institutional Review Board Statement: Not applicable.

Informed Consent Statement: Not applicable.

Data Availability Statement: Not applicable.

Conflicts of Interest: This manuscript has no financial or non-financial competing interests.

Abbreviations

HMW-GS	High molecular weight glutenin subunit
LMW-GS	Low molecular weight glutenin subunit
MATE	Multidrug and toxin efflux
PCR	Polymerase chain reaction
AS-PCR	Allele specific polymerase chain reaction
SNP	Single nucleotide polymorphism
ORF	Open reading frame
InDels	Insertion/Deletion variations
CTAB	Hexadecyl trimethyl ammonium bromide
RILs	Recombinant inbred lines
MYA	Million years ago
SDS-PAGE	Sodium dodecyl sulfate polyacryl amide gel electrophoresis
MALDI-TOF-MS	Matrix-assisted laser desorption ionization tandem time of flight mass spectrometry Molecular weight

References

1. Kumar, R.; Mamrutha, H.M.; Kaur, A.; Venkatesh, K.; Sharma, D.; Singh, G.P. Optimization of Agrobacterium-mediated transformation in spring bread wheat using mature and immature embryos. *Mol. Biol. Rep.* **2019**, *46*, 1845–1853. [CrossRef] [PubMed]
2. Spannagl, M.; Martis, M.M.; Pfeifer, M.; Nussbaumer, T.; Mayer, K.F. Analysing complex Triticeae genomes–concepts and strategies. *Plant Methods* **2013**, *9*, 1–9. [CrossRef] [PubMed]
3. Kumar, A.; Kapoor, P.; Chunduri, V.; Sharma, S.; Garg, M. Potential of *Aegilops sp.* for improvement of grain processing and nutritional quality in wheat (*Triticum aestivum*). *Front. Plant Sci.* **2019**, *10*, 308. [CrossRef]
4. Biesiekierski, J.R. What is gluten? *J. Gastroenterol. Hepatol.* **2017**, *32*, 78–81. [CrossRef] [PubMed]
5. Wrigley, C.W. Giant proteins with flour power. *Nature* **1996**, *381*, 738–739. [CrossRef]

6. Mioduszewski, Ł.; Cieplak, M. Viscoelastic properties of wheat gluten in a molecular dynamics study. *PLoS Comput. Biol.* **2021**, *17*, e1008840. [CrossRef] [PubMed]
7. Bonilla, J.C.; Erturk, M.Y.; Schaber, J.A.; Kokini, J.L. Distribution and function of LMW glutenins, HMW glutenins, and gliadins in wheat doughs analyzed with 'in situ' detection and quantitative imaging techniques. *J. Cereal Sci.* **2020**, *93*, 102931. [CrossRef]
8. Payne, P.I.; Holt, L.M.; Law, C.N. Structural and genetical studies on the high-molecular-weight subunits of wheat glutenin. *Theor. Appl. Genet.* **1981**, *60*, 229–236. [CrossRef]
9. Law, C.N.; Payne, P.I. Genetical aspects of breeding for improved grain protein content and type in wheat. *J. Cereal Sci.* **1983**, *1*, 79–93. [CrossRef]
10. Li, Y.; Fu, J.; Shen, Q.; Yang, D. High-molecular-weight glutenin subunits: Genetics, structures, and relation to end use qualities. *Int. J. Mol. Sci.* **2020**, *22*, 184. [CrossRef]
11. Branlard, G.; Dardevet, M. Diversity of grain protein and bread wheat quality: II. Correlation between high molecular weight subunits of glutenin and flour quality characteristics. *J. Cereal Sci.* **1985**, *3*, 345–354. [CrossRef]
12. Brönneke, V.; Zimmermann, G.; Killermann, B. Effect of high molecular weight glutenins and D-zone gliadins on bread-making quality in German wheat varieties. *Cereal Res. Commun.* **2000**, *28*, 187–194. [CrossRef]
13. Lama, S.; Kabir, M.R.; Akhond, M.A.Y. Biochemical and molecular characterization of Bangladeshi wheat varieties for bread-making quality. *Plant Tissue Cult. Biotechnol.* **2018**, *28*, 57–68. [CrossRef]
14. Pang, B.S.; Zhang, X.Y. Isolation and molecular characterization of high molecular weight glutenin subunit genes *1Bx13* and *1By16* from hexaploid wheat. *J. Integr. Plant Biol.* **2008**, *50*, 329–337. [CrossRef] [PubMed]
15. Shewry, P.R.; Halford, N.G.; Tatham, A.S. The high molecular weight subunits of wheat glutenin. *J. Cereal Sci.* **1992**, *15*, 105–120. [CrossRef]
16. Tatham. A.S.; Drake, A.F.; Shewry. P.R. Contormational studies of synthetic peptides corresponding to the repetitive region of the high molecular weight (HMW) glutenin subunits of wheat. *J. Cereal Sci.* **1990**, *11*, 189–200. [CrossRef]
17. Pézolet, M.; Bonenfant, S.; Dousseau, F.; Popineau, Y. Conformation of wheat gluten proteins Comparison between functional and solution states as determined by infrared spectroscopy. *FEBS Lett.* **1992**, *299*, 247–250. [CrossRef]
18. Lafiandra, D.; Shewry, P.R. Wheat Glutenin polymers 2, the role of wheat glutenin subunits in polymer formation and dough quality. *J. Cereal Sci.* **2022**, *106*, 103487. [CrossRef]
19. Huebner, F.R.; Bietz, J.A.; Wall, J.S. Disulfide bonds: Key to wheat protein functionality. *Adv. Exp. Med. Biol.* **1977**, *86A*, 67–88.
20. Tatham, A.S.; Miflin, B.J.; Shewry, P.R. The beta-turn conformation in wheat gluten proteins: Relationship to gluten elasticity. *Cereal Chem.* **1985**, *62*, 405–442.
21. Gao, J.H.; Yu, P.X.; Liang, H.R.; Fu, J.H.; Luo, Z.Y.; Yang, D. The wPDI Redox Cycle coupled conformational change of the repetitive domain of the HMW-GS 1Dx5-A computational study. *Molecules* **2020**, *25*, 4393. [CrossRef] [PubMed]
22. Cazalis, R.; Aussenac, T.; Rhazi, L.; Marin, A.; Gibrat, J.F. Homology modeling and molecular dynamics simulations of the N-terminal domain of wheat high molecular weight glutenin subunit 10. *Protein Sci.* **2003**, *12*, 34–43. [CrossRef]
23. Robert, A.; Barkoutsos, P.K.; Woerner, S.; Tavernelli, I. Resource-efficient quantum algorithm for protein folding. *npj Quantum Inform.* **2021**, *7*, 1–5. [CrossRef]
24. Magoshi, J.; Becker, M.; Han, Z.; Nakamura, S. Thermal properties of seed proteins. *J. Therm. Anal. Calorim.* **2002**, *70*, 833–839. [CrossRef]
25. Calucci, L.; Pinzino, C.; Capocchi, A.; Galleschi, L.; Ghiringhelli, S.; Saviozzi, F.; Zandomeneghi, M. Structure and dynamics of high molecular weight glutenin subunits of durum wheat (*Triticum durum*) in water and alcohol solutions studied by electron paramagnetic resonance and circular dichroism spectroscopies. *J. Agric. Food Chem.* **2001**, *49*, 359–365. [CrossRef] [PubMed]
26. Porath, J. From gel filtration to adsorptive size exclusion: Communications from the 11th international conference on methods in protein structure analysis. *J. Protein. Chem.* **1997**, *16*, 463–468. [CrossRef]
27. Bryant, P.; Pozzati, G.; Elofsson, A. Improved prediction of protein-protein interactions using alphafold2. *Nat. Commun.* **2022**, *13*, 1265. [CrossRef]
28. Jumper, J.; Evans, R.; Pritzel, A.; Green, T.; Hassabis, D. Highly accurate protein structure prediction with alphafold. *Nature* **2021**, *596*, 583–589. [CrossRef]
29. Zweckstetter, M. NMR hawk-eyed view of AlphaFold2 structures. *Protein Sci.* **2021**, *30*, 2333–2337. [CrossRef]
30. Duan, W.; Lu, F.; Cui, Y.; Zhang, J.; Du, X.; Hu, Y.; Yan, Y. Genome-wide identification and characterisation of wheat MATE genes reveals their roles in aluminium tolerance. *Int. J. Mol. Sci.* **2022**, *23*, 4418. [CrossRef]
31. Yan, Y.; Hsam, S.L.K.; Yu, J.Z.; Jiang, Y.; Ohtsuka, I.; Zeller, F.J. HMW and LMW glutenin alleles among putative tetraploid and hexaploid European spelt wheat (*Triticum spelta* L.) progenitors. *Theor. Appl. Genet.* **2003**, *107*, 1321–1330. [CrossRef] [PubMed]
32. An, X.; Li, Q.; Yan, Y.; Xiao, Y.; Hsam, S.L.K.; Zeller, F.J. Genetic diversity of European spelt wheat (*Triticum aestivum* ssp. *spelta* L. *em.* Thell.) revealed by glutenin subunit variations at the *Glu-1* and *Glu-3* loci. *Euphytica* **2005**, *146*, 193–201. [CrossRef]
33. Li, Q.Y.; Yan, Y.M.; Wang, A.L.; An, X.L.; Zhang, Y.Z.; Hsam, S.L.K.; Zeller, F.J. Detection of HMW glutenin subunit variations among 205 cultivated emmer accessions (*Triticum turgidum* ssp. *dicoccum*. *Plant Breed.* **2006**, *125*, 120–124. [CrossRef]
34. Yan, Y.; Hsam, S.L.K.; Yu, J.; Jiang, Y.; Zeller, F.J. Allelic variation of the HMW glutenin subunits in *Aegilops tauschii* accessions detected by sodium dodecyl sulphate (SDS-PAGE), acid polyacrylamide gel (A-PAGE) and capillary electrophoresis. *Euphytica* **2003**, *130*, 377–385. [CrossRef]

35. Zhang, Y.; Li, X.; Wang, A.; An, X.; Zhang, Q.; Pei, Y.; Gao, L.; Ma, W.; Appels, R.; Yan, Y. Novel x-type high-molecular-weight glutenin genes from *Aegilops tauschii* and their implications on the wheat origin and evolution mechanism of *Glu-D1-1* proteins. *Genetics* **2008**, *178*, 23–33. [CrossRef]
36. Cui, D.; Wang, J.; Li, M.; Lu, Y.; Yan, Y. Functional assessment and SNP-based molecular marker development of two 1Sl-encoded HMW glutenin subunits in *Aegilops longissima* L. *Mol. Breed.* **2019**, *39*, 1–15. [CrossRef]
37. Hu, J.; Wang, J.; Deng, X.; Yan, Y. Cloning and characterization of special HMW glutenin subunit genes from *Aegilops longissima* L. and their potential for wheat quality improvement. *3 Biotech.* **2019**, *9*, 1–13. [CrossRef] [PubMed]
38. Ma, C.; Yang, Y.; Li, X.; Ge, P.; Guo, G.; Subburaj, S.; Zeller, F.J.; Hsam, S.L.K.; Yan, Y. Molecular cloning and characterization of six novel HMW-GS genes from *Aegilops speltoides* and *Aegilops kotschyi*. *Plant Breed.* **2013**, *132*, 284–289. [CrossRef]
39. Pu, Z.E.; Liu, Y.X.; Li, W. Origin and genetics of spelt wheat (*Triticum spelta* L.) and its utilization in the genetic improvement of common wheat (*Triticum aestivum* L.). *J. Plant. Genet. Resour.* **2009**, *10*, 475–479.
40. Liu, M.; Zhao, Q.; Qi, F.; Stiller, J.; Tang, S.; Miao, J.; Liu, C. Sequence divergence between spelt and common wheat. *Theor. Appl. Genet.* **2018**, *131*, 1125–1132. [CrossRef]
41. Campbell, K.G. Spelt: Agronomy, genetics, and breeding. *Plant Breed. Rev.* **1997**, *15*, 187–213.
42. Frakolaki, G.; Giannou, V.; Topakas, E.; Tzia, C. Chemical characterization and breadmaking potential of spelt versus wheat flour. *J. Cereal Sci.* **2018**, *79*, 50–56.
43. Zeven, A.C.; Turkensteen, L.J.; Stubbs, R.W. Spelt (*Triticum spelta* L.) as a possible source of race-non-specific resistance to yellow rust (*Puccinia striiformis* West.). *Euphytica* **1968**, *17*, 381–384. [CrossRef]
44. Piergiovanni, A.R.; Rizzi, R.; Pannacciulli, E.; Gatta, C.D. Mineral composition in hulled wheat grains: A comparison between emmer (*Triticum dicoccon* Schrank) and spelt (*T. spelta* L.) accessions. *Int. J. Food Sci. Nutr.* **1997**, *48*, 381–386. [CrossRef]
45. Akel, W.; Thorwarth, P.; Mirdita, V.; Weissman, E.A.; Liu, G.; Würschum, T.; Longin, C.F.H. Can spelt wheat be used as heterotic group for hybrid wheat breeding? *Theor. Appl. Genet.* **2018**, *131*, 973–984. [CrossRef]
46. Halford, N.G.; Field, J.M.; Blair, H.; Urwin, P.; Moore, K.; Robert, L.; Thompson, R.; Flavell, R.B.; Shewry, P.R. Analysis of HMW glutenin subunits encoded by chromosome 1A of bread wheat (*Triticum aestivum* L.) indicates quantitative effects on grain quality. *Theor. Appl. Genet.* **1992**, *83*, 373–378. [CrossRef] [PubMed]
47. D'Ovidio, R.; Lafiandra, D.; Porceddu, E. Identification and molecular characterization of a large insertion within the repetitive domain of a high-molecular-weight glutenin subunit gene from hexaploid wheat. *Theor. Appl. Genet.* **1996**, *93*, 1048–1053. [CrossRef]
48. Yan, Y.M.; Jiang, Y.; An, X.L.; Pei, Y.H.; Li, X.H.; Zhang, Y.Z.; Wang, A.L.; He, Z.H.; Xia, X.C.; Bekes, F.; et al. Cloning, expression and functional analysis of HMW glutenin subunit *1By8* gene from Italy pasta wheat (*Triticum turgidum* L. ssp. *durum*). *J. Cereal Sci.* **2009**, *50*, 398–406. [CrossRef]
49. Liang, X.; Zhen, S.; Han, C.; Chang, W.; Yan, Y. Molecular characterization and marker development for hexaploid wheat-specific HMW glutenin subunit *1By18* gene. *Mol. Breed.* **2015**, *35*, 221.
50. Fevzioglu, M.; Ozturk, O.K.; Hamaker, B.R.; Campanella, O.H. Quantitative approach to study secondary structure of proteins by FT-IR spectroscopy, using a model wheat gluten system. *Int. J. Biol. Macromol.* **2020**, *164*, 2753–2760. [CrossRef]
51. Ruomei, W.; Junwei, Z.; Fei, L.; Nannan, L.; Prodanović, S.; Yueming, Y. Cloning and molecular characterization of two novel LMW-m type glutenin genes from *Triticum spelta* L. *Genetika* **2021**, *53*, 141–155.
52. Wang, Y.; Zhen, S.; Luo, N.; Han, C.; Lu, X.; Li, X.; Xia, X.; He, Z.; Yan, Y. Low molecular weight glutenin subunit gene *Glu-B3h* confers superior dough strength and breadmaking quality in wheat (*Triticum aestivum* L.). *Sci. Rep.* **2016**, *6*, 27182. [CrossRef] [PubMed]
53. Skolnick, J.; Gao, M.; Zhou, H.; Singh, S. Alphafold 2: Why it works and its implications for understanding the relationships of protein sequence, structure, and function. *J. Chem. Inf. Model.* **2021**, *61*, 4827–4831. [CrossRef] [PubMed]
54. Lutz, E.; Wieser, H.; Koehler, P. Identification of disulfide bonds in wheat gluten proteins by means of mass spectrometry/electron transfer dissociation. *J. Agric. Food Chem.* **2012**, *60*, 3708–3716. [CrossRef] [PubMed]
55. Shewry, P.; Gilbert, S.; Savage, A.; Tatham, A.S.; Wan, Y.F.; Belton, P.S.; Wellner, N.; D'Ovidio, R.; Békés, F.; Halford, N.G. Sequence and properties of HMW subunit IBx20 from pasta wheat (*Triticum durum*) which is associated with poor end use properties. *Theor. Appl. Genet.* **2003**, *106*, 744–750. [CrossRef] [PubMed]
56. Pirozi, M.R.; Margiotta, B.; Lafiandra, D.; MacRitchie, F. Composition of polymeric proteins and bread-making quality of wheat lines with allelic HMW-GS differing in number of cysteines. *J. Cereal Sci.* **2008**, *48*, 117–122. [CrossRef]
57. Shewry, P.R.; Halford, N.G. Cereal seed storage proteins: Structures, properties and role in grain utilization. *J. Exp. Bot.* **2002**, *53*, 947–958. [CrossRef]
58. Wang, N.; Ma, S.; Li, L.; Zheng, X. Aggregation characteristics of protein during wheat flour maturation. *J. Agric. Food Chem.* **2019**, *99*, 719–725. [CrossRef]
59. Andrews, J.L.; Skerritt, J.H. Quality-related epitopes of high Mr subunits of wheat glutenin. *J. Cereal Sci.* **1994**, *19*, 219–229. [CrossRef]
60. Ramesh, P.; Mallikarjuna, G.; Sameena, S.; Kumar, A.; Gurulakshmi, K.; Reddy, B.V.; Sekhar, A.C. Advancements in molecular marker technologies and their applications in diversity studies. *J. Biosci.* **2020**, *45*, 1–15. [CrossRef]
61. Ravel, C.; Faye, A.; Ben-Sadoun, S.; Ranoux, M.; Gérard Branlard. SNP markers for early identification of high molecular weight glutenin subunits (HMW-GSs) in bread wheat. *Theor. Appl. Genet.* **2020**, *133*, 751–770. [CrossRef] [PubMed]

62. Schwarz, G.; Sift, A.; Wenzel, G.; Mohler, V. DHPLC scoring of a SNP between promoter sequences of HMW glutenin x-type alleles at the Glu-D1 locus in wheat. *J. Agric. Food Chem.* **2003**, *51*, 4263–4267. [CrossRef] [PubMed]
63. Mackie, A.M.; Sharp, P.J.; Lagudah, E.S. The nucleotide and derived amino acid sequence of a HMW glutenin gene from *Triticum tauschii* and comparison with those from the D genome of bread wheat. *J. Cereal Sci.* **1996**, *24*, 73–78. [CrossRef]
64. Allaby, R.G.; Banerjee, M.; Brown, T.A. Evolution of the high molecular weight glutenin loci of the A, B, D, and G genomes of wheat. *Genome* **1999**, *42*, 296–307. [CrossRef] [PubMed]
65. Buchan, D.W.; Jones, D.T. The PSIPRED protein analysis workbench: 20 years on. *Nucleic Acids Res.* **2019**, *47*, W402–W407. [CrossRef]
66. Buchan, D.W.; Ward, S.M.; Lobley, A.E.; Nugent, T.C.O.; Bryson, K.; Jones, D.T. Protein annotation and modelling servers at University College London. *Nucleic Acids Res.* **2010**, *38*, W563–W568. [CrossRef]
67. Mirdita, M.; Schütze, K.; Moriwaki, Y.; Heo, L.; Ovchinnikov, S.; Steinegger, M. ColabFold: Making protein folding accessible to all. *Nat. Methods* **2022**, *19*, 679–682. [CrossRef]
68. Varadi, M.; Anyango, S.; Deshpande, M.; Nair, S.; Natassia, C.; Yordanova, G.; Velankar, S. AlphaFold Protein Structure Database: Massively expanding the structural coverage of protein-sequence space with high-accuracy models. *Nucleic Acids Res.* **2022**, *50*, D439–D444. [CrossRef]

International Journal of *Molecular Sciences*

MDPI

Review

Physiology and Molecular Breeding in Sustaining Wheat Grain Setting and Quality under Spring Cold Stress

Hui Su [1], Cheng Tan [1], Yonghua Liu [2], Xiang Chen [1], Xinrui Li [1], Ashley Jones [3], Yulei Zhu [1,*] and Youhong Song [1,4,*]

[1] School of Agronomy, Anhui Agricultural University, Hefei 230036, China
[2] School of Horticulture, Hainan University, Haikou 570228, China
[3] Research School of Biology, The Australian National University, Canberra, ACT 2601, Australia
[4] Centre for Crop Science, Queensland Alliance for Agriculture and Food Innovation, The University of Queensland, Brisbane, QLD 4072, Australia
* Correspondence: zhuyulei@ahau.edu.cn (Y.Z.); y.song@ahau.edu.cn (Y.S.)

Abstract: Spring cold stress (SCS) compromises the reproductive growth of wheat, being a major constraint in achieving high grain yield and quality in winter wheat. To sustain wheat productivity in SCS conditions, breeding cultivars conferring cold tolerance is key. In this review, we examine how grain setting and quality traits are affected by SCS, which may occur at the pre-anthesis stage. We have investigated the physiological and molecular mechanisms involved in floret and spikelet SCS tolerance. It includes the protective enzymes scavenging reactive oxygen species (ROS), hormonal adjustment, and carbohydrate metabolism. Lastly, we explored quantitative trait loci (QTLs) that regulate SCS for identifying candidate genes for breeding. The existing cultivars for SCS tolerance were primarily bred on agronomic and morphophysiological traits and lacked in molecular investigations. Therefore, breeding novel wheat cultivars based on QTLs and associated genes underlying the fundamental resistance mechanism is urgently needed to sustain grain setting and quality under SCS.

Keywords: *Triticum aestivum* L.; spring frost; spikelet development; grain set and quality; QTLs

check for updates

Citation: Su, H.; Tan, C.; Liu, Y.; Chen, X.; Li, X.; Jones, A.; Zhu, Y.; Song, Y. Physiology and Molecular Breeding in Sustaining Wheat Grain Setting and Quality under Spring Cold Stress. *Int. J. Mol. Sci.* **2022**, *23*, 14099. https://doi.org/10.3390/ijms232214099

Academic Editors: Jinsong Bao and Jianhong Xu

Received: 15 September 2022
Accepted: 12 November 2022
Published: 15 November 2022

Publisher's Note: MDPI stays neutral with regard to jurisdictional claims in published maps and institutional affiliations.

Copyright: © 2022 by the authors. Licensee MDPI, Basel, Switzerland. This article is an open access article distributed under the terms and conditions of the Creative Commons Attribution (CC BY) license (https://creativecommons.org/licenses/by/4.0/).

1. Introduction

Wheat provides approximately 20% of the food energy and protein produced for human consumption [1], and grain quality is an important indicator due to market value and consumer acceptance [2,3]. Wheat grain quality is a complex combination of various traits, mainly controlled by genotypic and environmental factors [4]. Climate change is causing a temperature shift and ecological landscapes that negatively impact wheat yield and quality [5]. During the last several decades, it has been reported that spring cold stress (SCS) has caused severe losses in wheat production and grain quality. For example, in Australia, the SCS events that frequently occurred at wheat reproductive stage typically resulted in yield losses of 10%, and it's more than 85% in various farmlands [6,7]. Nearly 85% of China's total area planted with winter wheat experiences widespread SCS [8,9]. Reports from North America and Europe indicated that late frost spells are one of the most economically damaging agricultural climate hazards, causing substantial economic losses in 2017 [10,11]. Consequently, the abiotic stress of SCS threatens the safety of crop production systems worldwide. Wheat growth and development have been subjected to more frequent cold stress as climate change continues [12].

The SCS events often occur during the reproductive development in winter wheat [13]. The reproductive development is composed of floral initiation, pollen grain and embryo development, pollination, fertilization and grain setting, etc. [14]. When wheat suffers from frost during the reproductive growth period, it causes the wheat spike cells to lose water and wither, affecting the young spike's normal development and increasing the young spike's mortality [15,16]. Malfunctions and irreversible abortion of male and female reproductive organs and gametophytes are the main reasons for cold-induced male and female infertility [17]. During SCS, the anthers display irregular hypertrophy and vacuolation of the tapetum, an unusual accumulation of starch and protein in the plastids, and poor pollen tube development [18,19]. Zhang et al. (2021) stated that low-temperature stress significantly reduced the expression and activity of the sucrose invertase (CWINV) coding gene in young ears at the booting stage, inhibited the transport of sucrose to pollen sac, and then hindered the normal development of pollens [20]. Occurrence of SCS at late reproductive growth resulted in smaller dark-colored seeds with a wrinkled epidermis, poor seed setting and quality [21].

Wheat responds to cold stress by regulating key physiological, biochemical, and molecular mechanisms [22]. Under cold stress, a wide range of chemicals or protective proteins are produced, including soluble carbohydrates, proline, and cold-resistance proteins [23], which are involved in regulating osmotic potential, preventing ice crystal formation, the stability of cell membranes and reactive oxygen species (ROS) scavenging [24]. At the molecular level, estimates of phenotypic plasticity were used to identify loci associated with stress tolerance. Candidate genes involved in phytohormone-mediated processes for stress tolerance were proved to be involved in cold stress responses [25]. Cold acquisition of freezing tolerance requires the orchestration of disparate physiological and biochemical changes, and these changes are mainly mediated through the differential expression of genes [26,27]. Some of these genes encode effector molecules directly involved in stress mitigation, and others encode proteins for signal transduction or transcription factors that control gene pool expression [26]. Genes involved in plant metabolism were differentially expressed to avoid injury and damage associated with SCS; it includes the encodings of Ca^{2+} binding proteins, protein kinases, and inorganic pyrophosphatase [28].

Understanding the potential regulatory mechanisms behind SCS tolerance is necessary to create wheat breeding varieties with improved grain setting and quality under cold stress. In this review study, we further summarized the consequences of SCS and explored the potential mechanisms to sustain wheat grain setting and quality under SCS. The objectives of this study are to (i) make clear the physiological and molecular mechanism in controlling grain setting and quality under SCS, and (ii) propose breeding strategies in combatting SCS during reproductive stage.

2. Effects of SCS on Grain Number and Quality in Wheat

Under varying climatic conditions, the SCS events have become more frequent, intense, and prolonged. The SCS events often occur during the reproductive stage of winter wheat, which is critical for the establishment of the panicle [29]. The SCS compromises the development of young spike and floret; nutrient distribution is altered, and floret stunting (or sterility) occurs, resulting in poor grain set and quality (Figure 1).

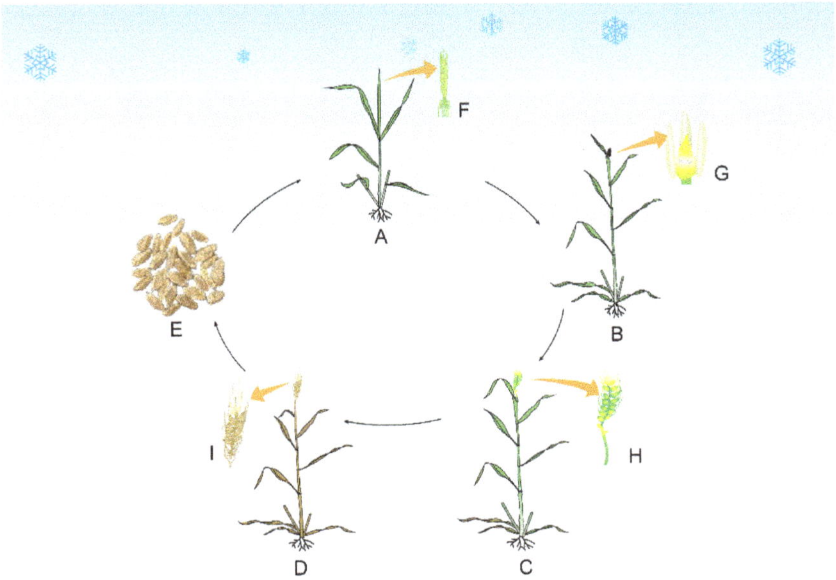

Figure 1. A schematic diagram visually demonstrating the impacts of SCS initiated at tetrad stage (**A**) on subsequent wheat growth and development at booting stage (**B**), anthesis stage (**C**) and maturation stage (**D**). (**F**) Indicates young spikelet development at tetrad stage. (**G**) Indicates tapetum degeneration and pollen sterility in the developing anthers at booting stage. (**H**) Indicates reduced pollen viability and thus spikelet fertility. (**I**) Indicates reduced grain-filling rate and period and enhanced grain abortion, and thus less grain number and quality (**E**).

2.1. Grain Number

Grain number is a significant factor in determining wheat grain yield [30]. The stages from jointing to flowering are critical to prevent florets from degenerating and increase the grain setting rate [31–33]. Under SCS conditions, the lower spike number per plant and grain number per spike were primarily responsible for reduced grain production (Table 1; Figure 2) [34]. Compared with spring wheat cultivars, semi-winter wheat has stronger cold resistance. For example, under low temperature of −2–6 °C for 3 days at the jointing stage, the grain number per spike was lowered by 1.3–4.4% in Yangmai16 (spring wheat), while decreased by 0.6–1.0% in Xumai30 (semi-winter wheat) [35]. Meanwhile, cold stress led to different yields of different genotypes at the reproductive stage [36]. Compared with the control, low temperature led to zero harvest of diploid genotypes, and the yield of tetraploid genotypes decreased significantly, while hexaploid genotypes acquired relatively high maintenance rate of grain yield among three species [36]. Additionally, the yield loss caused by SCS also depends on the intensity of the low temperature and its duration [37]. Ji et al. (2017) exposed two wheat cultivars at the booting stage to freezing temperature at 2, −2, −4 and −6 °C for 2–6 d in a convective freezing chamber, causing 13.9–85.2% grain yield reduction in spring wheat, while resulting 3.2–85.9% grain yield loss in semi-winter wheat [35]. With the temperature declined to −5 °C and −7 °C at the vegetative growth stage, the grain yield decreased by 10–100% [38]. In each case, the SCS events during the reproductive development significantly affected the growth and development of younger spikes and florets, causing pollen infertility and poor grain setting [39], thereby resulting in a decrease in the number of grains.

Table 1. Effects of SCS treatment at different stages on grain number per spike in wheat.

Cultivars	Period	Duration	Low Temperature	Grain Number	Drop Percentage (%)	Reference
Xinong979	Booting stage	12 h	15 °C/20 °C −3.5 °C/20 °C −5.5 °C/20 °C	42.2 24.8 13.1	- 41.2 67.0	[40]
Changhan58	Booting stage	12 h	15 °C/20 °C −3.5 °C/20 °C −5.5 °C/20 °C	40.7 32.3 17.5	- 20.6 57.0	
Yangmai15	Stem elongation stage	3 d	5 °C/10 °C −3 °C/0 °C	40.9 34.3	- 16.1	[41]
Yangmai16	Jointing stage	3 d	6 °C/16 °C/11 °C −2 °C/8 °C/2 °C −4 °C/6 °C/1 °C −6 °C/4 °C/−1 °C	40.7 40.2 39.2 38.9	- 1.3 3.6 4.4	[35]
	Booting stage	3 d	6 °C/16 °C/11 °C −2 °C/8 °C/2 °C −4 °C/6 °C/1 °C −6 °C/4 °C/−1 °C	40.0 37.2 36.4 14.4	- 7.0 8.9 63.9	
Xumai30	Jointing stage	3 d	6 °C/16 °C/11 °C −2 °C/8 °C/2 °C −4 °C/6 °C/1 °C −6 °C/4 °C/−1 °C	36.4 36.2 36.2 36.1	- 0.6 0.6 1.0	
	Booting stage	3 d	6 °C/16 °C/11 °C −2 °C/8 °C/2 °C −4 °C/6 °C/1 °C −6 °C/4 °C/−1 °C	36.4 32.5 31.1 21.1	- 10.7 14.6 42.1	
XM21	Jointing stage	5 d	Approximately 8 °C lower than the ambient temperature	-	4.6–5.9	[42]
XZ24	Jointing stage	5 d	Approximately 8 °C lower than the ambient temperature	-	12.3–13.9	
DM22	Jointing stage	39 d	15 °C/20 °C 5 °C/15 °C	14.0 8.4	- 40.0	[36]
DM31	Jointing stage	39 d	15 °C/20 °C 5 °C/15 °C	21.0 4.0	0 81.0	
L8275	Jointing stage	39 d	15 °C/20 °C 5 °C/15 °C	19.0 17.0	0 10.5	
MO1	Jointing stage	44 d	15 °C/20 °C 5 °C/15 °C	10.4 0	0 100	
MO2	Jointing stage	44 d	15 °C/20 °C 5 °C/15 °C	13.6 0	0 100	

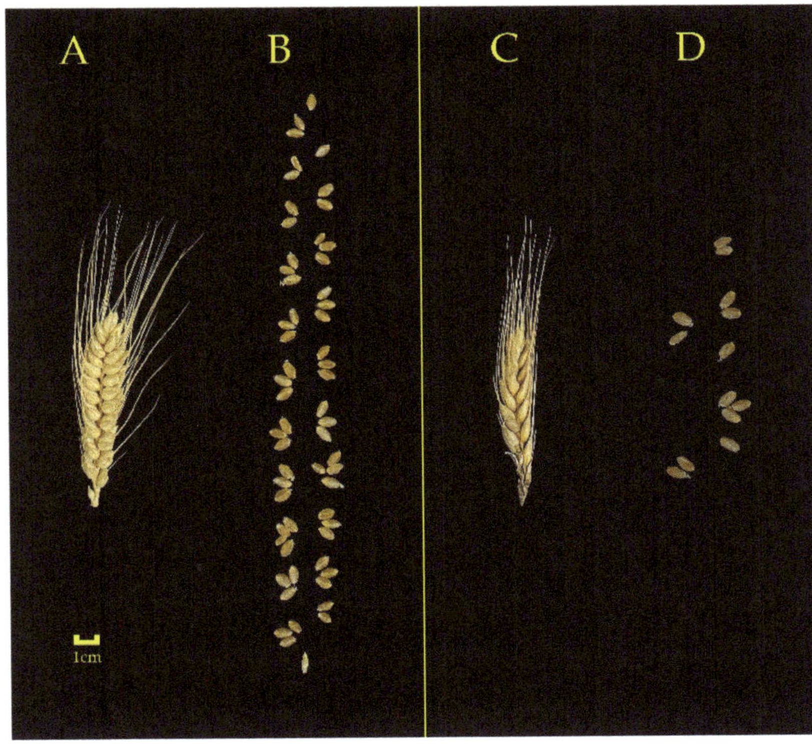

Figure 2. The photos of spikes (**A**,**C**) and grains (**B**,**D**) in wheat under the control (**A**,**B**) and spring cold stress (**C**,**D**) −2 °C for 6 h. The photo visually shows the effect of cold stress on the size and color of spike and grain number.

2.2. *Grain Quality*

Grain quality is primarily based on appearance and nutritional quality [43]. It is well known that mostly spring cold stress events are often encountered during the reproductive period in wheat, which seriously affects the absorption and distribution of nutrients [39]. Grain quality relative to its appearance refers to external morphological characteristics, including grain length, width, and aspect ratio [44]. For example, wheat responds to SCS (−4 °C for 12 h) at the jointing stage by increasing the ratio between grain length and width (L/W) for 0.4–14.2% while decreasing the equivalent diameter in 0.9–11.0% and grain area in 1.6–20.2% [45]. Compared to the cold-tolerant genotype, the grain width and L/W of the sensitive wheat genotype were more susceptible to low temperatures [46]. It also reported that the grain width is more sensitive to low temperatures than the grain length [46].

In addition to affecting morphological appearance, the quality of grain nutrition is adversely affected by SCS [47]. For wheat grain nutrition quality, protein content is of key significance [48]. It has been noted that SCS limits the production of nitrogen compounds and nonstructural carbohydrates, which decreases the transit of protein and total soluble sugar from stems into grains, resulting in a decline in wheat quality [36]. Under low temperatures at the booting stage, the mean accumulation of total protein decreased by 4.8–6.9%, albumin by 5.8–9.6%, globulin by 8.4–15.4%, gliadin by 13.2–18.4%, and glutenin by 17.8–29.1% [49]. In addition to this, reductions in the concentrations of amylose, amylopectin and total starch were also observed under different low-temperature levels [46]. According to a recent report, the total starch in wheat grains, as well as the rate of accumulation of straight-chain and branched-chain starch, were closely related to the activities of starch branching enzyme (SBE), soluble starch synthase (SSS), granule-bound starch synthase (GBSS) and adenosine diphosphate glucose pyrophosphorylase (AGPase) [50], while the activity of essential starch synthesis enzymes is particularly sensitive to SCS during grain development [51]. The low temperature during the reproductive stage decreased the activities of crucial starch synthesis enzymes (AGPase, SSS, GBSS, and SBE) in the grain, thereby reducing the accumulation of starch, resulting in a decreasing grain quality [52].

3. Physiological Mechanism of Controlling Wheat Resistance to Cold Stress

3.1. Protective Enzymes for Oxidation

Cold stress often leads to excess accumulation of reactive oxygen species (ROS) such as superoxide radical (O^{2-}) and hydrogen peroxide (H_2O_2), which causes oxidative damage to DNA, proteins, and lipids, leading to the inhibition of wheat seed development [53,54]. Hence, the balanced ROS production level was achieved at the intracellular level which promotes the normal growth, development, and cellular metabolism (Figure 3) [55].

The activation of subcellular antioxidant mechanisms can provide some resistance to SCS in wheat while also decreasing oxidative burst in the photosynthesis machinery [56]. Activities of antioxidant enzymes, such as peroxidase (POD), superoxide dismutase (SOD), and catalase (CAT), play an essential role in protecting plants from oxidative damage by ROS scavenging [57,58]. Several studies have reported that alterations in the activity of numerous antioxidant defense system enzymes help plants to handle oxidative stress in wheat [59,60]. For example, cold stress (4 °C and −4 °C) increased the activity of SOD by 6.8–68.3%, POD by 16.6–69.4%, CAT by 6.0–53.8% in a wheat spikelet, compared to optimum temperature (16 °C) [61]. Furthermore, antioxidant chemicals, including proline, glutathione (GSH) and ascorbic acid (AsA), also play critical roles in protecting plants from ROS damage caused by cold stress [62]. Under SCS, the accumulation of proline eliminates oxygen free radicals, which balances the osmotic pattern in the cell, and maintains the normal state of the membrane [63]. For example, the application of exogenous proline improved wheat's cold tolerance, due to the increased accumulation of free proline and sucrose, by coordinating carbon and nitrogen metabolism [64]. It is noted that the AsA–GSH cycle, including ascorbate peroxidase (APX), monodehydroascorbate reductase (MDHAR), dehydroascorbate reductase (DHAR), and glutathione reductase (GR), are very effective in improving wheat cold tolerance, particularly to ROS stress [65]. For example, AsA could induce the up-regulation of diverse antioxidants (super oxide dismutase (SOD), peroxidase (POD), and catalase (CAT)), thus offsetting the adverse effects of cold stress at early and reproductive stages of wheat [66].

Another key mechanism in plant cold stress responses is the regulation of transcription by endogenous hormones and ROS [67]. Once induced by cold stress, hormones change the ROS levels due to increasing transcription or talking about post-translational modification/activation of proteins and transforming ROS signaling [68]. For instance, it has been demonstrated that the ROS generated by RBOHs mediates an interaction between ABA and BRs, enhancing cold tolerance in *Arabidopsis* [69]. According to a recent study, the application of exogenous BRs increased antioxidant capability, directing the reduction of oxidative damage caused by ROS bursts [70].

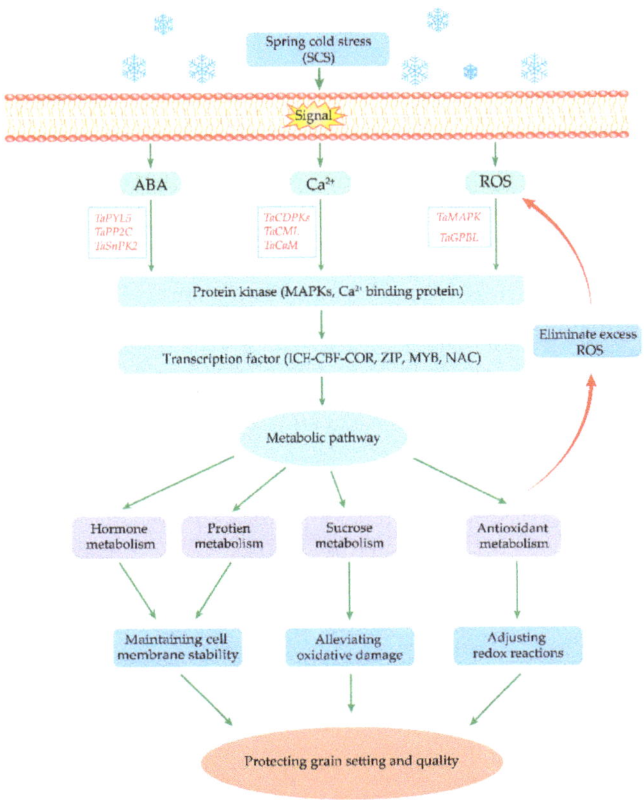

Figure 3. Overview of wheat responses to spring cold stress, which induces several protective measures to regulate grain setting and quality. Firstly, cold stress triggers multiple channel activation leading to the increased ABA, Ca^{2+} and ROS concentrations in the cytosol. The main components in the core ABA signaling transduction pathway include ABA receptor *TaPYL5, TaPP2C, TaSnRKs*, and the Ca^{2+} signaling transduction pathway include *TaCDPKs, TaCML, TaCaM*, which have a positive regulation of cold stress. Secondly, component changes in the MAPK cascade pathway were influenced by the activation of the ABA, ROS and Ca^{2+} pathway. Thirdly, cold stress response-induced signal transduction leads to the activation of multiple transcription factors, thereby regulating the metabolic hormone, protein, sucrose and antioxidant pathway. These alterations mitigate cell membrane damage and regulate intracellular osmotic balance, preventing the loss of grain yield and quality.

3.2. Carbohydrate Metabolism

Carbohydrate metabolism plays an essential role in energy availability for plant development and also has a role in temperature acclimation [71]. In plants, several soluble sugars, such as sucrose, glucose, sucrose, fructose, raffinose and trehalose, act as biofilm protectors by interacting with the lipid bilayer. This interaction has a role in reducing membrane damage, as the sugars function as osmoprotectants and provide adaption to the cold environment [72,73].

The soluble sugars sucrose, glucose, trehalose, and fructose start accumulating in response to cold stress, enhancing cold tolerance during the reproductive stage of crops [74]. For instance, the buildup of soluble sugars under SCS can raise the amount of proline, which controls osmotic pressure, scavenges reactive oxygen species, and stabilizes biomolecule structure, reducing low-temperature damage [75,76]. Fructans, which are highly water soluble, act as osmoregulatory substances to prevent the formation of ice crystals in the cytoplasm and improve biofilm stability, enhancing crop cold tolerance [77]. Recent research has confirmed a high correlation between fructan accumulation and cold tolerance due to increasing transcript levels of the Cor (cold-responsive)/Lea (late-embryogenesis-abundant), C-repeat-binding factor (CBF), and fructan biosynthesis-related genes in the wheat family [77]. Trehalose has been found to act as an osmoprotectant, and stabilizes protein integrity in plants [78]. Importantly, exogenous trehalose prevented floret degeneration under low-temperature conditions and increased floret fertility in young spikelets, minimized any loss in grain number per spike [43].

Recently, the Sugars Will Eventually be Exported Transporters (SWEETs) have been reported to regulate abiotic stress tolerance, sugar transport, plant growth and development [79]. The SWEETs also play vital roles in oxidative and osmotic stress tolerance [80]. In wheat, the genome-wide analysis revealed 105 SWEETs, and 59% exhibited significant expression changes under abiotic stresses [81]. Importantly, *AtSWEET16* and *AtSWEET17* are two bidirectional vesicular fructose transporters that maintain glycan homeostasis and promote the accumulation of fructose in vacuoles, which may be beneficial in stress tolerance responses [82,83]. A further understanding of sugar metabolism and transport will be key in reducing any sugar starvation in crop reproductive development and enhancing seed setting rate.

3.3. *Hormones and Ca^{2+} Signals*

Plants adapt to environmental changes in low-temperature settings by a sequence of cellular reactions triggered by signaling molecules (e.g., hormone signals, Ca^{2+} signal), which result in plant defense and adaptability to adverse conditions [84,85]. Plant hormones, such as abscisic acid (ABA) [86], jasmonic acid (JA) [87], and salicylic acid (SA) [88], have been reported to play a significant role in regulating grain quality. Past findings revealed that many plants experience higher endogenous ABA levels in response to cold stress [89,90]. In wheat, the application of exogenous ABA is reported to enhance cold tolerance by increasing the activities of antioxidant enzymes and reducing H_2O_2 contents under cold stress [91]. In particular, ABA-dependent gene expression, which includes the ABA receptors, protein phosphatases type-2C (PP2Cs), Snf1-related kinase 2s (SnRK2s), and AREB/ABF regulon, controlled by the raised ABA levels, helped plants adapt to abiotic stress cold stress [92]. According to Zhang et al. (2018), the significant up-regulation of the *SnRK2.11*, serine/threonine-protein kinase and serine/threonine-protein phosphatase PP1-like was considered to be a significant reason for improving cold tolerance in wheat during the reproductive stage [28]. These genes were believed to function in ABA signaling in guard cells.

Additionally, JA also plays a mediating role in synthesis and signaling to mediate low-temperature tolerance [93]. For instance, endogenous JA levels were found to be increased in wheat [94], rice [95], and Arabidopsis [96], enhancing the frost resistance of crops. JA functions as an upstream signal of the ICE-CBF pathway, positively modulating freezing responses [97]. *JAZ1* and *JAZ4* are JA signaling negative regulators interacting with ICE1 and ICE2 to repress their transcriptional activity [98]. Subsequently, they regulate the expression of CBF and other low-temperature responsive genes, thus affecting wheat cold resistance [97].

It is well known that SA plays a vital role in responding to abiotic stresses, apart from regulating crop growth, ripening and development [98,99]. SA activates the active oxygen species before low-temperature exposure; it promotes an increase in antioxidant enzyme activity and higher mRNA content of *TaFeSOD*, *TaMnSOD*, *TaCAT* gene transcripts, and free Proline after SCS [100]. Freezing stress during the reproductive stage shows salicylic acid-primed wheat up-regulated the expression level of the WRKY gene (*WRKY19*), heat shock transcription factor (*HSF3*), mitochondrial alternative oxidase (AOX1a), and heat shock protein (HSP70), which contributes to increasing of antioxidant capacity and protection of photosystem in parallel with lower malonaldehyde content, superoxide radical production as compared with non-primed wheat [101]. Further research has demonstrated that SA treatment reduces ice nucleate and induces anti-freezing protein, which inhibits the formation of ice crystals in plant cells [88].

Ca^{2+} is an essential secondary messenger in plants in response to cold stress [102]. Ca^{2+} sensors such as calmodulins (CaMs), CaM-like proteins (CMLs), Ca^{2+}-dependent protein kinases (CPKs/CDPKs), and calcineurin B-like proteins (CBLs) are the primary transmitters of the Ca^{2+} signal that is induced by cold stress [103–105]. For example, *OsCPK27*, *OsCPK25*, and *OsCPK17* activated MAPK, ROS, and nitric oxide pathways in response to cold stress [85,106]. Recently, genome-wide identification and expression analysis also show that 18 *TaCaM* and 230 *TaCML* gene members were identified in the wheat genome, and *TaCML17*, *21*, *30*, *50*, *59* and *75* were identified related with responses to cold stress in wheat [107].

3.4. Transcription Factors

Wheat genomes contain a large number of transcription factors that play important roles in cold-stress biological processes, including CBF [108], basic leucine zipper (bZIP) [109], MYB [110], and NAC [111].

The ICE-CBF-COR signaling pathway is widely recognized as essential for cold adaptation [112]. The receptor protein detects cold stress and initiates signal transduction, activating and regulating the ICE gene, which up-regulates the transcription and expression of the CBF gene [113]. In wheat, five ICE genes, 37 CBF genes and 11 COR genes were discovered in the wheat genome database [114]. Wheat CBF genes have been demonstrated to improve cold tolerance in other plants, as shown with transgenic barley containing *TaCBF14* and *TaCBF15* genes [115]. A vast variety of transcription factors are also important, such as CBF1, CBF2, and CBF3 [116] and C-type repeats (CTR) [117], which play crucial roles in the biological processes of abiotic stressors in wheat. Previous studies reported that cold-regulated transcriptional activator CBF3 positively regulates cold stress responses in wheat [118]. The RNA-seq data and qRT-PCR revealed that the ICE, CBF, and COR genes have varying expression patterns in different wheat organs, with ICE genes mainly up-regulated in the grain, CBF in the root and stem, and COR in the leaf and grain [114]. All these results show that the ICE–CBF–COR cascade plays a crucial role in the response of wheat to cold stress (Figure 3).

The bZIP genes are involved in important regulatory processes of plant growth and physiological metabolisms, such as promoting anthocyanin accumulation [119] and other signals [120]. Similarly, the bZIP gene also has a variety of biological functions under abiotic stress, and 187 bZIP genes have been predicted in wheat [121]. And the majority of bZIPs linked to frost tolerance in plants are positive regulators [122]. For example, phenotypic analysis and related physiological indicators of cold resistance showed that overexpression of *TaABI5* could enhance cold resistance [109]. In recent years, 15 bZIP genes with variable expression were found in early wheat spikes, and most showed an increase in in expression in response to SCS [123]. Furthermore, the bZIP genes are involved in ABA signaling and play a role in responding to freezing stress in the later stage of wheat [109]. Similarly, MYB and NAC are crucial in controlling plant growth and cold stress responses [110,124].

4. Breeding Strategies to Develop SCS-Resistant Wheat

Superior wheat genotypes are needed for SCS resistance, which will be made possible by breeding cold-resistant cultivars that maintain yield stability and high quality [125]. Appropriate measures need to be taken to cope with the consequences of SCS in wheat during the reproductive stage, to improve crop yield and quality. Strategies to strengthen SCS resistance include selecting cold-tolerance cultivars, identifying QTL/genes, and exploiting closely linked markers in wheat.

4.1. QTLs Associated with Cold Resistance

Genetic components such as QTLs have great potential to accelerate traditional breeding processes [126]. QTLs related to cold tolerance and the underlying molecular mechanisms have been thoroughly studied in wheat [127,128]. There are loci for cold resistance on 1B, 1D, 2B, 2D, 4D, 5A, 5D, and 7A, with 5A and 5D suspected to carry significant genes of interest [129,130]. Wheat chromosome 5A plays a key role in cold acclimation and frost tolerance [119]. Three key genes responsible for SCS tolerance, *Fr-1* (e.g., *Fr-A1*, *Fr-B1*) and *Fr-2* (e.g., *Fr-D1*), were located on chromosomes 5A, 5B, and 5D [131,132], with two loci being mapped within a distance of approximately 30 cM [118]. The *Fr-1* maps close to the vernalization locus Vernalization-1 (*Vrn-1*), so they showed highly homologous [133]. The *Vrn1* acts as a positive regulator of vernalization and regulates the transition from vegetative to reproductive growth in wheat [134]. The *Fr-A*^{m}*2* locus is made up of a group of eleven CBF genes that are activated during vernalization, which in turn activate the COR genes necessary for enhanced cold tolerance of wheat [135,136].

Genome-wide association studies (GWAS) of traits related to wheat resistance and tolerance are essential to understanding their genetic structure for improving breeding selection efficiency [137]. 23 QTL regions located on 11 chromosomes (1A, 1B, 2A, 2B, 2D, 3A, 3D, 4A, 5A, 5B and 7D) were detected for frost tolerance in 276 winter wheat genotypes by GWAS, eight novel QTLs were discovered on chromosomes 1B, 2D, 3A, 3D, 4A and 7D [129]. Eighty SNP loci distributed in all the 21 chromosomes were associated with the resistance of SCS with 16.6–36.2% phenotypic variation by GWAS, six loci of these were stable loci with more than two traits, and multiple superior alleles were obtained from the associated loci related to SCS traits [138]. Nevertheless, the majority of the QTL intervals for low-temperature tolerance reported by GWAS are still huge, and there are too many candidate genes; the causal genes for cold tolerance are still challenging to find.

Of the different genome editing approaches, CRISPR/Cas9 genome editing module has evolved as a successful tool in modulating genes essential for developing high-stress resistance of crops [139]. Meanwhile, CRISPR/Cas9 allows the manipulation of the wheat genome for improved agronomic performance, resistance to biotic and abiotic stresses, higher yields, and better grain quality [140]. For example, Tian et al. (2013) [141] cloned *TaSnRK2.3*, then further determined its expression patterns under freezing stresses in wheat emerging and characterized its function in Arabidopsis. Overexpression of *TaSnRK2.3* significantly enhanced tolerance to freezing stress, enhancing the expression of cold stress-responsive genes and ameliorating physiological indices [141]. Additionally, it showed that overexpressing *TaFBA-A10* led to the increased activity of FBA, as well as regulating key enzymes in the Calvin cycle and the glycolysis rate to enhance cold tolerance of wheat [142]. Therefore, acceptance and utilization of new plant breeding technologies involving genome editing confer opportunities for sustainable agriculture and ensure global food security.

4.2. Cultivars for SCS Resistance Based on Agronomic Traits

Wheat yield is associated with several agronomic traits which have been used to make better cultivars, increasing the yield and quality of wheat [143]. Given the high heritability of the traits and the relevance of wheat yield, agronomic traits can be used as selection criteria in breeding and cultivar development (Table 2) [144]. Cold stress affects agronomic traits at every developmental stage, but the reproductive stages are relatively more sensitive [145]. Specifically, cold stress affects the development of young spikes and flowers, grain characteristics and quality [146,147]. Some researchers have classified inversions into five major categories based on the degree of damage to the spikelet: grade 1 for no apparent frost damage, grade 2 for frost damage less than 1/3, grade 3 for frost damage between 1/3 and 1/2, grade 4 for frost damage greater than 1/2, and grade 5 for all young spikes that died from freezing [148]. Similarly, frost damage also impaired stem development, resulting in lower plant height and a decrease number of spikes [149]. For example, using the dead stem rate to classify 120 wheat cultivars into five classes of very strong, moderate, weak and very weak, and determining the criteria for categorizing wheat spring frost resistance evaluation classes [150].

Table 2. Tolerant and sensitive wheat genotypes and their performances in response to spring cold stress.

Selected Indicators	Cold Stress Method	Growth Phase	Tolerant Genotypes	Performance of Tolerant Genotypes	Sensitive Genotypes	Performance of Sensitive Genotypes	References
Agronomic traits	Cryogenic incubator and solar thermal chamber	The anther seperation stage	Shannong 7859, Beijing 841, Jinmai 47, Xinmai 9, Yumai 49.	TSR $\geq$ 0.90	Neixiang 188, Zhengmai 7698, Xinong 889	TSR < 0.70	[138]
Agronomic traits	Field nature identification	1th to 15th in March	Yannong 5158, Huaimai 28, Huaimai 33, Jinan 17, Fanmai 5, Yannong 19, Xumai 35	Higher plant height (PH), larger grain number of main stem spike (GNMSS), GYPP, heavier grain yield per plant, stronger cold resistance, and better comprehensive agronomic traits	Jimai22, Huaimai22, Jinan17, Guomai9 Liangxing66, Zhoumai27, SXM208	Fewer GNMSS, lighter TKW, lower GYPP, weak cold resistance, and poor comprehensive agronomic traits	[151]
Agronomic traits	Field nature identification	5th to 7th in April	Bainong 207, Xinong 511	Low frozen spikelet rate, and high rate of seed setting of frozen spikelet	Zhengmai 366, Fengdecunmai 5	High frozen spikelet rate, and low rate of seed setting of frozen spikelet	[148]
Agronomic traits	Artificial chamber	3th in April	Yannong19	The correlation between GBSS activity, the starch content and the thousand kernel weight was highly significant	Yangmai18	The correlation between GBSS activity, the starch content and the thousand kernel weight was not significant	[52]
Agronomic traits	Artificial chamber	From pistil and stamen primordia differentiation stage to anthesis stage	Jimai22, Yannong19	Low dead stem rate and few residual spikes	Zhengmai8329, Wanmai50, Zhengmai366, Xian8	High rate of dead stems and many residual spikes	[147]
Physiological traits	Intelligent biochemical incubator	Jointing and booting stage	Taishan 6426	Photosynthesis rate (Pn), Transpiration rate (Tr) and Stomatal conductance (Gs) were decreased, and Internal CO_2 concentration (Ci) increased	Taishan 4033, Jimai22	Pn, Tr and Gs decreased, and Ci were significantly decreased, and Ci increased significantly overall, and Fv/Fm decreased significantly	[152]
Physiological traits	Cryogenic incubator and solar thermal chamber	The young microspore stage	Young	Control the unsaturated lipid levels to maintain membrane fluidity	Wyalkatchem		[13]

Moreover, biomass accumulation is also a significant source of grain yield and a growth process sensitive to cold stress [150]. The SCS has adverse effects on several wheat metrics, including the mean leaf area index (MLAI), mean net assimilation rate (MNAR), harvest index (HI), biomass per plant (BPPM), and grain yield per plant (GYPP) [35]. These metrics can be utilized in wheat breeding programs to assist in developing cold-tolerant varieties.

4.3. Cultivars for SCS Resistance Based on Molecular Traits

It is critical for breeding to understand the physiological features linked to genetic improvements in yield and quality [153]. When SCS harms wheat, a variety of complicated physiological and biochemical changes take place inside the plant that has an impact on yield and quality. Reactive oxygen [154], MDA content, antioxidant enzyme activity [56], carbohydrates [155], osmoregulatory substances [87], hormone content [91], starch content [156], and photosynthesis [157] are often used as physiological and biochemical indicators for wheat inversion identification (Table 2). According to Zhang et al. (2019), the quantity of wheat-bearing grain can be considered to determine POD activity, SOD activity, and MDA level as indices of wheat cold resistance [158]. To determine the extent of freezing damage, Wang et al. (2022) used principal component-affiliate function-stepwise regression analysis to screen seven important physiological indicators: chlorophyll a, leaf water content, proline, Fv/Fm, soluble protein, MDA, and SOD. The equation coefficient of determination between the predicted value of the integrated index of freezing damage and yield established from this reached 0.898 [159]. Following an abrupt temperature drop, it was discovered that in cold-tolerant wheat cultivars, the expression of genes encoding antioxidant enzymes increased, antioxidant enzyme activity was improved, and ROS content was decreased, whereas ROS content was higher, and some leaves died in cold-sensitive wheat cultivars [160].

To enhance wheat tolerance to SCS and improve sustainability, many researchers focus on understanding the key molecular targets, regulatory pathways and signaling designed for genotype–environment interactions [161,162]. As an important research tool for functional genes, transcriptome sequencing has been employed in regulatory network investigations of plants under abiotic stress [163]. In wheat, 450 genes were found to have altered transcript abundance following 14 low-temperature treatments, including 130 candidates for transcription factors, protein kinases, ubiquitin ligases, GTP, RNA, and Ca^{2+} binding proteins genes [164]. Transcriptome sequencing of cold stress during reproductive stages in wheat identified 562 up-regulated, and 314 down-regulated differentially expressed genes, and these genes were mainly involved in photosynthesis, lipid and carbohydrate synthesis, amino acid and protein accumulation [165]. According to transcriptomics and metabolomics analysis, the ABA/JA phytohormone signaling and proline biosynthesis pathways play an important role in regulating cold tolerance in wheat [94]. Transcription is only part of the response; many researchers also employ proteomics for in-depth analysis of protein changes, offering global analysis of protein accumulation [166]. Proteomic analysis has been carried out in wheat under SCS [167], with various proteins being identified as having a role in cold tolerance, providing protection against cold damage [168]. For instance, the proteomic analysis of wheat under low temperatures revealed an upregulation of the expression of proteins involved in signal transduction, carbohydrate metabolism, stress and defense responses, and phenylpropane biosynthesis [169].

5. Conclusions and Future Perspectives

SCS incidents more often occur under changing climatic conditions, causing a serious threat to wheat reproductive tissues and grain production. The SCS is detrimental to the development of the floret and spikelet in wheat; thus, compromising the grain number and quality. A premium cultivar tolerating SCS is a prerequisite for sustaining wheat farming. The review shows that the protection of young, tender spikelet issues in wheat from cold stress impacts was mainly dependent on the collective contribution of antioxidant enzyme activity, carbohydrate accumulation, hormone signaling and transcriptional regulation. The effort of breeding cultivars with simple agronomic and morpho-physiological traits has been made in coping with cold stress, which should be improved by identifying novel SCS-tolerant QTLs or genes with regards to floret and spikelet development in new breeding strategies which embrace fundamental mechanisms. Further studies on multi-omics, from genomics to phenomics, to identify the genes regulating cold tolerance will be necessary for future breeding programs.

Author Contributions: Y.S. conceived this review; H.S., Y.Z., C.T., X.C. and X.L. collected information and drafted this review; H.S., A.J., Y.L., Y.Z. and Y.S. finalized the paper. All authors have read and agreed to the published version of the manuscript.

Funding: The work was supported by grants from the National Key Research and Development Plan Program of China (No. 2017YFD0301307), Graduate Innovation Found of Anhui Agricultural University (2021yjs-2), and the National Natural Science Foundation of China (No. 31901540).

Institutional Review Board Statement: Not applicable.

Informed Consent Statement: Not applicable.

Data Availability Statement: All of the data generated or analyzed during this study are included in this published article.

Acknowledgments: We thank Muhammad A. Hassan for providing valuable suggestions for the article.

Conflicts of Interest: We declare no conflict of interest.

References

1. Wang, G.Y.; Li, X.Y.; Dong, J.; Jin, R.; Yang, L.; Huo, L.H.; Chen, L.M.; Zhao, W.C.; Gao, X. Proteomics unravels new candidate genes of *Dasypyrum villosum* for improving wheat quality. *J. Proteom.* **2021**, *245*, 104292. [CrossRef] [PubMed]
2. Zhong, Y.X.; Yang, M.T.; Cai, J.; Wang, X.; Zhou, Q.; Cao, W.X.; Dai, T.B.; Jiang, D. Nitrogen topdressing timing influences the spatial distribution patterns of protein components and quality traits of flours from different pearling fractions of wheat (*Triticum aestivum* L.) grains. *Field Crops Res.* **2018**, *216*, 120–128. [CrossRef]
3. Nuttall, J.G.; O'Leary, G.J.; Panozzo, J.F.; Waller, C.K.; Barlow, K.M.; Fitzgerald, G.J. Models of grain quality in wheat—A review. *Field Crops Res.* **2017**, *202*, 136–145. [CrossRef]
4. Castroverde, C.D.M.; Dina, D. Temperature regulation of plant hormone signaling during stress and development. *J. Exp. Bot.* **2021**, *72*, 7436–7458. [CrossRef]
5. Kawasaki, K.; Uchida, S. Quality matters more than quantity: Asymmetric temperature effects on crop yield and quality grade. *Am. J. Agric. Econ.* **2016**, *98*, 1195–1209. [CrossRef]
6. Crimp, S.J.; Zheng, B.; Khimashia, N.; Gobbett, D.L.; Chapman, S.; Howden, M.; Nicholls, N. Recent changes in southern Australian frost occurrence: Implications for wheat production risk. *Crop Pasture Sci.* **2016**, *67*, 801–811. [CrossRef]
7. Ferrante, A.; Cullis, B.R.; Smith, A.B.; Able, J.A. A multi-environment trial analysis of frost susceptibility in wheat and barley under australian frost-prone field conditions. *Front. Plant Sci.* **2021**, *12*, 722637. [CrossRef]
8. Xiao, L.J.; Liu, L.L.; Asseng, S.; Xia, Y.M.; Tang, L.; Liu, B.; Cao, W.X.; Zhu, Y. Estimating spring frost and its impact on yield across winter wheat in China. *Agric. For. Meteorol.* **2018**, *260*, 154–164. [CrossRef]
9. Yue, Y.J.; Zhou, Y.; Wang, J.A.; Ye, X.Y. Assessing wheat frost risk with the support of GIS: An approach coupling a growing season meteorological index and a hybrid fuzzy neural network model. *Sustainability* **2016**, *8*, 1308. [CrossRef]
10. Ma, Q.Q.; Huang, J.G.; Hanninen, H.; Berninger, F. Divergent trends in the risk of spring frost damage to trees in Europe with recent warming. *Glob. Change Biol.* **2019**, *25*, 351–360. [CrossRef]
11. Papagiannaki, K.; Lagouvardos, K.; Kotroni, V.; Papagiannakis, G. Agricultural losses related to frost events: Use of the 850 hPa level temperature as an explanatory variable of the damage cost. *Nat. Hazards Earth Syst. Sci.* **2014**, *14*, 2375–2386. [CrossRef]
12. Chen, J.H.; Liu, J.B.; Zhang, X.J.; Chen, S.Q.; Huang, W.; Chen, J.; Zhang, S.R.; Zhou, A.F.; Chen, F.H. Unstable little ice age climate revealed by high-resolution proxy records from northwestern China. *Clim. Dyn.* **2019**, *53*, 1517–1526. [CrossRef]
13. Cheong, B.E.; Onyemaobi, O.; Ho, W.W.H.; Ben, B.T.; Rupasinghe, T.W.T.; Roessner, U.; Dolferus, R. Phenotyping the chilling and freezing responses of young microspore stage wheat spikes using targeted metabolome and lipidome profiling. *Cells* **2020**, *9*, 1309. [CrossRef]
14. Hu, J.; Ren, B.; Dong, S.; Liu, P.; Zhao, B.; Zhang, J. Poor development of spike differentiation triggered by lower photosynthesis and carbon partitioning reduces summer maize yield after waterlogging. *Crop J.* **2021**, *10*, 478–489. [CrossRef]
15. Cheong, B.E.; Ho, W.W.H.; Biddulph, B.; Wallace, X.; Rathjen, T.; Rupasinghe, T.W.T.; Roessner, U.; Dolferus, R. Phenotyping reproductive stage chilling and frost tolerance in wheat using targeted metabolome and lipidome profiling. *Metabolomic* **2019**, *15*, 144. [CrossRef]

16. Thakur, P.; Kumar, S.; Malik, J.A.; Berger, J.D.; Nayyar, H. Cold stress effects on reproductive development in grain crops: An overview. *Environ. Exp. Bot.* **2010**, *67*, 429–443. [CrossRef]
17. Kiran, A.; Kumar, S.; Nayyar, H.; Sharma, K.D. Low temperature-induced aberrations in male and female reproductive organ development cause flower abortion in chickpea. *Plant Cell Environ.* **2019**, *42*, 2075–2089. [CrossRef]
18. Gao, Y.; Zhang, Y.X.; Ma, Q.; Su, S.N.; Li, C.Y.; Ding, J.F.; Zhu, M.; Zhu, X.K.; Guo, W.S. Effects of low temperature in spring on fertility of pollen and formation of grain number in wheat. *Acta Agron. Sin.* **2021**, *47*, 104–115. (In Chinese) [CrossRef]
19. Tang, Z.H.; Zhang, L.P.; Yang, D.; Zhao, C.P.; Zheng, Y.L. Cold stress contributes to aberrant cytokinesis during male meiosis I in a wheat thermosensitive genic male sterile line. *Plant Cell Environ.* **2011**, *34*, 389–405. [CrossRef]
20. Zhang, W.J.; Wang, J.Q.; Huang, Z.L.; Mi, L.; Xu, K.F.; Wu, J.J.; Fan, Y.H.; Ma, S.Y.; Jiang, D.G. Effects of low temperature at booting stage on sucrose metabolism and endogenous hormone contents in winter wheat spikelet. *Front. Plant Sci.* **2019**, *10*, 498. [CrossRef]
21. Cromey, D.W.; Wright, H.W. Effects of frost during grain filling on wheat yield and grain structure. *N. Z. J. Crop Hortic. Sci.* **1998**, *26*, 279–291. [CrossRef]
22. Hassan, M.A.; Xiang, C.; Farooq, M.; Muhammad, N.; Yan, Z.; Hui, X.; Ke, Y.Y.; Bruno, A.K.; Zhang, L.L.; Li, J.C. Cold stress in wheat: Plant acclimation responses and management strategies. *Front. Plant Sci.* **2021**, *12*, 676884. [CrossRef] [PubMed]
23. Ding, Y.L.; Shi, Y.T.; Yang, S.H. Advances and challenges in uncovering cold tolerance regulatory mechanisms in plants. *New Phytol.* **2019**, *222*, 1690–1704. [CrossRef] [PubMed]
24. Dong, C.H.; Zolman, B.K.; Bartel, B.; Lee, B.; Stevenson, B.; Agarwal, M.; Zhu, J.K. Disruption of *Arabidopsis CHY1* reveals an important role of metabolic status in plant cold stress signaling. *Mol. Plant* **2009**, *2*, 59–72. [CrossRef] [PubMed]
25. Kusmec, A.; Srinivasan, S.; Nettleton, D.; Schnable, P.S. Distinct genetic architectures for phenotype means and plasticities in Zea mays. *Nat. Plants* **2017**, *3*, 715–723. [CrossRef]
26. Winfield, M.O.; Lu, C.G.; Wilson, I.D.; Coghill, J.A.; Edwards, K.J. Plant responses to cold: Transcriptome analysis of wheat. *Plant Biotechnol. J.* **2010**, *8*, 749–771. [CrossRef]
27. Kosmala, A.; Bocian, A.; Rapacz, M.; Jurczyk, B.; Zwierzykowski, Z. Identification of leaf proteins differentially accumulated during cold acclimation between Festuca pratensis plants with distinct levels of frost tolerance. *J. Exp. Bot.* **2009**, *60*, 3595–3609. [CrossRef]
28. Zhang, S.; Song, G.; Gao, J.; Li, Y.; Guo, D.; Fan, Q.; Sui, X.; Chu, X.; Huang, C.; Liu, J.; et al. Transcriptome characterization and differential expression analysis of cold-responsive genes in young spikes of common wheat. *J. Biotechnol.* **2014**, *189*, 48–57. [CrossRef]
29. Zhang, Z.; Chapman, S.C.; Christopher, J.T.; Frederiks, T.M.; Chenu, K. Frost trends and their estimated impact on yield in the Australian wheatbelt. *J. Exp. Bot.* **2015**, *66*, 3611–3623. [CrossRef]
30. Zhang, H.; Richards, R.; Riffkin, P.; Berger, J.; Christy, B.; O'Leary, G.; Acuña, T.B.; Merry, A. Wheat grain number and yield: The relative importance of physiological traits and source-sink balance in southern Australia. *Eur. J. Agron.* **2019**, *110*, 125935. [CrossRef]
31. Martino, D.L.; Abbate, P.E. Frost damage on grain number in wheat at different spike developmental stages and its modelling. *Eur. J. Agron.* **2019**, *103*, 13–23. [CrossRef]
32. Powell, N.; Ji, X.M.; Ravash, R.; Edlington, J.; Dolferus, R. Yield stability for cereals in a changing climate. *Funct. Plant Biol.* **2012**, *39*, 539–552. [CrossRef] [PubMed]
33. Frederiks, T.M.; Christopher, J.T.; Harvey, G.L.; Sutherland, M.W.; Borrell, A.K. Current and emerging screening methods to identify post-head-emergence frost adaptation in wheat and barley. *J. Exp. Bot.* **2012**, *63*, 5405–5416. [CrossRef]
34. Liu, L.; Ji, H.; An, J.; Shi, K.; Ma, J.; Liu, B.; Tang, L.; Cao, W.; Zhu, Y. Response of biomass accumulation in wheat to low-temperature stress at jointing and booting stages. *Environ. Exp. Bot.* **2019**, *157*, 46–57. [CrossRef]
35. Ji, H.T.; Xiao, L.J.; Xia, Y.M.; Song, H.; Liu, B.; Tang, L.; Cao, W.X.; Zhu, Y.; Liu, L.L. Effects of jointing and booting low temperature stresses on grain yield and yield components in wheat. *Agric. For. Meteorol.* **2017**, *243*, 33–42. [CrossRef]
36. Li, P.F.; Ma, B.L.; Xiong, Y.C.; Zhang, W.Y. Morphological and physiological responses of different wheat genotypes to chilling stress: A cue to explain yield loss. *J. Sci. Food Agric.* **2017**, *97*, 4036–4045. [CrossRef]
37. Liu, L.L.; Xia, Y.; Liu, B.; Chang, C.; Xiao, L.; Shen, J.; Tang, L.; Cao, W.; Zhu, Y. Individual and combined effects of jointing and booting low-temperature stress on wheat yield. *Eur. J. Agron.* **2020**, *113*, 125989. [CrossRef]
38. Fuller, M.P.; Fuller, A.M.; Kaniouras, S.; Christophers, J.; Fredericks, T. The freezing characteristics of wheat at ear emergence. *Eur. J. Agron.* **2007**, *26*, 435–441. [CrossRef]
39. Livingston, D.P.; Tuong, T.D.; Isleib, T.G.; Murphy, J.P. Differences between wheat genotypes in damage from freezing temperatures during reproductive growth. *Eur. J. Agron.* **2016**, *74*, 164–172. [CrossRef]
40. Liang, Z.M.; Luo, J.; Wei, B.; Liao, Y.C.; Liu, Y. Trehalose can alleviate decreases in grain number per spike caused by low-temperature stress at the booting stage by promoting floret fertility in wheat. *J. Agron. Crop Sci.* **2021**, *207*, 717–732. [CrossRef]
41. Yu, X.; Hao, D.; Yang, J.; Ran, L.; Xiong, F. Effects of low temperature at stem elongation stage on the development, morphology, and physicochemical properties of wheat starch. *PeerJ* **2020**, *8*, e9672. [CrossRef]
42. Li, X.; Cai, J.; Liu, F.; Dai, T.; Cao, W.; Jiang, D. Spring freeze effect on wheat yield is modulated by winter temperature fluctuations: Evidence from meta-analysis and simulating experiment. *J. Agron. Crop Sci.* **2015**, *201*, 288–300. [CrossRef]

43. Zhou, H.; Xia, D.; He, Y. Rice grain quality—traditional traits for high quality rice and health-plus substances. *Mol. Breed.* **2020**, *40*, 1. [CrossRef]
44. Zhao, D.S.; Liu, Q.Q.; Zhang, C.Q.; Li, Q.F. Genetic control of grain appearance quality in rice. *Biotechnol. Adv.* **2022**, *60*, 108014. [CrossRef] [PubMed]
45. Wang, S.L.; Wang, Z.L.; Wang, P.; Wang, H.W.; Wu, Y.G.; Yin, Y.P. Freeze resistance analysis of different wheat cultivars based on the relationships between physiological indices and grain yield. *Chin. J. Ecol.* **2011**, *22*, 1477–1484. (In Chinese)
46. Liu, L.L.; Song, H.; Shi, K.J.; Liu, B.; Zhang, Y.; Tang, L.; Cao, W.X.; Zhu, Y. Response of wheat grain quality to low temperature during jointing and booting stages-On the importance of considering canopy temperature. *Agric. For. Meteorol.* **2019**, *278*, 107658. [CrossRef]
47. Maningat, C.C.; Seib, P.A.; Bassi, S.D.; Woo, K.S.; Lasater, G.D. Chapter 10—Wheat starch: Production, properties, modification and uses. In *Starch*, 3rd ed.; BeMiller, J., Whistler, R., Eds.; Academic Press: San Diego, CA, USA, 2009; pp. 441–510.
48. Gebbing, T.; Schnyder, H. Pre-anthesis reserve utilization for protein and carbohydrate synthesis in grains of wheat. *Plant Physiol.* **1999**, *121*, 871–878. [CrossRef]
49. Zhang, C.; Gu, K.; Gu, D.; Zhang, S.; Wu, J. Quantifying the effect of low-temperature events on the grain quality formation of wheat. *J. Cereal Sci.* **2021**, *100*, 103257. [CrossRef]
50. Wang, Z.B.; Li, W.H.; Qi, J.C.; Shi, P.C.; Yin, Y.A. Starch accumulation, activities of key enzyme and gene expression in starch synthesis of wheat endosperm with different starch contents. *J. Food Sci. Technol.* **2014**, *51*, 419–429. [CrossRef]
51. Sharma, D.; Tiwari, R.; Gupta, V.K.; Rane, J.; Singh, R. Genotype and ambient temperature during growth can determine the quality of starch from wheat. *J. Cereal Sci.* **2018**, *79*, 240–246. [CrossRef]
52. Zhang, W.J.; Zhao, Y.; Li, L.Y.; Xu, X.; Yang, L.; Luo, Z.; Wang, B.B.; Ma, S.Y.; Fan, Y.H.; Huang, Z.L. The effects of short-term exposure to low temperatures during the booting stage on starch synthesis and yields in wheat grain. *Front. Plant Sci.* **2021**, *12*, 684784. [CrossRef] [PubMed]
53. Caverzan, A.; Casassola, A.; Brammer, S.P. Antioxidant responses of wheat plants under stress. *Genet. Mol. Biol.* **2016**, *39*, 1–6. [CrossRef] [PubMed]
54. Dreyer, A.; Dietz, K.J. Reactive oxygen species and the redox-regulatory network in cold stress acclimation. *Antioxidants* **2018**, *7*, 169. [CrossRef] [PubMed]
55. Meena, K.K.; Sorty, A.M.; Bitla, U.M.; Choudhary, K.; Gupta, P.; Pareek, A.; Singh, D.P.; Prabha, R.; Sahu, P.K.; Gupta, V.K.; et al. Abiotic stress responses and microbe-mediated mitigation in plants: The omics strategies. *Front. Plant Sci.* **2017**, *8*, 172. [CrossRef] [PubMed]
56. Li, X.N.; Cai, J.; Liu, F.L.; Dai, T.B.; Cao, W.X.; Jiang, D. Cold priming drives the sub-cellular antioxidant systems to protect photosynthetic electron transport against subsequent low temperature stress in winter wheat. *Plant Physiol. Biochem.* **2014**, *82*, 34–43. [CrossRef]
57. Liu, Y.H.; Li, J.J.; Zhu, Y.L.; Jones, A.; Rose, R.J.; Song, Y.H. Heat stress in legume seed setting: Effects, causes, and future prospects. *Front. Plant Sci.* **2019**, *10*, 938. [CrossRef]
58. Pasquariello, M.S.; Di Patre, D.; Mastrobuoni, F.; Zampella, L.; Scortichini, M.; Petriccione, M. Influence of postharvest chitosan treatment on enzymatic browning and antioxidant enzyme activity in sweet cherry fruit. *Postharvest Biol. Technol.* **2015**, *109*, 45–56. [CrossRef]
59. Bao, Y.; Xing, J.; Liang, Y.; Ren, Z.; Fu, L.; Yu, J.; Wang, D.; Zhang, D.; Xu, Q.; Jing, C. Analysis of overwintering indexes of winter wheat in alpine regions and establishment of a cold resistance model. *Field Crops Res.* **2022**, *275*, 108347. [CrossRef]
60. Turk, H.; Erdal, S.; Genisel, M.; Atici, O.; Demir, Y.; Yanmis, D. The regulatory effect of melatonin on physiological, biochemical and molecular parameters in cold-stressed wheat seedlings. *Plant Growth Regul.* **2014**, *74*, 139–152. [CrossRef]
61. Jiang, G.; Hassan, M.A.; Muhammad, N.; Arshad, M.; Chen, X.; Xu, Y.; Xu, X.; Ni, Q.; Liu, B.; Yang, W.K.; et al. Comparative physiology and transcriptome analysis of young spikes in response to late spring coldness in wheat (*Triticum aestivum* L.). *Front. Plant Sci.* **2022**, *13*, 811884. [CrossRef]
62. Zita, K.; Simon-Sarkadi, L.; Csongor, S.; Klára, K.; Gábor, C.; Gábor, K. Differential effects of cold acclimation and abscisic acid on free amino acid composition in wheat. *Plant Sci.* **2011**, *180*, 61–68.
63. Liu, K.; Wang, L.; Xu, Y.; Chen, N.; Ma, Q.; Li, F.; Chong, K. Overexpression of *OsCOIN*, a putative cold inducible zinc finger protein, increased tolerance to chilling, salt and drought, and enhanced proline level in rice. *Planta* **2007**, *226*, 1007–1016. [CrossRef] [PubMed]
64. Wang, W.L.; Wang, X.; Lv, Z.S.; Khanzada, A.; Huang, M.; Cai, J.; Zhou, Q.; Huo, Z.Y.; Jiang, D. Effects of cold and salicylic acid priming on free proline and sucrose accumulation in winter wheat under freezing stress. *J. Plant Growth Regul.* **2022**, *41*, 2171–2184. [CrossRef]
65. Hasanuzzaman, H.; Bhuyan, M.H.M.B.; Anee, T.I.; Parvin, K.; Nahar, K.; Al-Mahmud, J.; Fujita, M. Regulation of ascorbate-glutathione pathway in mitigating oxidative damage in plants under abiotic stress. *Antioxidants* **2019**, *9*, 384. [CrossRef] [PubMed]
66. Traiq, S.; Sadia, L.; Hamad, K.; Fazal, M.; Nie, L.X. Ascorbic acid priming enhances seed germination and seedling growth of winter wheat under low temperature due to late sowing in Pakistan. *Agronomy* **2019**, *9*, 757.
67. Devireddy, A.R.; Tschaplinski, T.J.; Tuskan, G.A.; Muchero, W.; Chen, J.G. Role of reactive oxygen species and hormones in plant responses to temperature changes. *Int. J. Mol. Sci.* **2021**, *22*, 8843. [CrossRef]

68. Chinnusamy, V.; Zhu, J.K.; Zhu, J.K. Cold stress regulation of gene expression in plants. *Trends Plant Sci.* **2007**, *12*, 444–451. [CrossRef]
69. Lv, B.; Tian, H.; Zhang, F.; Liu, J.; Lu, S.; Bai, M.; Li, C.; Ding, Z. Brassinosteroids regulate root growth by controlling reactive oxygen species homeostasis and dual effect on ethylene synthesis in *Arabidopsis*. *PLoS Genet.* **2018**, *14*, e1007144. [CrossRef]
70. Fang, P.; Yan, M.; Chi, C.; Wang, M.; Zhou, Y.; Zhou, J.; Shi, K.; Xia, X.; Foyer, C.H.; Yu, J. Brassinosteroids act as a positive regulator of photoprotection in response to chilling stress. *Plant Physiol.* **2019**, *180*, 2061–2076. [CrossRef]
71. Liu, S.; Li, Z.; Wu, S.; Wan, X. The essential roles of sugar metabolism for pollen development and male fertility in plants. *Crop J.* **2021**, *9*, 1223–1236. [CrossRef]
72. Furtauer, L.; Weiszmann, J.; Weckwerth, W.; Nagele, T. Dynamics of plant metabolism during cold acclimation. *Int. J. Mol. Sci.* **2019**, *20*, 5411. [CrossRef] [PubMed]
73. Sami, F.; Yusuf, M.; Faizan, M.; Faraz, A.; Hayat, S. Role of sugars under abiotic stress. *Plant Physiol. Biol.* **2016**, *109*, 54–61. [CrossRef] [PubMed]
74. Zulfiqar, F.; Akram, N.A.; Ashraf, M. Osmoprotection in plants under abiotic stresses: New insights into a classical phenomenon. *Planta* **2020**, *251*, 3. [CrossRef] [PubMed]
75. Saddhe, A.A.; Manuka, R.; Penna, S. Plant sugars: Homeostasis and transport under abiotic stress in plants. *Physiol. Plant.* **2021**, *171*, 739–755. [CrossRef]
76. Salmon, Y.; Lintunen, A.; Dayet, A.; Chan, T.; Dewar, R.; Vesala, T.; Holtta, T. Leaf carbon and water status control stomatal and nonstomatal limitations of photosynthesis in trees. *New Phytol.* **2020**, *226*, 690–703. [CrossRef]
77. Krasensky, J.; Jonak, C. Drought, salt, and temperature stress-induced metabolic rearrangements and regulatory networks. *J. Exp. Bot.* **2012**, *63*, 1593–1608. [CrossRef]
78. Yokota, H.; Iehisa, J.C.; Shimosaka, E.; Takumi, S. Line differences in Cor/Lea and fructan biosynthesis-related gene transcript accumulation are related to distinct freezing tolerance levels in synthetic wheat hexaploids. *J. Plant Physiol.* **2015**, *176*, 78–88. [CrossRef]
79. Gupta, P.; Deswal, R. Antifreeze proteins enable plants to survive in freezing conditions. *J. Biosci.* **2014**, *39*, 931–944. [CrossRef]
80. Jeena, G.S.; Kumar, S.; Shukla, R.K. Structure, evolution and diverse physiological roles of sweet sugar transporters in plants. *Plant Mol. Biol.* **2019**, *100*, 351–365. [CrossRef]
81. Qin, J.X.; Jiang, Y.J.; Lu, Y.Z.; Zhao, P.; Wu, B.J.; Li, H.X.; Wang, Y.; Xu, S.B.; Sun, Q.X.; Liu, Z.S. Genome-wide identification and transcriptome profiling reveal great expansion of sweet gene family and their wide-spread responses to abiotic stress in wheat (*Triticum aestivum* L.). *J. Integr. Agric.* **2020**, *19*, 1704–1720. [CrossRef]
82. Klemens, P.A.W.; Patzke, K.; Deitmer, J.; Spinner, L.; Le, H.R.; Bellini, C.; Bedu, M.; Chardon, F.; Krapp, A.; Neuhaus, H.E. Overexpression of the vacuolar sugar carrier AtSWEET16 modifies germination, growth, and stress tolerance in *Arabidopsis*. *Plant Physiol.* **2013**, *163*, 1338–1352. [CrossRef] [PubMed]
83. Chardon, F.; Bedu, M.; Calenge, F.; Klemens, P.A.W.; Spinner, L.; Clement, G.; Chietera, G.; Leran, S.; Ferrand, M.; Lacombe, B.; et al. Leaf fructose content is controlled by the vacuolar transporter *SWEET17* in *Arabidopsis*. *Curr. Biol.* **2013**, *23*, 697–702. [CrossRef] [PubMed]
84. Rocha, F.R.; Papini-Terzi, F.S.; Nishiyama, M.Y.; Vencio, R.Z.N.; Vicentini, R.; Duarte, R.D.C.; Rosa, V.E.; Vinagre, F.; Barsalobres, C.; Medeiros, A.H.; et al. Signal transduction-related responses to phytohormones and environmental challenges in sugarcane. *BMC Genom.* **2007**, *8*, 71. [CrossRef] [PubMed]
85. Almadanim, M.C.; Alexandre, B.M.; Rosa, M.T.G.; Sapeta, H.; Leitão, A.E.; Ramalho, J.C.; Lam, T.T.; Negrão, S.; Abreu, I.A.; Oliveira, M.M. Rice calcium-dependent protein kinase OsCPK17 targets plasma membrane intrinsic protein and sucrose-phosphate synthase and is required for a proper cold stress response. *Plant Cell Environ.* **2017**, *40*, 1197–1213. [CrossRef]
86. Wilkinson, S.; Kudoyarova, G.R.; Veselov, D.S.; Arkhipova, T.N.; Davies, W.J. Plant hormone interactions: Innovative targets for crop breeding and management. *J. Exp. Bot.* **2012**, *63*, 3499–3509. [CrossRef]
87. An, J.P.; Wang, X.F.; Zhang, X.W.; You, C.X.; Hao, Y.J. Apple B-box protein BBX37 regulates jasmonic acid mediated cold tolerance through the JAZ-BBX37-ICE1-CBF pathway and undergoes MIEL1-mediated ubiquitination and degradation. *New Phytol.* **2021**, *229*, 2707–2729. [CrossRef]
88. Zhao, Y.; Song, C.; Brummell, D.A.; Shuning, Q.I.; Duan, Y. Salicylic acid treatment mitigates chilling injury in peach fruit by regulation of sucrose metabolism and soluble sugar content. *Food Chem.* **2021**, *358*, 129867. [CrossRef] [PubMed]
89. Huang, X.B.; Shi, H.Y.; Hu, Z.R.; Liu, A.; Amombo, E.; Chen, L.; Fu, J.M. ABA is involved in regulation of cold stress response in bermudagrass. *Front. Plant Sci.* **2017**, *8*, 1613. [CrossRef]
90. Zhang, N.; Li, L.J.; Zhang, L.R.; Li, J.; Fang, Y.H.; Zhao, L.; Ren, Y.; Chen, F. Abscisic acid enhances tolerance to spring freeze stress and regulates the expression of ascorbate-glutathione biosynthesis-related genes and stress-responsive genes in common wheat. *Mol. Breed.* **2020**, *40*, 108. [CrossRef]
91. Yu, J.; Cang, J.; Lu, Q.W.; Fan, B.; Xu, Q.H.; Li, W.N.; Wang, X.T. ABA enhanced cold tolerance of wheat 'dn1' via increasing ROS scavenging system. *Plant Signal. Behav.* **2020**, *15*, 1780403. [CrossRef]
92. Nakashima, K.; Yamaguchi-Shinozaki, K. ABA signaling in stress-response and seed development. *Plant Cell Rep.* **2013**, *32*, 959–970. [CrossRef] [PubMed]
93. Ghorbel, M.; Brini, F.; Sharma, A.; Landi, M. Role of jasmonic acid in plants: The molecular point of view. *Plant Cell Rep.* **2021**, *40*, 1471–1494. [CrossRef] [PubMed]

94. Zhao, Y.; Zhou, M.; Xu, K.; Li, J.H.; Li, S.S.; Zhang, S.H.; Yang, X.J. Integrated transcriptomics and metabolomics analyses provide insights into cold stress response in wheat. *Crop J.* **2019**, *7*, 857–866. [CrossRef]
95. Zeng, X.; Jiang, J.; Wang, F.; Liu, W.; Zhang, S.; Du, J.; Yang, C. Rice *OsClo5*, a caleosin protein, negatively regulates cold tolerance through the jasmonate signalling pathway. *Plant Biol.* **2022**, *24*, 52–61. [CrossRef] [PubMed]
96. Wei, X.S.; Liu, S.; Sun, C.; Xie, G.S.; Wang, L.Q. Convergence and divergence: Signal perception and transduction mechanisms of cold stress in *Arabidopsis* and rice. *Plants* **2021**, *10*, 1864. [CrossRef] [PubMed]
97. Hu, Y.R.; Jiang, L.Q.; Wang, F.; Yu, D.Q. Jasmonate regulates the inducer of CBF expression-C-repeat binding factor/DRE binding factor1 cascade and freezing tolerance in *Arabidopsis*. *Plant Cell* **2013**, *25*, 2907–2924. [CrossRef] [PubMed]
98. Devi, J.; Bhatia, S.; Alam, M.S.; Dhillon, T.S. Effect of calcium and salicylic acid on quality retention in relation to antioxidative enzymes in radish stored under refrigerated conditions. *J. Food Sci. Technol.* **2018**, *55*, 1116–1126. [CrossRef]
99. Liu, J.L.; Qiu, G.Y.; Liu, C.; Li, H.; Chen, X.D.; Fu, Q.L.; Lin, Y.C.; Guo, B. Salicylic acid, a multifaceted hormone, combats abiotic stresses in plants. *Life* **2022**, *12*, 886. [CrossRef]
100. Ignatenko, A.; Talanova, V.; Repkina, N.; Titov, A. Exogenous salicylic acid treatment induces cold tolerance in wheat through promotion of antioxidant enzyme activity and proline accumulation. *Acta Physiol. Plant* **2019**, *41*, 80. [CrossRef]
101. Wang, W.L.; Wang, X.; Zhang, J.; Huang, M.; Cai, J.; Zhou, Q.; Dai, T.B.; Jiang, D. Salicylic acid and cold priming induce late-spring freezing tolerance by maintaining cellular redox homeostasis and protecting photosynthetic apparatus in wheat. *Plant Growth Regul.* **2019**, *91*, 109–121. [CrossRef]
102. Yuan, P.G.; Yang, T.B.; Poovaiah, B.W. Calcium signaling-mediated plant response to cold stress. *Int. J. Mol. Sci.* **2018**, *19*, 3896. [CrossRef] [PubMed]
103. Zipfel, C.; Oldroyd, G.E.D. Plant signalling in symbiosis and immunity. *Nature* **2017**, *543*, 328–336. [CrossRef] [PubMed]
104. Yuan, P.G.; Du, L.Q.; Poovaiah, B.W. Ca^{2+}/Calmodulin-Dependent *AtSR1/CAMTA3* plays critical roles in balancing plant growth and immunity. *Int. J. Mol. Sci.* **2018**, *19*, 1764. [CrossRef] [PubMed]
105. Mori, K.; Renhu, N.; Naito, M.; Nakamura, A.; Miura, K. Ca^{2+}-permeable mechanosensitive channels MCA1 and MCA2 mediate cold-induced cytosolic Ca^{2+} increase and cold tolerance in *Arabidopsis*. *Sci. Rep.* **2018**, *8*, 550. [CrossRef] [PubMed]
106. Lv, X.Z.; Li, H.Z.; Chen, X.X.; Xiang, X.; Guo, Z.X.; Yu, J.Q.; Zhou, Y.H. The role of calcium-dependent protein kinase in hydrogen peroxide, nitric oxide and ABA-dependent cold acclimation. *J. Exp. Bot.* **2018**, *69*, 4127–4139. [CrossRef]
107. Liu, Y.W.; Chen, W.Y.; Liu, L.B.; Su, Y.H.; Li, Y.; Jia, W.Z.; Jiao, B.; Wang, J.; Yang, F.; Dong, F.S.; et al. Genome-wide identification and expression analysis of calmodulin and calmodulin-like genes in wheat (*Triticum aestivum L.*). *Plant Signal. Behav.* **2022**, *17*, 2013646. [CrossRef]
108. Tweneboah, S.; Oh, S.-K. Biological roles of NAC transcription factors in the regulation of biotic and abiotic stress responses in solanaceous crops. *J. Plant Biotechnol.* **2017**, *44*, 248. [CrossRef]
109. Liang, Y.; Xia, J.Q.; Jiang, Y.S.; Bao, Y.Z.; Chen, H.C.; Wang, D.J.; Zhang, D.; Yu, J.; Cang, J. Genome-wide identification and analysis of bZIP gene family and resistance of *TaABI5 (TabZIP96)* under freezing stress in wheat (*Triticum aestivum L.*). *Int. J. Mol. Sci.* **2022**, *23*, 2351. [CrossRef]
110. Han, Z.L.; Shang, X.W.; Shao, L.X.; Wang, Y.; Zhu, X.J.; Fang, W.P.; Ma, Y.C. Meta-analysis of the effect of expression of MYB transcription factor genes on abiotic stress. *PeerJ* **2021**, *9*, e11268. [CrossRef]
111. Zuo, Z.F.; Kang, H.G.; Park, M.Y.; Jeong, H.; Sun, H.J.; Song, P.S.; Lee, H.Y. Zoysia japonica MYC type transcription factor *ZjICE1* regulates cold tolerance in transgenic *Arabidopsis*. *Plant Sci.* **2019**, *289*, 110254. [CrossRef]
112. Hwarari, D.; Guan, Y.L.; Ahmad, B.; Movahedi, A.L.; Min, T.; Hao, Z.D.; Lu, Y.; Chen, J.H.; Yang, L.M. CE-CBF-COR signaling cascade and its regulation in plants responding to cold stress. *Int. J. Mol. Sci.* **2022**, *23*, 1549. [CrossRef] [PubMed]
113. Li, X.; Liu, C.; Zhao, Z.; Ma, D.; Zhang, J.; Yang, Y.; Liu, Y.; Liu, H. *COR27* and *COR28* are novel regulators of the COP1-HY5 regulatory hub and photomorphogenesis in *Arabidopsis*. *Plant Cell* **2020**, *32*, 3139–3154. [CrossRef] [PubMed]
114. Guo, J.; Ren, Y.K.; Tang, Z.H.; Shi, W.P.; Zhou, M.X. Characterization and expression profiling of the ICE-CBF-COR genes in wheat. *PeerJ* **2019**, *7*, e8190. [CrossRef]
115. Soltesz, A.; Smedley, M.; Vashegyi, I.; Galiba, G.; Harwood, W.; Vagujfalvi, A. Transgenic barley lines prove the involvement of *TaCBF14* and *TaCBF15* in the cold acclimation process and in frost tolerance. *J. Exp. Bot.* **2013**, *64*, 1849–1862. [CrossRef] [PubMed]
116. Babben, S.; Schliephake, E.; Janitza, P.; Berner, T.; Keilwagen, J.; Koch, M.; Arana-Ceballos, F.A.; Templer, S.E.; Chesnokov, Y.; Pshenichnikova, T.; et al. Association genetics studies on frost tolerance in wheat (*Triticum aestivum* L.) reveal new highly conserved amino acid substitutions in CBF-A3, CBF-A15, VRN3 and PPD1 genes. *BMC Genom.* **2018**, *19*, 409. [CrossRef] [PubMed]
117. Takumi, S.; Koike, A.; Nakata, M.; Kume, S.; Ohno, R.; Nakamura, C. Cold-specific and light-stimulated expression of a wheat (*Triticum aestivum L.*) Cor gene *Wcor15* encoding a chloroplast-targeted protein. *J. Exp. Bot.* **2003**, *54*, 2265–2274. [CrossRef]
118. Vagujfalvi, A.; Galiba, G.; Cattivelli, L.; Dubcovsky, J. The cold-regulated transcriptional activator Cbf3 is linked to the frost-tolerance locus *Fr-A2* on wheat chromosome 5A. *Mol. Genet. Genom.* **2003**, *269*, 60–67. [CrossRef]
119. Luo, Q.J.; Mittal, A.; Jia, F.; Rock, C.D. An autoregulatory feedback loop involving PAP1 and TAS4 in response to sugars in *Arabidopsis*. *Plant Mol. Biol.* **2012**, *80*, 117–129. [CrossRef]
120. Kim, S.; Kang, J.Y.; Cho, D.I.; Park, J.H.; Kim, S.Y. ABF2, an ABRE-binding bZIP factor, is an essential component of glucose signaling and its overexpression affects multiple stress tolerance. *Plant J.* **2004**, *40*, 75–87. [CrossRef]

121. Li, X.Y.; Gao, S.Q.; Tang, Y.M.; Li, L.; Zhang, F.J.; Feng, B.N.; Fang, Z.F.; Ma, L.J.; Zhao, C.P. Genome-wide identification and evolutionary analyses of bZIP transcription factors in wheat and its relatives and expression profiles of anther development related *TabZIP* genes. *BMC Genom.* **2015**, *16*, 976. [CrossRef]
122. Cai, W.T.; Yang, Y.L.; Wang, W.W.; Guo, G.Y.; Liu, W.; Bi, C.L. Overexpression of a wheat (*Triticum aestivum* L.) bZIP transcription factor gene, *TabZIP6*, decreased the freezing tolerance of transgenic *Arabidopsis* seedlings by down-regulating the expression of CBFs. *Plant Physiol. Biochem.* **2018**, *124*, 100–111. [CrossRef] [PubMed]
123. Yu, X.R.; Jiang, Y.F.; Yao, H.H.; Ran, L.P.; Zang, Y.; Xiong, F. Cytological and molecular characteristics of delayed spike development in wheat under low temperature in early spring. *Crop J.* **2022**, *10*, 840–852. [CrossRef]
124. Chen, J.; Gong, Y.; Gao, Y.; Zhou, Y.B.; Chen, M.; Xu, Z.S.; Guo, C.H.; Ma, Y.Z. *TaNAC48* positively regulates drought tolerance and ABA responses in wheat (*Triticum aestivum L.*). *Crop J.* **2021**, *9*, 785–793. [CrossRef]
125. Mondal, S.; Sallam, A.; Sehgal, D.; Biswal, A.; Nehela, Y. Advances in breeding for abiotic stress tolerance in wheat. In *Genomic Designing for Abiotic Stress Resistant Cereal Crops*; Springer: Cham, Switzerland, 2021; pp. 71–103.
126. Jha, U.C.; Bohra, A.; Jha, R. Breeding approaches and genomics technologies to increase crop yield under low-temperature stress. *Plant Cell Rep.* **2017**, *36*, 1–35. [CrossRef]
127. Mickelbart, M.V.; Hasegawa, P.M.; Bailey-Serres, J. Genetic mechanisms of abiotic stress tolerance that translate to crop yield stability. *Nat. Rev. Genet.* **2015**, *16*, 237–251. [CrossRef]
128. Zhu, J.; Pearce, S.; Burke, A.; See, D.R.; Skinner, D.Z.; Dubcovsky, J.; Garland-Campbell, K. Copy number and haplotype variation at the *VRN-A1* and central *FR-A2* loci are associated with frost tolerance in hexaploid wheat. *Theor. Appl. Genet.* **2014**, *127*, 1183–1197. [CrossRef]
129. Soleimani, B.; Lehnert, H.; Babben, S.; Keilwagen, J.; Koch, M.; Arana-Ceballos, F.A.; Chesnokov, Y.; Pshenichnikova, T.; Schondelmaier, J.; Ordon, F.; et al. Genome wide association study of frost tolerance in wheat. *Sci. Rep.* **2022**, *12*, 5275. [CrossRef]
130. Li, Q.; Byrns, B.; Badawi, M.A.; Diallo, A.B.; Danyluk, J.; Sarhan, F.; Laudencia-Chingcuanco, D.; Zou, J.T.; Fowler, D.B. Transcriptomic insights into phenological development and cold tolerance of wheat grown in the field. *Plant Physiol.* **2018**, *176*, 2376–2394. [CrossRef]
131. Sthapit, B.R.; Witcombe, J.R. Inheritance of tolerance to chilling stress in rice during germination and plumule greening. *Crop Sci.* **1998**, *38*, 660–665. [CrossRef]
132. Toth, B.; Galiba, G.; Feher, E.; Sutka, J.; Snape, J.W. Mapping genes affecting flowering time and frost resistance on chromosome 5B of wheat. *Theor. Appl. Genet.* **2003**, *107*, 509–514. [CrossRef]
133. Dhillon, T.; Pearce, S.P.; Stockinger, E.J.; Distelfeld, A.; Li, C.X.; Knox, A.K.; Vashegyi, I.; Vagujfalvi, A.; Galiba, G.; Dubcovsky, J. Regulation of freezing tolerance and flowering in temperate cereals: The VRN-1 connection. *Plant Physiol.* **2010**, *153*, 1846–1858. [CrossRef] [PubMed]
134. Chen, A.; Dubcovsky, J. Wheat tilling mutants show that the vernalization gene vrn1 down-regulates the flowering repressor vrn2 in leaves but is not essential for flowering. *PLoS Genet.* **2012**, *8*, e1003134. [CrossRef] [PubMed]
135. Knox, A.K.; Li, C.X.; Vágújfalvi, A.; Galiba, G.; Stockinger, E.J.; Dubcovsky, J. Identification of candidate CBF genes for the frost tolerance locus *Fr-A^{m}2* in *Triticum monococcum*. *Plant Mol. Biol.* **2008**, *67*, 257–270. [CrossRef] [PubMed]
136. Sieber, A.N.; Longin, C.F.H.; Leiser, W.L.; Wurschum, T. Copy number variation of CBF-A14 at the *Fr-A2* locus determines frost tolerance in winter durum wheat. *Theor. Appl. Genet.* **2016**, *129*, 1087–1097. [CrossRef]
137. Pang, Y.L.; Wu, Y.Y.; Liu, C.X.; Li, W.H.; St Amand, P.; Bernardo, A.; Wang, D.F.; Dong, L.; Yuan, X.F.; Zhang, H.R.; et al. High-resolution genome-wide association study and genomic prediction for disease resistance and cold tolerance in wheat. *Theor. Appl. Genet.* **2021**, *134*, 2857–2873. [CrossRef]
138. Xue, H.; Yu., K.; Ma, X.D.; Liu, X.D.; Liu, Y.H.; Zhu, B.L.; Liu, D.C.; Zhamg, A.M.; Zhan, K.H. Evaluation and correlation analysis of wheat cultivars in Huanghuai wheat area that are tolerant to spring cold. *J. Triticeae Crops* **2018**, *38*, 1174–1188. (In Chinese)
139. Bhat, M.A.; Mir, R.A.; Kumar, V.; Shah, A.A.; Zargar, S.M.; Rahman, S.; Jan, A.T. Mechanistic insights of CRISPR/Cas-mediated genome editing towards enhancing abiotic stress tolerance in plants. *Physiol. Plant.* **2022**, *172*, 1255–1268. [CrossRef]
140. Kim, D.; Alptekin, B.; Budak, H. CRISPR/Cas9 genome editing in wheat. *Funct. Integr. Genom.* **2018**, *18*, 31–41. [CrossRef] [PubMed]
141. Tian, S.J.; Mao, X.G.; Zhang, H.Y.; Chen, S.S.; Zhai, C.C.; Yang, S.M.; Jing, R.L. Cloning and characterization of *TaSnRK*2.3, a novel SnRK2 gene in common wheat. *J. Exp. Bot.* **2013**, *64*, 2063–2080. [CrossRef]
142. Peng, K.K.; Tian, Y.; Cang, J.; Yu, J.; Wang, D.J.; He, F.X.; Jiao, H.R.; Tan, Y.G. Overexpression of *TaFBA-A10* from winter wheat enhances freezing tolerance in *Arabidopsis thaliana*. *J. Plant Growth Regul.* **2022**, *41*, 314–326. [CrossRef]
143. Nkhathutsheleni, M.T.; Jacob, M.; Hussein, S.; Alfred, O. Agronomic and physiological traits, and associated quantitative trait loci (qtl) affecting yield response in wheat (*Triticum aestivum* L.): A review. *Front. Plant Sci.* **2019**, *10*, 1428.
144. Gao, F.M.; Ma, D.Y.; Yin, G.H.; Rasheed, A.; Dong, Y.; Xiao, Y.G.; Xia, X.C.; Wu, X.X.; He, Z.H. Genetic progress in grain yield and physiological traits in chinese wheat cultivars of southern yellow and huai valley since 1950. *Crop Sci.* **2017**, *57*, 760–773. [CrossRef]
145. Manasa, S.L.; Panigrahy, M.; Panigrahi, K.C.S.; Rout, G.R. Overview of cold stress regulation in plants. *Bot. Rev.* **2022**, *88*, 359–387. [CrossRef]
146. Venzhik, Y.V.; Titov, A.F.; Koteyeva, N.K.; Miroslavov, E.A.; Talanova, V.V. The effect of low temperature on wheat roots causes rapid changes in chloroplast ultrastructure in wheat leaves. In *Doklady Biological Sciences: Proceedings of the Academy of Sciences of the USSR, Biological Sciences Sections*; Springer: Cham, Switzerland, 2012; Volume 445, pp. 230–234.

147. Hu, X.X.; Huang, J.Y.; Xiao, S.J. Influence of late frost on different wheat cultivars the first report of investigation on late frost injury to wheat in 1998. *Chin. J. Agric.* **2001**, *20*, 28–30. (In Chinese)
148. Ou, X.Q.; Wang, Y. Preliminary study on wheat breeding for late spring coldness tolerance in south of Huanghuai region. *J. Triticeae Crops* **2019**, *39*, 560–566. (In Chinese)
149. Valluru, R.; Link, J.; Claupein, W. Consequence of early chilling stress in two Triticum sprcirs: Plastic responses and adaptive significance. *Plant Biol.* **2012**, *14*, 641–651. [CrossRef]
150. Sassenrath, G.F.; Ort, D.R. The relationship between inhibition of photosynthesis at low temperature and the inhibition of photosynthesis after rewarming in chill-sensitive tomato. *Plant Physiol. Biochem.* **1990**, *28*, 457–465.
151. Zheng, J.C.; Liu, T.; Zheng, Q.X.; Li, J.Q.; Qian, Y.C.; Li, J.C.; Zhan, Q.W. Identification of cold tolerance and analysis of genetic diversity for major wheat cultivars in jiang huai region of China. *Pak. J. Bot.* **2020**, *52*, 839–849. [CrossRef]
152. Wang, R.X.; Yan, C.S.; Zhang, X.Y.; Sun, G.Z.; Qian, Z.G.; Qi, X.L.; Mou, Q.H.; Xiao, S.H. Effects of spring low temperature on yield and photosynthetic characteristics of wheat. *Acta Agron. Sin.* **2018**, *44*, 288–296. (In Chinese) [CrossRef]
153. Zhang, Y.; Xu, W.; Wang, W.; Dong, H.; Qi, X.; Zhao, M.; Fang, Y.H.; Gao, C.; Hu, L. Progress in genetic improvement of grain yield and related physiological traits of Chinese wheat in Henan Province. *Field Crops Res.* **2016**, *199*, 117–128. [CrossRef]
154. Al-Issawi, M.; Rihan, H.Z.; Al-Shmgani, H.; Fuller, M.P. Molybdenum application enhances antioxidant enzyme activity and COR15a protein expression under cold stress in wheat. *J. Plant Interact.* **2016**, *11*, 5–10. [CrossRef]
155. Shahryar, N.; Maali-Amiri, R. Metabolic acclimation of tetraploid and hexaploid wheats by cold stress-induced carbohydrate accumulation. *J. Plant Physiol.* **2016**, *204*, 44–53. [CrossRef] [PubMed]
156. Shi, K.J.; Yin, T.W.; Zhu, Y.; Liu, B.; Tang, L.; Cao, W.X.; Liu, L.L. Estimating the effect of low-temperature stress on the spatial distribution patterns of protein in wheat grains. *J. Cereal Sci.* **2022**, *105*, 103461. [CrossRef]
157. Venzhik, Y.; Talanova, V.; Titov, A. The effect of abscisic acid on cold tolerance and chloroplasts ultrastructure in wheat under optimal and cold stress conditions. *Acta Physiol. Plant.* **2016**, *38*, 63. [CrossRef]
158. Zhang, Z.Y.; Wang, Z.; Wang, B.; Wang, Z.W.; Zhu, Q.D.; Huo, Y.F.; Ru, Z.G.; Liu, M.J. Effects of low temperature treatment on grain yield and physiological characteristics of young ear in different wheat cultivars. *North China Agric. J.* **2019**, *34*, 130–139. (In Chinese)
159. Wang, Y.Y.; Liu, W.; He, L.; Ren, D.C.; Duan, J.Z.; Hu, X.; Guo, T.C.; Wang, Y.H.; Feng, W. Evaluation of freezing damage and water effect of wheat based on multivariate statistical analysis. *Sci. Agricul. Sin.* **2022**, *55*, 1301–1318. (In Chinese)
160. Xu, J.; Li, Y.; Sun, J.; Du, L.; Zhang, Y.; Yu, Q.; Liu, X. Comparative physiological and proteomic response to abrupt low temperature stress between two winter wheat cultivars differing in low temperature tolerance. *Plant Biol.* **2013**, *15*, 292–303. [CrossRef] [PubMed]
161. Bhat, K.A.; Mahajan, R.; Rakhtoon, M.M.; Urwat, U.; Bashir, Z.; Shah, A.A.; Agrawal, A.; Bhat, B.; Sofi, P.A.; Masi, A.; et al. Low temperature stress tolerance: An insight into the omics approaches for legume crops. *Front. Plant Sci.* **2022**, *13*, 888710. [CrossRef]
162. Parida, A. Metabolomics-guided elucidation of abiotic stress tolerance mechanisms in plants. In *Plant Metabolites and Regulation Under Environmental Stress*; Academic Press: New York, NY, USA, 2018; pp. 89–131.
163. Chen, R.; Wei, Q.; Liu, Y.; Wei, X.; Xie, T. Transcriptome sequencing and functional characterization of new sesquiterpene synthases from Curcuma wenyujin. *Arch. Biochem. Biophys.* **2021**, *709*, 108986. [CrossRef]
164. Monroy, A.F.; Dryanova, A.; Malette, B.; Oren, D.H.; Farajalla, M.R.; Liu, W.; Danyluk, J.; Ubayasena, L.W.C.; Kane, K.; Scoles, G.J. Regulatory gene candidates and gene expression analysis of cold acclimation in winter and spring wheat. *Plant Mol. Biol.* **2007**, *64*, 409–423. [CrossRef]
165. Diaz, M.L.; Soresi, D.S.; Basualdo, J.; Cuppari, S.J.; Carrera, A. Transcriptomic response of durum wheat to cold stress at reproductive stage. *Mol. Biol. Rep.* **2019**, *46*, 2427–2445. [CrossRef]
166. Jan, N.; Rather, A.; John, R.; Chaturvedi, P.; Ghatak, A.; Weckwerth, W.; Zargar, S.M.; Mir, R.A.; Khan, M.A.; Mir, R.R. Proteomics for abiotic stresses in legumes: Present status and future directions. *Crit. Rev. Biotechnol.* **2022**, 1–20. [CrossRef]
167. Zhang, N.; Zhang, L.R.; Zhao, L.; Ren, Y.; Cui, D.Q.; Chen, J.H.; Wang, Y.Y.; Yu, P.B.; Chen, F. iTRAQ and virus-induced gene silencing revealed three proteins involved in cold response in bread wheat. *Sci. Rep.* **2017**, *7*, 7524. [CrossRef] [PubMed]
168. Hao, Y.C.; Hao, M.; Cui, Y.J.; Kong, L.R.; Wang, H.W. Genome-wide survey of the dehydrin genes in bread wheat (*Triticum aestivum* L.) and its relatives: Identification, evolution and expression profiling under various abiotic stresses. *BMC Genom.* **2022**, *23*, 73. [CrossRef] [PubMed]
169. Xu, K.; Zhao, Y.; Gu, J.; Zhou, M.; Gao, L.; Sun, R.X.; Wang, W.W.; Zhang, S.H.; Yang, X.J. Proteomic analysis reveals the molecular mechanism underlying the cold acclimation and freezing tolerance of wheat (*Triticum aestivum* L.). *Plant Sci.* **2022**, *318*, 111242. [CrossRef] [PubMed]

MDPI
St. Alban-Anlage 66
4052 Basel
Switzerland
www.mdpi.com

International Journal of Molecular Sciences Editorial Office
E-mail: ijms@mdpi.com
www.mdpi.com/journal/ijms

Disclaimer/Publisher's Note: The statements, opinions and data contained in all publications are solely those of the individual author(s) and contributor(s) and not of MDPI and/or the editor(s). MDPI and/or the editor(s) disclaim responsibility for any injury to people or property resulting from any ideas, methods, instructions or products referred to in the content.

www.ingramcontent.com/pod-product-compliance
Lightning Source LLC
Chambersburg PA
CBHW041017270926
51850CB00024B/1373
9783036590554